AUTOMOTIVE CHASSIS SYSTEMS
Brakes, Steering, Suspension, and Alignment

James D. Halderman

Sinclair Community College

Prentice Hall
Englewood Cliffs, New Jersey   Columbus, Ohio

Library of Congress Cataloging-in-Publication Data

Halderman, James D.
    Automotive chassis systems : brakes, steering, suspension, and
alignment / James D. Halderman.
        p.   cm.
    Includes index.
    ISBN 0-13-052317-8
    1.   Automobiles—Chassis.   I.   Title.
TL255.H33   1995
629.24'028'8—dc20   95-31841
                                                                              CIP

Editor: Ed Francis
Editorial/Production Supervision: WordCrafters Editorial Services, Inc.
Design Coordinator: Jill E. Bonar
Production Manager: Pamela D. Bennett
Marketing Manager: Debbie Yarnell

This book was set in Century Book by Carlisle Communications, Ltd. and
was printed and bound by The Banta Company. The cover was printed by
Phoenix Color Corp.

© 1996 by Prentice-Hall, Inc.
A Simon & Schuster Company
Englewood Cliffs, New Jersey 07632

Printed in the United States of America

10   9   8   7   6   5   4

ISBN: 0-13-052317-8

Prentice-Hall International (UK) Limited, *London*
Prentice-Hall of Australia Pty. Limited, *Sydney*
Prentice-Hall of Canada, Inc., *Toronto*
Prentice-Hall Hispanoamericana, S.A., *Mexico*
Prentice-Hall of India Private Limited, *New Delhi*
Prentice-Hall of Japan, Inc., *Tokyo*
Simon & Schuster Asia Pte. Ltd., *Singapore*
Editora Prentice-Hall do Brasil, Ltda., *Rio de Janeiro*

CONTENTS

TECH TIPS

PREFACE

In this *Automotive Chassis Systems* textbook, all theory and servicing on a particular topic such as disc brakes, are included in the same chapter. This was done at the suggestion of teachers and automotive instructors from throughout the United States and Canada. Putting theory and service together eliminates the need to find servicing and troubleshooting procedures in another chapter or in another book. The purpose, function, and parts are discussed first, then the servicing and diagnosis are presented all in one chapter. This makes learning easier since the operation and the service procedures are closely linked. The machining of drums and rotors is together in a separate chapter. Since both items require the same basic steps and equipment, this eliminates repetition of similar steps and precautions, and makes teaching easier and more effective. Tech tips and diagnostic stories are included throughout the text. These help the reader by emphasizing important chassis system information.

Each chapter includes:

- Specific Objectives
- The Purpose, Function, and Parts Description
- Servicing and Testing Procedures

- Diagnosis and Troubleshooting
- Tech Tips
- Troubleshooting Guide (selected chapters)
- Chapter Summary
- Review, Discussion-Type, Questions
- ASE Certification-Type Multiple Choice Questions

All terms are printed in **bold** type and defined at first use *and* in the detailed glossary. A combination of line drawings and photographs makes the subject of chassis systems come alive. This text was written by an automotive instructor for use by other automotive instructors.

A sample ASE test for Brakes is included in Appendix 1 and a sample ASE test for Steering, Suspension, and Alignment is included in Appendix 2. Also included in the appendix is a wheel lug nut tightening torque chart.

At the request of many automotive instructors, a complete chart of drum and rotor "machine to" and "discard" dimension information is presented in Appendix 8 at the end of the text for easy access.

ACKNOWLEDGMENTS

The author gratefully acknowledges the following companies for technical information and the use of their illustrations:

Allied Signal, Inc.
Ammco Tools, Inc.
Arrow Automotive Industries
Bear Automotive
Chrysler Corporation
CR Services
DANA Corporation
DURALCAN USA a Division of
 Alcan Aluminum Corporation
Ferodo America
FMC Corporation
Ford Motor Company
Friction Materials Standards Institute, Inc.
General Motors Corporation
Gibson Products, A Division of Rolero, Inc.
Hunter Engineering Company
Lee Manufacturing Company
Monroe Shock Absorbers
MOOG Automotive, Inc., A Subsidiary of Cooper
 Industries

Northstar Manufacturing Company, Inc.
Nuturn Corporation
Oldsmobile Motor Division
SHIMCO International, Inc.
SKF, USA, Inc.
Specialty Product Company
Tire and Rim Association, Inc.
Toyota Motor Sales, USA, Inc.
TRW Inc.
Wagner Division, Cooper Industries, Inc.
Wurth USA, Inc.

Portions of materials contained herein have been reprinted with permission of General Motors Corporation, Service Technology Group.

I also wish to thank my colleagues and students at Sinclair Community College in Dayton, Ohio, for their comments and suggestions. A special thank you to Herbert Ellinger for the unrestricted use of his many illustrations.

Most of all, I wish to thank my wife, Michelle, for her assistance in all phases of manuscript preparation.

James D. Halderman

◀ Chapter 1 ▶

CHASSIS CONSTRUCTION, FASTENERS, AND SAFETY

OBJECTIVES

After studying Chapter 1, the reader will be able to:

1. Describe the various types of frames and construction used on vehicles.
2. Explain how the body is constructed and how to locate the lift points under a vehicle.
3. Explain the safe use of tools, automotive chemicals, and hoisting methods.
4. Discuss hazardous materials and how to handle them.

The chassis is the framework of any vehicle. The suspension, steering, braking components, and drivetrain components are mounted to the chassis. The chassis has to be a rigid and strong platform to support the suspension components. The suspension system allows the wheels and tires to follow the contour of the road. The connections between the chassis, the suspension, and the drivetrain must be made of rubber to dampen noise, vibration, and harshness (NVH). The construction of today's vehicles requires the use of many different materials.

FRAME CONSTRUCTION

Frame construction usually consists of channel-shaped steel beams welded and/or fastened together (see Figure 1–1).

There are many terms used to label or describe the frame of a vehicle including:

- **Ladder frame.** This is a common name for a type of perimeter frame where the transverse connecting members are straight across as in Figure 1–1.
- **Perimeter frame.** A perimeter frame consists of welded or riveted frame members around the entire perimeter of the body (see Figure 1–2).
- **Stub frame.** A stub frame is a partial frame often used on unit-body vehicles to support the power train and suspension components. Also called a "cradle" on many front-wheel-drive vehicles (see Figure 1–3).

FIGURE 1-1 Typical frame of a vehicle.

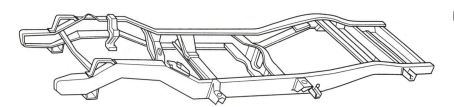

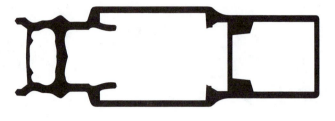

FIGURE 1-2 Perimeter frame.

◄ **TECH TIP** ►

GVW

Gross vehicle weight (GVW) is the weight of the vehicle plus the weight of all passengers the vehicle is designed to carry [× 150 lb (68 kg) each], plus the maximum allowable payload or luggage load. *Curb weight* is the weight of a vehicle "wet," that is, with a full tank of fuel and all fluids filled but without passengers or cargo (luggage). *Model weight* is the weight of a vehicle "wet" and with passengers.

The GVW is found stamped on a plate fastened to the doorjamb of most vehicles. A high GVW rating does not mean that the vehicle itself weighs a lot more than other vehicles. For example, a light truck with a GVW of 6000 lb (2700 kg) will not "ride" like an old 6000-lb luxury car.

Trucks are often overloaded with cargo and the braking system may not be able to handle the extra weight. The service technician should always check with the driver of the truck to see if the vehicle has been overloaded before attempting to "correct" a braking problem when loaded. No vehicle should be loaded beyond its rated GVW rating.

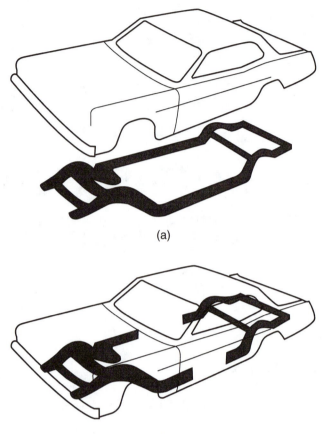

(a)

(b)

FIGURE 1–3 (a) Separate body and frame construction; (b) unitized construction: the small frame members are for support of the engine and suspension components. Many vehicles attach the suspension components directly to the reinforced sections of the body and do not require the rear frame section.

NOTE: A typical vehicle contains about 10,000 individual parts.

UNIT-BODY CONSTRUCTION

Unit-body construction (sometimes called *unibody*) is a design that combines the body with the structure of the frame. The body is composed of many individual stamped-steel panels welded together. The strength of this type of construction lies in the *shape* of the assembly. The typical vehicle uses 300 separate stamped steel panels that are spot-welded together to form a vehicle's body (see Figure 1–4).

SPACE FRAME CONSTRUCTION

Space frame construction consists of formed sheet steel used to construct a framework of the entire vehicle. The vehicle is driveable without the body, which uses plastic or steel panels to cover the steel framework (see Figure 1–5).

FIGURE 1–4 Note the ribbing and the many different pieces of sheet metal used in the construction of this body.

FIGURE 1–5 Space frame for a GM van, showing it without the exterior body panels. The framework surrounding the vehicle is a three-dimensional measuring system capable of accurate measurement of the vehicle.

FIGURE 1–6 Hollander interchange manuals are available for both domestic and imported vehicles.

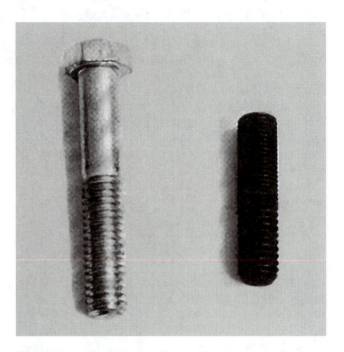

FIGURE 1–7 Typical bolt on the left and stud on the right. Note the different thread pitch on the top and bottom portions of the stud.

◀ **TECH TIP** ▶

HOLLANDER INTERCHANGE MANUAL

Most salvage businesses that deal with wrecked vehicles use a reference book entitled the *Hollander Interchange Manual* (see Figure 1–6). In this yearly publication, every vehicle part is given a number. If a part from another vehicle has the same Hollander number, the parts are interchangeable.

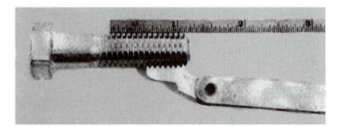

FIGURE 1–8 Thread pitch gauge used to measure the pitch of the thread. This is a ½-in.-diameter bolt with 13 threads to the inch (½ × 13).

THREADED FASTENERS

Most of the threaded fasteners used on engines are cap screws. They are called **cap screws** when they are threaded into a casting. Automotive service technicians usually refer to these fasteners as **bolts,** regardless of how they are used. In this chapter, they are called bolts. Sometimes, studs are used for threaded fasteners. A **stud** is a short rod with threads on both ends. Often, a stud will have coarse threads on one end and fine threads on the other end. The end of the stud with coarse threads is screwed into the casting. A nut is used on the opposite end to hold the parts together (see Figure 1–7).

The fastener threads *must* match the threads in the casting or nut. The threads may be either measured in fractions of an inch (called *fractional*) or metric. The size is measured across the outside of the threads. This is called the **crest** of the thread.

Fractional threads are either coarse or fine. Coarse threads are called *UNC* (Unified National Coarse) and

fine threads are called *UNF* (Unified National Fine). Standard combinations of sizes and number of threads per inch (called **pitch**) are used. Pitch can be measured with a thread gauge (see Figure 1–8). Fractional threads are specified by giving the diameter in fractions of an inch and the number of threads per inch. Typical UNC thread sizes would be ⁵⁄₁₆–18 and ½–13. Similar UNF thread sizes would be ⁵⁄₁₆–24 and ½–20.

The size of a metric bolt is specified by the letter M followed by the diameter in millimeters across the outside (crest) of the threads. Typical metric sizes would be M8 and M12. Fine metric threads are specified by the thread diameter followed by an × and the distance between the threads measured in millimeters (M8 × 1.5).

Bolts are identified by their diameter and length as measured from below the head, as shown in Figure 1–9. Bolts are made from many different types of steel. For this reason, some are stronger than others. The strength or classification of a bolt is called the **grade**. The bolt heads are marked to indicate their grade strength. Fractional bolts have lines on the head to indicate the grade, as shown in Figure 1–10. The actual grade of these bolts is two more than the number of lines on the bolt head. Metric bolts have a decimal number to indicate the grade. More lines or a higher grade number indicate a stronger bolt. In some cases, nuts and machine screws have similar grade markings. Figure 1–10 also shows head markings for common metric bolts.

CAUTION: *Never* use hardware store (nongraded) bolts, studs, or nuts on any vehicle steering, suspension, or brake component. Always use the exact size and grade of hardware that is specified and used by the vehicle manufacturer.

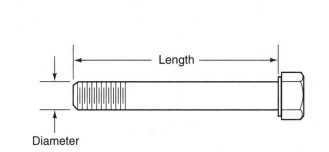

FIGURE 1–9 Bolt size identification.

Nuts. Most nuts used on cap screws have the same hex size as the cap screw head. Some inexpensive nuts use a hex size larger than the cap screw head. Metric nuts are often marked with dimples to show their strength.

◀ **TECH TIP** ▶

A ½-IN. WRENCH DOES NOT FIT A ½-IN. BOLT

A common mistake made by persons new to the automotive field is that the size of a bolt or nut is not the size of the head. The size (outside diameter of threads) is usually smaller than the size of the wrench or socket that fits the head of the bolt or nut. For example:

Wrench size	Thread size
⁷⁄₁₆-in.	¼-in.
½-in.	⁵⁄₁₆-in.
⁹⁄₁₆-in.	⅜-in.
⅝-in.	⁷⁄₁₆-in.
¾-in.	½-in.
10 mm	6 mm
12 or 13 mm[a]	8 mm
14 or 17 mm[a]	10 mm

[a]European (SI) metric.

FIGURE 1–10 Typical bolt (cap screw) grade markings and approximate strength.

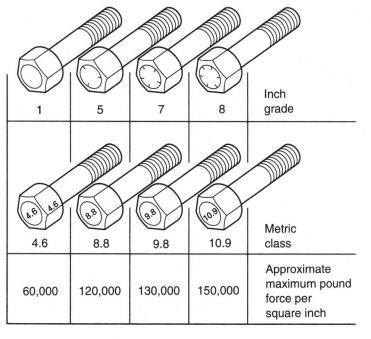

	1	5	7	8	Inch grade
	4.6	8.8	9.8	10.9	Metric class
	60,000	120,000	130,000	150,000	Approximate maximum pound force per square inch

FIGURE 1–11 Types of lock nuts: on the left, a nylon ring; in the center, a distorted shape; and on the right, a castle for use with a cotter key.

More dimples indicate stronger nuts. Some nuts and cap screws use interference-fit threads to keep them from loosening accidentally. This is done by slightly distorting the shape of the nut or by deforming part of the threads. Nuts can also be kept from loosening with a nylon washer fastened in the nut, or with a nylon patch or strip on the threads (see Figure 1–11).

NOTE: Most locking nuts are grouped together and are commonly referred to as *prevailing torque nuts*. This means that the nut will "hold" its tightness or torque and not loosen with movement or vibration. Most prevailing torque nuts should be replaced whenever removed to be assured that the nut will not loosen during service. Always follow the manufacturer's recommendations.

◀ TECH TIP ▶

IT JUST TAKES A SECOND

Whenever removing any automotive component, it is wise to put the bolts back into the holes a couple of threads with your hand. This assures that the right bolt will be used in its original location when the component or part is put back on the vehicle. Often, a fastener of the same diameter is used on a component, but the length of the bolt may vary. Spending just a couple of seconds to put the bolts and nuts back where they belong when the part is removed can save a lot of time when the part is being installed. Besides making certain that the right fastener is being installed in the right place, this method helps prevent bolts and nuts from being lost or kicked away. How much time have you wasted looking for that lost bolt or nut?

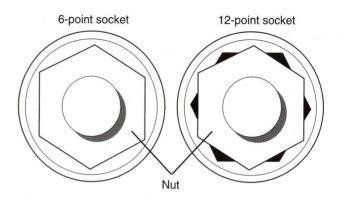

FIGURE 1–12 A six-point socket fits the head of the bolt or nut on all sides. A 12-point socket can round off the head of a bolt or nut if a lot of force is applied.

Anaerobic sealers such as LOCTITE are used on the threads when the nut or cap screw must be both locked and sealed.

Washers. Washers are often used under cap screw heads and under nuts. Plain flat washers are used to provide an even clamping load around the fastener. Lock washers are added to prevent accidental loosening. In some accessories, the washers are locked onto the nut to provide easy assembly.

SAFETY TIPS FOR USING HAND TOOLS

1. Always *pull* a wrench toward you for best control and safety. Never push a wrench.
2. Keep wrenches and all hand tools clean to help prevent rust and for a better, firmer grip.
3. Always use a six-point socket or a box end wrench to break loose a tight bolt or nut (see Figure 1–12).
4. Use a box end wrench for torque and an open end wrench for speed.
5. Never use a pipe extension or other types of "cheater bars" on a wrench or ratchet handle. If more force is required, use a larger tool or use penetrating oil and/or heat on the frozen fastener. (If heat is used on a bolt or nut to remove it, always replace it with a new part.)
6. Always use the proper tool for the job. If a specialized tool is required, use the proper tool and do not try to use another tool improperly.
7. Never expose any tool to excessive heat. High temperatures can reduce the strength ("draw the temper") of metal tools.

8. Never use a hammer on any wrench or socket handle unless you are using a special "staking-face" wrench designed to be used with a hammer.

9. Replace any tools that are damaged or worn.

SAFETY TIPS FOR TECHNICIANS

Safety is not just a buzzword on a poster in the work area. Safe work habits can reduce accidents, injuries, ease the workload, and keep employees pain-free. Suggested safety tips include:

1. **Safety glasses should be worn at all times while servicing any vehicle.**
2. Watch your toes—always keep your toes protected with steel-toed safety shoes. If safety shoes are not available, leather-topped shoes offer more protection than canvas or cloth.
3. Wear gloves to protect your hands from rough or sharp surfaces. It is recommended that thin rubber gloves be worn when working around automotive liquids such as engine oil, antifreeze, transmission fluid, or any other liquids that may be hazardous.
4. A service technician working under a vehicle should wear a **bump cap** to protect the head against under-vehicle objects and the pads of the lift.
5. Remove jewelry that may get caught on something or act as a conductor to an exposed electrical circuit.
6. Avoid loose or dangling clothing.
7. When lifting any object, get a secure grip with solid footing. Keep the load close to your body to minimize the strain. Lift with your legs and arms, not your back.
8. Do not twist your body when carrying a load. Instead, pivot your feet to help prevent strain to the spine.
9. Ask for help when moving or lifting heavy objects.
10. Push a heavy object rather than pulling it. [This is the opposite to how you should work with tools: Never push a wrench! If a bolt or nut loosens, your entire weight is used to propel your hand(s) forward. This usually results in cuts, bruises, or other painful injury.]
11. Work with objects, parts, and tools that are between chest high and waist high while standing. If seated, work at tasks that are at elbow height.

SAFETY IN LIFTING (HOISTING) A VEHICLE

Many chassis and under-body service procedures require that the vehicle be hoisted or lifted off the ground. The simplest methods involve the use of drive-on ramps or a floor jack and safety (jack) stands, but in-ground or surface-mounted lifts provide greater access.

Setting the Pads. All automobile and light truck service manuals include recommended locations to be used when hoisting (lifting) a vehicle. Newer vehicles carry a decal on the driver's door indicating the recommended lift points. The recommended standard for the lift points is given in the Society of Automotive Engineers, (SAE) Standard JRP-2184. These locations typically include the following:

1. The vehicle should be centered on the lift or hoist so as not to overload one side or put too much force either too far forward or rearward.
2. The pads of the lift should be spread as far apart as possible to provide a stable platform.
3. The pad should be placed under a portion of the vehicle that is strong and capable of supporting the weight of the vehicle, including:
   a. Pinch welds at the bottom edge of the body are generally considered to be strong.

CAUTION: Even though pinch weld seams are the recommended location for hoisting many vehicles with unitized body, care should be taken not to place the pad(s) too far forward or rearward. Incorrect placement of the vehicle on the lift could cause imbalance of the vehicle on the lift and the vehicle could fall. This is exactly what happened to the vehicle in Figure 1–13.

---

   b. Boxed areas of the body are the best places to place the pads on a vehicle without a frame. Be careful to note whether the arms of the lift may come in contact with other parts of the vehicle before the pad touches the intended location. Commonly damaged areas include:
      (1) Rocker panel moldings
      (2) Exhaust system (including catalytic converter)
      (3) Tires, especially if the edges of the pads or arms are sharp
   c. Frame or boxed perimeter frame members of the body (see Figure 1–14).

FIGURE 1–13   This car fell from the lift because the pads were not set correctly. A technician had just removed the left front strut assembly when the car fell. No one was hurt, but the car was a total loss.

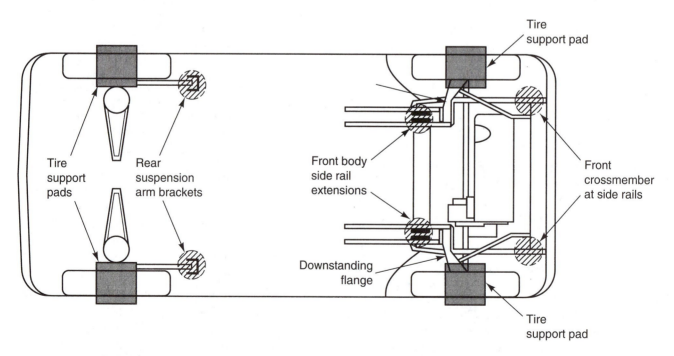

FIGURE 1–14   Typical hoist setting illustration from a service manual. (Courtesy of Ford Motor Company)

4. The vehicle should be raised about 1 ft (30 cm) off the floor, then stopped. The vehicle should be shaken to check for stability. If the vehicle seems to be stable when checked a short distance from the floor, continue raising the vehicle and continue to view the vehicle until it has reached the desired height.

CAUTION: Do not look away from the vehicle while it is being raised (or lowered) on a hoist. Often, one side or one end of the hoist can stop or fail, resulting in the vehicle being slanted enough to slip or fall, creating physical damage not only to the vehicle and/or hoist but also to the technician or others who may be near.

HINT: Most hoists can be safely placed at any desired height. For ease while working, the area that you are working on should be at chest level. When working on brakes or suspension components, it is not necessary to work on them down near the floor or over your head; raise the hoist so that the components are at chest level.

---

5. Before lowering the hoist, the safety latch(es) must be released and the direction of the controls reversed to lower the hoist. For additional safety, the speed downward is often adjusted to be as slow as possible.

HAZARDOUS MATERIALS

The EPA (Environmental Protection Agency) regulates and controls the handling of hazardous materials in the United States. A material is considered hazardous if it meets one or more of the following conditions:

1. Contains over 1000 parts per million (ppm) of halogenated compounds (halogenated compounds are chemicals containing chlorine, fluorine, bromine, or iodine). Common items that contain these solvents include:
   - Carburetor cleaner
   - Silicone spray
   - Aerosols
   - Adhesives
   - Stoddard solvent
   - Trichloromethane
   - Gear oils
   - Brake cleaner
   - A/C compressor oils
   - Floor cleaners
   - Anything else that contains *chlor* or *fluor* in its ingredient name
2. Has a flash point below 140°F (60°C).
3. Is corrosive (a pH of 2 or less or 12. 5 or higher).
4. Contains toxic metals or organic compounds. Volatile organic compounds (VOCs) must also be limited and controlled. This classification greatly affects the painting and finishing aspects of the automobile industry.

Always follow recommended procedures for handling of any chemicals and dispose of all used engine oil and other possible waste products according to local, province, state, or federal laws.

To help safeguard workers and the environment, the following is recommended:

1. A technician's hands should always be washed thoroughly after touching used engine oils, transmission fluids, and greases. Dispose of all waste oil according to established standards and laws in your area.

NOTE: The Environmental Protection Agency (EPA) current standard permits less than 1000 parts per million (ppm) of total halogens (chlorinated solvents) that used engine oil can contain and still be able to be recycled. Oil containing greater amounts of halogens must be considered hazardous waste (see Figure 1–15).

FIGURE 1–15 Waste products such as used engine oil, transmission oil, and antifreeze should be kept separated and disposed of or recycled according to established standards and laws.

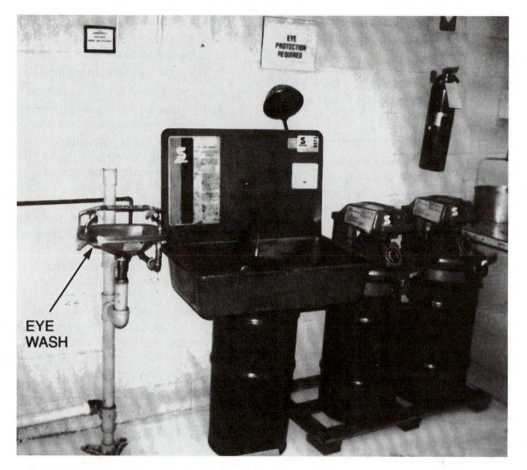

FIGURE 1–16 Eyewash station.

2. Asbestos and products that contain asbestos are known cancer-causing agents. Even though most brake linings and clutch facing materials no longer contain asbestos from the factory, millions of vehicles are being serviced every day that *may* contain asbestos. The general procedure for handling asbestos is to put the used parts into a sealed plastic bag and return the parts as cores for rebuilding or to be disposed of according to current laws and regulations.

3. Eyewash stations should be readily accessible near the work area or near where solvents or other contaminants could get into the eyes (see Figure 1–16).

MATERIAL SAFETY DATA SHEETS

Businesses and schools in the United States are required to provide a detailed data sheet on each chemi-

cal and other material that a person may be exposed to in the area. These sheets of information on each material that *may* be harmful are called **material safety data sheets** (MSDS).

SUMMARY

1. Vehicle chassis designs include frame, unit-body, and space-frame construction. A full-frame vehicle is often stronger and quieter and permits the towing of heavier loads. Unit-body and space-frame designs are often lighter and more fuel efficient.

2. Bolts, studs, and nuts are commonly used as fasteners in the chassis. The sizes for fractional and metric threads are different and are not interchangeable. Grade is the rating of the strength of a fastener.

3. Whenever a vehicle is raised above the ground, it must be supported at a substantial section of the body or frame.

4. Hazardous materials include common automotive chemicals, liquids, and lubricates, especially those whose ingredients contain *chlor* or *fluor* in the name. Asbestos fibers should be avoided and removed according to current laws and regulations.

REVIEW QUESTIONS

1. List two advantages and two disadvantages of frame and unit-body chassis construction.

2. List three precautions necessary to perform whenever hoisting (lifting) a vehicle.

3. List five common automotive chemicals or products that may be considered hazardous materials.

4. List five precautions every technician should adhere to when working with automotive products and chemicals.

MULTIPLE-CHOICE QUESTIONS

1. Two technicians are discussing the hoisting of a vehicle. Technician A says to put the pads of a lift under a notch at the pinch weld of a unit-body vehicle. Technician B says to place the pads on four corners of the frame of a full-frame vehicle. Which technician is correct?
   a. A only
   b. B only
   c. Both A and B
   d. Neither A nor B

2. The correct location for hoisting or jacking the vehicle can often be found in the
   a. service manual.
   b. shop manual.
   c. owner's manual.
   d. all of the above.

3. Hazardous materials include all *except*
   a. engine oil.
   b. asbestos.
   c. water.
   d. brake cleaner.

4. To check if a product or substance being used is hazardous, consult
   a. a dictionary.
   b. MSDS.
   c. SAE.
   d. EPA.

5. For best heavy-duty towing, the tow vehicle should have what type of construction?
   a. Space frame
   b. Unit body
   c. Frame
   d. Body

6. For the best working position, the work should be at
   a. neck or head level.
   b. knee or ankle level.
   c. overhead about 1 ft.
   d. chest or elbow level.

7. When working with hand tools, always
   a. push the wrench—don't pull toward you.
   b. pull a wrench—don't push a wrench.

8. A high-strength bolt is identified by
   a. the UNC symbol.
   b. lines on the head.
   c. strength letter codes.
   d. the coarse threads.

◀ Chapter 2 ▶

BRAKING SYSTEM PRINCIPLES, COMPONENTS, AND OPERATION

OBJECTIVES

After studying Chapter 2, the reader will be able to:

1. State the operating principles of the braking system.
2. List the parts and terms for disc and drum brakes.
3. Describe how the hydraulic brake system can be used to transfer pressure to each wheel.
4. Discuss the coefficient of friction and describe how the friction between the lining material and the drum or rotor stops wheels.
5. List the types of brake fluids and their application.
6. Discuss how ABS units help in braking and vehicle stability when stopping on slippery surfaces.

Brakes are by far the most important mechanism on any vehicle because the safety and lives of those riding in the vehicle depend on proper operation of the braking system. It has been estimated that the brakes on the average vehicle are applied 50,000 times a year!

"Brakes stop wheels—not vehicles." This basic fact means that the best brakes in the world only stop the rotation of the tire/wheel assembly. It is the friction between the tire and the pavement that accomplishes the stopping or slowing of a vehicle. A vehicle being driven on an icy street will take an extremely long time to stop. Even with a vehicle equipped with an antilock braking system (ABS), there still has to be **friction** (**traction**) between the tire and the road for the vehicle to stop.

HOW BRAKES STOP VEHICLES

Brakes are an energy-absorbing mechanism that converts vehicle movement into heat while stopping the rotation of the wheels. All braking systems are designed to reduce the speed and stop a moving vehicle and to keep it from moving if the vehicle is stationary. **Service brakes** are the main driver-operated vehicle brakes (see Figure 2–1). Service brakes are also called **base brakes** or **foundation brakes.**

Most vehicles built since the late 1920s use a brake on each wheel. To stop a wheel, the driver exerts a force on a brake pedal. This force on the pedal pressurizes brake fluid in a master cylinder. This hydraulic force (liquid under pressure) is transferred through steel lines to a wheel cylinder or caliper at each wheel. Hydraulic pressure to each wheel cylinder or caliper is used to force friction materials against the brake drum or rotor. The friction between the stationary friction material and the rotating drum or rotor (disc) causes the rotating part to slow and eventually stop. Since the wheels are

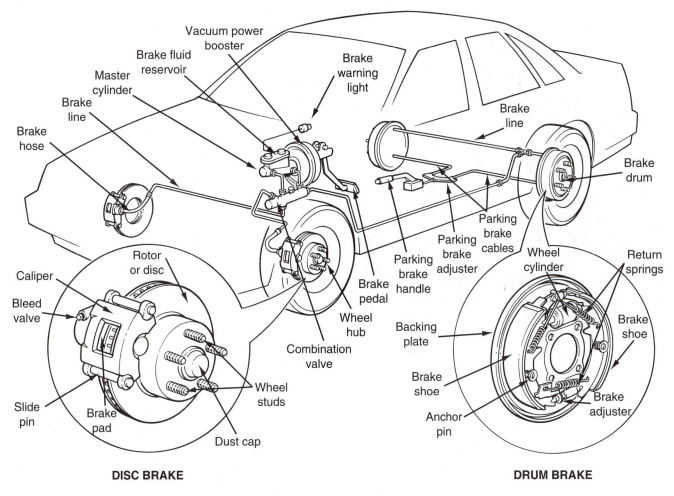

FIGURE 2–1   Typical vehicle brake system showing all typical components.

attached to the drums or rotors, the wheels of the vehicles also stop.

The heavier the vehicle and the higher the speed, the more heat the brakes have to be able to absorb. Long, steep hills can cause the brakes to overheat, reducing the friction necessary to slow and stop a vehicle (see Figures 2–2 and 2–3).

DRUM BRAKES

Drum brakes are used on the rear of many rear-wheel-drive, front-wheel-drive, and four-wheel-drive vehicles. Since the early 1970s, few vehicles have used drum brakes on the front wheels. When drum brakes are applied, brake shoes are moved outward against a rotating brake drum. The wheel studs for the wheels are attached to the drum. When the drum slows and stops, the wheels also slow and stop.

FIGURE 2–2   Brakes change the energy of the moving vehicle into heat. Too much heat and brakes fail as indicated on this sign coming down from Pike's Peak in Colorado, USA, at 14,000 ft (4300 m).

FIGURE 2–3 When driving down long, steep grades, select a lower transmission gear to allow the engine compression to help maintain vehicle speed.

Drum brakes are economical to manufacture, service, and repair. Parts for drum brakes are generally readily available and reasonably priced (see Figure 2–4). On some vehicles, an additional drum brake is used as a parking brake on vehicles equipped with rear disc brakes .

DISC BRAKES

Disc brakes are used on the front of most vehicles built since the early 1970s and on the rear wheels of many vehicles. A disc brake operates by squeezing brake pads on both sides of a rotor or disc that is attached to the wheel (see Figure 2–5).

Type of brake	Rotating part	Friction part
Drum brakes	Brake drum	Brake "shoes"
Disc brakes	Rotor or disc	Brake "pads"

Due to the friction between the road surface and the tires, the vehicle stops. To summarize, the events necessary to stop a vehicle include:

1. The driver presses on the brake pedal.
2. The brake pedal pressure is transferred hydraulically to a wheel cylinder or caliper at each wheel.
3. Hydraulic pressure inside the wheel cylinder or caliper presses friction materials (brake shoes or pads) against rotating brake drums or rotors.
4. The friction slows and stops the drum or rotor. Since the drum or rotor is bolted to the wheel of the vehicle, the wheel also stops.
5. When the wheels of the vehicle slow and stop, the tires must have friction (traction) with the road surface to stop the vehicle.

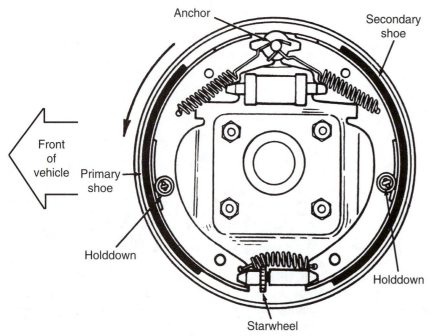

FIGURE 2–4 Typical drum brake assembly. (Wagner Division, Cooper Industries Inc.)

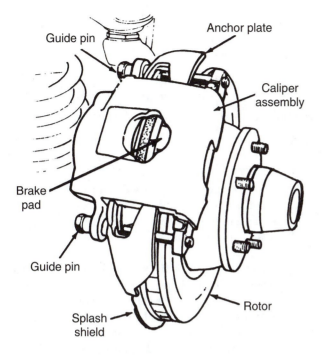

FIGURE 2–5 Typical disc brake assembly. (Wagner Division, Cooper Industries Inc.)

WEIGHT TRANSFER DURING BRAKING

Whenever the brakes are applied on a vehicle in motion, the weight of the vehicle is transferred forward while driving forward. Most vehicles can be observed "nosing downward" when the brakes are applied. Greater braking power is required for the front brakes because of this weight transfer (see Figure 2–6). It is estimated that the front brakes handle about 60 to 70 percent of the braking power on rear-wheel-drive vehicles and 80 percent on front-wheel-drive vehicles (see Figure 2-7).

FRICTION

Friction is the force that resists the motion between two objects that are in contact (see Figures 2–8). The amount of friction between two surfaces, called the **coefficient of friction**, is represented by a number from zero to 1. The Greek lowercase letter **mu (μ)** is used to represent this factor.

$\mu = 0$ (no friction between surfaces)

$\mu = 1$ (maximum friction between surfaces)

To stop a vehicle there must be friction between the brake lining material and the rotating brake part.

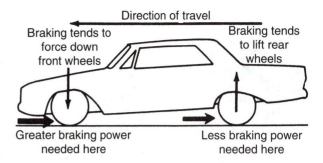

FIGURE 2–6 When brakes are applied, vehicle weight is transferred toward the front. This weight transfer requires that more braking power be available on the front wheels and less on the rear wheels to avoid rear-wheel lockup. (Courtesy of EIS Brake Parts)

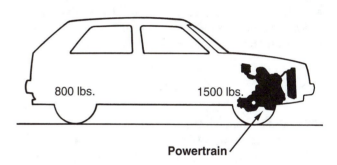

FIGURE 2–7 Front-wheel-drive vehicles have much of their vehicle weight over the front wheels. This means that most of the braking power occurs with the front brakes. (Courtesy of Hunter Engineering Company)

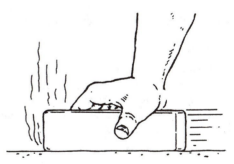

FIGURE 2–8 A brick being pushed against the floor creates heat from the friction between the two surfaces. (Courtesy of EIS Brake Parts)

as the brake drum or rotor. Road surfaces also provide varying amounts of traction (friction) for the tires. On ice, tires cannot "grip the road" because even if the brakes are working perfectly, the vehicle will not stop quickly because of the lack of friction between the tire and the ice (see Figure 2–9). Braking is especially

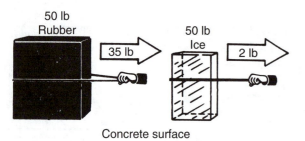

FIGURE 2–9 Friction varies with the type of material. A slippery surface has less friction (traction) than do other surfaces. (Wagner Division, Cooper Industries Inc.)

dangerous when one side of the vehicle is traveling over normal high-friction road surfaces while the other wheel on the other side of the vehicle is running on gravel or ice. Brake engineers call these **split-mu** conditions. Antilock braking systems are especially important during braking under these conditions.

BRAKE FLUID TYPES

Brake fluid is made from a combination of various types of glycol, a non-petroleum-based fluid. Brake fluid is a polyalkylene glycol–ether mixture, called **polyglycol** for short. **All polyglycol brake fluid is clear to amber in color.** Brake fluid has to have the following characteristics:

1. High boiling point
2. Low freezing point
3. Will not damage rubber parts in the brake system

CAUTION: DOT 3 brake fluid is a very strong solvent and can remove paint! Care is required when working with DOT 3 brake fluid to avoid contact with a vehicle's painted surfaces. It also takes the color out of leather shoes.

BRAKE FLUID SPECIFICATIONS

All automotive brake fluid must meet Federal Motor Vehicle Safety Standard 116. The Society of Automotive Engineers (SAE) and the Department of Transportation (DOT) have established brake fluid specification standards.

	DOT 3	DOT 4	DOT 5
Dry boiling point			
°F	401	446	500
°C	205	230	260
Wet boiling point			
°F	284	311	356
°C	140	155	180

The wet boiling point is often referred to as **equilibrium reflux boiling point** (ERBP), the name given to the method in the specification (SAE J1703) for how the fluid is exposed to moisture and tested.

DOT 3. This brake fluid is the type most often used. DOT 3 absorbs moisture that in it is *hygroscopic.* According to SAE, DOT 3 can absorb 2 percent of its volume in water per year. Moisture is absorbed by the brake fluid through microscopic seams in the brake system and around seals. Over time, the water will corrode the system and thicken the brake fluid. The moisture can also cause a spongy brake pedal due to reduced vapor-lock temperature (see Figure 2–10). DOT 3 must be used from a sealed (capped) container. If allowed to remain open for any length of time, DOT 3 will absorb moisture from the surrounding air.

DOT 4. This brake fluid is formulated for use by all vehicles, imported or domestic. It is commonly called LMA (low moisture absorption). DOT 4 does not absorb water as fast as DOT 3 but is still affected by moisture and should be used only from a sealed container. DOT 4 costs approximately double the cost of DOT 3. *DOT 4 can be used wherever DOT 3 is specified* (see Figure 2–11).

DOT 5. This brake fluid, commonly called **silicone brake fluid**, is made from polydimethylsiloxanes. It does not absorb water; that is, it is *nonhygroscopic.*

NOTE: Even though DOT 5 does not normally absorb water, it is still tested in a humidity chamber using standardized SAE procedures. After a fixed amount of time, the brake fluid is measured for boiling point. Since it has had a *chance* to absorb moisture, the boiling point after this sequence is called the *minimum wet boiling point.*

DOT 5 brake fluid is purple (violet) in color to distinguish it from DOT 3 or DOT 4 brake fluid. Silicones have about three times the amount of dissolved air as do glycol fluids (about 15 percent of dissolved air versus only about 5 percent for standard glycol brake fluid). It is this

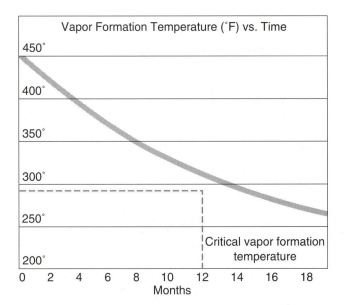

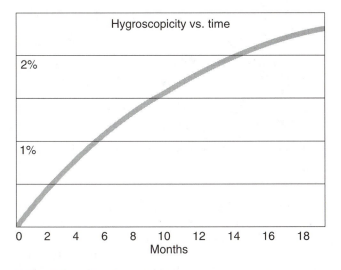

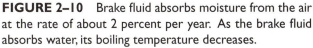

FIGURE 2–10   Brake fluid absorbs moisture from the air at the rate of about 2 percent per year. As the brake fluid absorbs water, its boiling temperature decreases.

characteristic of silicone brake fluid that makes it more difficult to bleed the hydraulic system of trapped air.

NOTE: The characteristics of DOT 5 silicone brake fluid to absorb air is one of the major reasons why it is *not* recommended for use with an antilock braking system (ABS). In an ABS, valves and pumps are used which can aerate the brake fluid. Brake fluid filled with air bubbles cannot lubricate the ABS components properly and will cause a low, soft brake pedal.

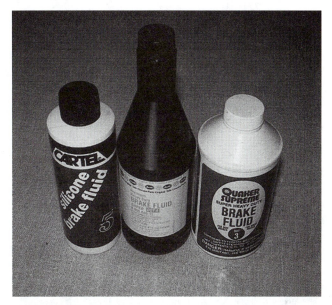

FIGURE 2–11   Brake fluid should be purchased in small containers to help prevent contamination and the possibility of moisture absorption after the container has been opened. On the left is silicone (DOT 5) brake fluid with DOT 4 in the center and DOT 3 on the right.

DOT 5 brake fluid should not be mixed with any other type of brake fluid. Therefore, the entire braking system must be flushed completely and refilled with DOT 5. DOT 5 is approximately four times the cost of DOT 3 brake fluid.

HYDRAULIC SYSTEM MINERAL OIL

Some French-built Citroen and British-designed Rolls-Royce vehicles use hydraulic system mineral oil (HSMO) as part of their hydraulic control systems. The systems in these vehicles use a hydraulic pump to pressurize hydraulic oil for use in the suspension leveling and braking systems.

CAUTION: Mineral hydraulic oil should never be used in a braking system that requires DOT 3 or DOT 4 polyglycol-based brake fluid. If *any* mineral oil, such as engine oil, transmission oil, or automatic transmission fluid (ATF), gets into a braking system that requires glycol brake fluid, *every* rubber part in the entire braking system *must* be replaced. Mineral oil causes the rubber compounds that are used in glycol brake fluid systems to swell (see Figure 2–12).

FIGURE 2–12   Both rubber sealing cups were of exactly the same size. The cup on the left was exposed to mineral oil. Notice how the seal greatly expanded.

◀ DIAGNOSTIC STORY ▶

PIKE'S PEAK BRAKE INSPECTION

All vehicles must stop about halfway down Pike's Peak Mountain in Colorado [14,110 ft (4300 m)] for a "brake inspection." When this author stopped at the inspection station, a uniformed inspector simply looked at the right front wheel and waved us on. I pulled over and asked the inspector what he was checking. He said that when linings and drums/rotors get hot, the vehicle loses brake effectiveness. But if the brake fluid boils, the vehicle loses its brakes entirely. The inspector was *listening* for boiling brake fluid at the front wheel and feeling for heat about 1 ft (30 cm) from the wheel. If the inspector felt heat 1 ft away from the brakes, the brakes were definitely too hot to continue and you would be instructed to pull over and wait for the brakes to cool. The inspector recommended placing the transmission into a lower gear which uses the engine to slow the vehicle during the descent without having to rely entirely on the brakes.

To help prevent hydraulic system mineral oil from being mixed with glycol brake fluid, *hydraulic mineral oils are green.*

BRAKE FLUID DISPOSAL

Current EPA laws permit used brake fluid to be disposed of with used engine oil. Brake fluid spilled on

◀ DIAGNOSTIC STORY ▶

SINKING BRAKE PEDAL

This author experienced what happens when the brake fluid is not changed regularly. Just as many technicians will tell you, we do not always do what we know should be done to our vehicles.

While driving a four-year-old vehicle on vacation in very hot weather in mountainous country, the brake pedal sank to the floor. When the vehicle was cold, the brakes were fine. But after several brake applications, the pedal became soft and spongy and sank slowly to the floor if pressure was maintained on the brake pedal. Because the brakes were okay when cold, I knew it had to be boiling brake fluid. Old brake fluid (four years old) often has a boiling point under 300°F (150°C). With the air temperature near 100°F (38°C), it does not take much more heat to start boiling the brake fluid. After bleeding over a quart (1 liter) of new brake fluid through the system, the brakes worked normally. I'll never again forget to replace the brake fluid as recommended by the vehicle manufacturers.

the floor should be cleaned up using absorbent material and the material disposed of in the regular trash. Brake fluid becomes a hazardous waste if spilled onto open ground where it can seep into groundwater. The disposal requirements for brake fluid spilled onto open ground vary with the exact amount spilled and other factors. Refer to local EPA guidelines and requirements for the exact rules and regulations in your area.

RUBBER TYPES

Vehicles use a wide variety of rubber in the braking system, suspension system, as well as the steering system and the engine. Rubber products are called **elastomers**. Some are oil- and grease-resistant elastomers and can be harmed by brake fluid, while others are brake fluid resistant and can swell or expand if they come in contact with oil or grease. See the rubber compatibility chart for types and compatible fluids.

Rubber Compatibility

Name	Abbreviation	OK	Not OK	Uses
Ethylene propylene diene (developed in 1963)	EPM, EPDM, EPR	Brake fluid, silicone fluids	Petroleum fluids	Most brake system seals and parts
Styrene butadiene (developed in 1920s)	SBR, Buna S, GRS	Brake fluid, silicone fluids, alcohols	Petroleum fluids	Some drum brake seals, O-rings
Nitrile (nitrile butadiene rubber)	NBR, Buna N	Petroleum fluids, ethylene glycol (antifreeze)	Brake fluid	Engine seals, O-rings
Neoprene (polychloroprene)	CR	Refrigerants [(Freon) R-12 and R-134a] petroleum fluids	Brake fluids	Refrigerant, O-rings
Polyacrylate	ACM	Petroleum fluids, automatic transmission fluids	Brake fluids	Automatic transmission and engine seals
Viton (fluorocarbon)	FKM	Petroleum fluids	Brake fluid, R-134a refrigerant	Engine seals, fuel system parts
Natural rubber	NR	Water, brake fluid	Petroleum fluids	Tires

BRAKE LINING COMPOSITION

Friction materials such as disc brake pads or drum brake shoes contain a mixture of ingredients. These materials include a binder such as a thermosetting resin, fibers for reinforcement, and friction modifiers to obtain a desired coefficient of friction (mu).

The various ingredients in brake lining are mixed and molded into the shape of the finished product. The fibers in the material are the only thing holding this mixture together. A large press is used to force the ingredients together to form a **brake block** that eventually becomes the brake lining. Many disc brake pads are **integrally molded** (see Figure 2–13).

Typical Asbestos (Organic) Lining Composition

Ingredient	Formula range (%)
Phenolic resin (binder)	9–15
Asbestos fiber	30–50
Organic friction modifiers (rubber scrap)	8–19
Inorganic friction modifiers (barytes, talc, whiting)	12–26
Abrasive particles (alumina)	4–20
Carbon	4–20

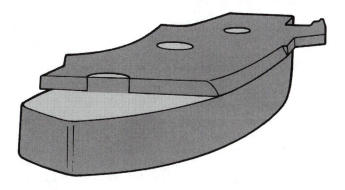

FIGURE 2–13   Cutaway shows friction material integrally molded into steel backing plate.

FIGURE 2–14 Poor-quality semimetallic disc brake pad. The screwdriver is pointing to large chunks of steel embedded in the lining.

Semimetallic Friction Material Composition. The term **semi-metallic** (also called **semi-mets**) refers to brake lining material that uses metal rather than asbestos in its formulation. It still uses resins and binders and is, therefore, not 100 percent metal but rather, semi-metallic. The metal in most metallic linings is made from metal particles that have been fused together without melting. This process, called **sintering,** is used to produce **sintered metal linings** (see Figure 2–14).

Ingredient	Formula range(%)
Phenolic resin	15–40
Graphite or carbon particles	15–40
Steel fibers	0–25
Ceramic powders	2–10
Steel, copper, and brass metal powders	15–40
Other modifiers (rubber scrap)	0–20

Most semi-metallic linings do not contain asbestos. Semi-metallic linings require a very smooth finish on the rotor because the metal in the lining does not conform to the surface of the rotor as do asbestos linings.

Nonasbestos Friction Material. Brake pads and linings that use synthetic material such as aramid fibers instead of steel are usually referred to as **nonasbestos, nonasbestos organic** (NAO), or **nonasbestos synthetic** (NAS). Linings are called *synthetic* because syn-

thetic (human-made) fibers are used. These linings use **aramid fiber** instead of metal as the base material. Aramid is the generic name for aromatic polyamide fibers. **Kevlar** is the Dupont brand name of aramid and a registered trademark of E.I. Dupont de Nemours and Company. Nonasbestos linings are often quieter than semimetallics and do not cause as much wear to brake rotors as do semi-metallic pads.

Carbon Fiber Friction Material. Carbon fiber brake lining is the newest and most expensive of the lining materials. Carbon fiber material is often called **carbon-fiber-reinforced carbon** (CFRC).

LINING EDGE CODES

The edge coding contains three groups of letters and numbers:

- *First group:* a series of letters that identify the manufacturer of the lining.
- *Second group:* a series of numbers, letters, or both that identify the lining compound or formula. This code is usually known to the manufacturer of the lining material and helps them identify the lining after manufacture.
- *Third group:* two letters that identify the coefficient of friction (see Figure 2–15).

The coefficient of friction is a pure number that indicates the amount of friction between two surfaces. For example, a material with a 0.55 coefficient of friction has more friction than a material with a coefficient of friction of 0.39. These codes were established by the Society of Automotive Engineers (SAE):

Code C = 0.00 to 0.15
Code D = 0.15 to 0.25
Code E = 0.25 to 0.35
Code F = 0.35 to 0.45
Code G = 0.45 to 0.55
Code H = 0.55 and above
Code Z = ungraded

The first letter, which is printed on the side of most linings, indicates its coefficient of friction when brakes are cold, and the second letter indicates the coefficient of friction of the brake lining when the brakes are hot. (For example, FF indicates that the brake lining mater-

FIGURE 2–15 Typical drum brake lining edge code.

ial has a coefficient of friction between 0.35 and 0.45 when both cold and hot.)

The code letters should not be interpreted as indicating the relative quality of the lining material. Lining wear, fade resistance, tensile strength, heat recovery rate, wet friction, noise, and coefficient of friction must be considered when purchasing good-quality linings. Unfortunately, there are no standards that a purchaser can check regarding all of these other considerations. For best brake performance, always purchase the best-quality name brand linings that you can afford.

NOTE: Brake squeal (noise) is usually caused by movement of the friction material more than the composition of the brake lining material.

HINT: While many brands of replacement brake lining provide acceptable stopping power and long life, purchasing factory brake lining from a dealer is usually the best opportunity to get lining material that meets all vehicle requirements. Aftermarket linings are *not* required by federal law to meet performance or wear standards that are required of original factory brake linings.

DANGERS OF EXPOSURE TO ASBESTOS

Friction materials such as brake and clutch linings often contain asbestos. While asbestos has been eliminated from most original-equipment friction materials, the automotive service technician cannot know whether or not

◀ TECH TIP ▶

COMPETITIVELY PRICED BRAKES

The term *competitively priced* means lower in cost. Most brake manufacturers offer "premium" as well as lower-priced linings to remain competitive with other manufacturers or importers of brake lining material produced overseas or in foreign countries. Organic asbestos brake lining is inexpensive to manufacture. In fact, according to warehouse distributors and importers, the box often costs more than the brake lining inside!

Professional brake service technicians should only install brake linings and pads that will give braking performance equal to that of the original factory brakes. This means that "competitive" asbestos linings should *never* be substituted for semimetallic or NAO original linings or pads. For best results, always purchase good-quality brake parts from a known brand name manufacturer.

the vehicle being serviced is or is not equipped with friction materials containing asbestos. It is important that *all* friction material be handled as if it does contain asbestos.

Asbestos exposure can cause scar tissue to form in the lungs. This condition, called **asbestosis**, causes increasing shortness of breath and permanent scarring of the lungs.

Even low exposures to asbestos can cause **mesothelioma**, a type of fatal cancer of the lining of the chest or abdominal cavity. Asbestos exposure can also increase the risk of **lung cancer** as well as cancer of the voice box, stomach, and large intestines. It usually takes 15 to 30 years or more for cancer or asbestos lung scarring to show up after exposure.

OSHA STANDARDS

The **Occupational Safety and Health Administration** (OSHA) has established acceptable levels of asbestos exposure. Any vehicle service establishment that does either brake or clutch work must limit employee exposure to asbestos to less than 0.1 fiber per cubic centimeter (cc) as determined by an air sample. If the level of exposure to employees is greater than specified, corrective measures must be performed and a large fine may be imposed.

EPA REGULATION OF ASBESTOS

The federal **Environmental Protection Agency** (EPA) has established procedures for the removal and disposal of asbestos. The EPA procedures require that products containing asbestos be "wetted" to prevent the asbestos fibers from becoming airborne. According to the EPA, asbestos-containing materials can be disposed of as regular waste—only when asbestos becomes airborne is it considered to be hazardous.

ASBESTOS HANDLING GUIDELINES

The air in the shop area can be tested by a testing laboratory, but this can be expensive. Tests have determined that asbestos levels can easily be kept below the recommended levels by using a solvent or a special vacuum.

NOTE: Even though asbestos is being removed from brake lining materials, the service technician cannot tell whether or not the old brake pads or shoes contain asbestos. Therefore, to be safe, the technician should *assume* that all brake pads or shoes contain asbestos.

HEPA Vacuum. A special **high-efficiency particulate air** (HEPA) vacuum system has been proven to be effective in keeping asbestos exposure levels below 0.1 fiber per cubic centimeter.

Solvent Spray. Many technicians use an aerosol can of brake cleaning solvent to wet the brake dust and prevent it from becoming airborne. Commercial brake cleaners are available that use a concentrated cleaner that is mixed with water (see Figure 2–16). The waste liquid is filtered, and when dry, the filter can be disposed of as solid waste.

CAUTION: Never use compressed air to blow brake dust. The fine talc-like brake dust can create a health hazard even if asbestos is not present or is present in dust rather than fiber form.

Disposal of Brake Dust and Brake Shoes. The hazard of asbestos occurs when asbestos fibers are air-

FIGURE 2–16 Portable brake wash units are popular because the brakes can be wetted down before removal of the brake drum or caliper reducing the chances of airborne brake dust. (Courtesy of Clayton)

borne. Old brake shoes and pads should be enclosed, preferably in a plastic bag, to help prevent any of the brake material from becoming airborne. *Always follow current federal and local laws considering disposal of all waste.*

ANTILOCK BRAKE SYSTEM OPERATION

The purpose of an **antilock brake system** (ABS) is to prevent the wheels from locking during braking, especially on low-friction surfaces such as wet, icy, or snowy roads. Remember, it is the friction between the tire tread and the road that does the actual stopping of the vehicle. Therefore, ABS does not mean that a vehicle can stop quickly on all road surfaces. ABS uses sensors at the wheels to measure the wheel speed. If a wheel is rotating slower than the others, indicating pos-

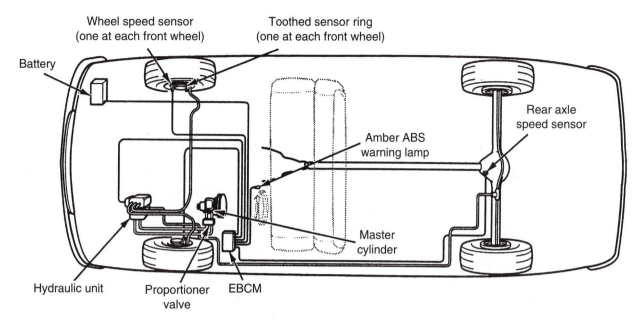

FIGURE 2–17 Typical components of an antilock braking system (ABS) used on a rear-wheel-drive vehicle. (Courtesy of General Motors)

sible lockup (for example, on an icy spot), the ABS computer will control the brake fluid pressure to that wheel for a fraction of a second. *A locked wheel has less traction to the road surface than that of a rotating wheel.*

The ABS computer can reapply the pressure from the master cylinder to the wheel a fraction of a second later. Therefore, if a wheel *starts* to lockup, the purpose of the ABS system is to *pulse* the brakes on and off to maintain directional stability with maximum braking force. Many ABS units will cause the brake pedal to pulse if the unit is working in the ABS mode. The pulsating brake pedal is a cause for concern for some drivers; however, the pulsing brake pedal informs the driver that the ABS is being activated. Some ABS units use an isolator valve in the ABS unit to prevent brake pedal pulsations during ABS operation. With these types of systems, it is often difficult for the driver to know if and when the ABS unit is working to control a locking wheel. See Figure 2–17 for an overview of a typical ABS on a rear-wheel-drive vehicle.

Another symptom of normal ABS unit operation is the activation of the hydraulic pressure pump used by many ABS units. In some ABS units, the hydraulic pump is run every time the vehicle is started and moved. Other types of units operate randomly or whenever the pressure in the system calls for the pump to operate. See Chapter 10 for additional details on antilock braking systems.

SUMMARY

1. Brakes stop wheels—not vehicles.
2. Hydraulic pressure is sent to each wheel when the driver pushes on the brake pedal and creates a pressure buildup in the master brake cylinder according to Pascal's law.
3. During a typical brake application, only about 1 teaspoon of brake fluid is moved from the master cylinder to create the pressure increase throughout the entire brake system.
4. All brake fluid is specified by DOT and SAE. DOT 3 brake fluid is the most commonly recommended brake fluid for all types of vehicles.
5. An antilock braking system pulses the hydraulic force to the wheels to prevent the tires from locking up. A locked tire has lower friction than that of a rolling tire.

REVIEW QUESTIONS

1. Explain how hydraulic pressure is used to stop the rotation of the wheels.

2. List the coefficient of friction edge codes.

3. Describe brake fluid and how it should be used and handled.

4. Explain how ABS units prevent wheel lockup.

MULTIPLE-CHOICE QUESTIONS

1. Disc brakes use replaceable friction material called
   a. linings.
   b. pads.

2. A wheel that locks up during braking
   a. has less friction than a rolling tire.
   b. results in straighter stops.
   c. results in shorter stops.
   d. has more friction to the road than that of a rolling tire.

3. Coefficient of friction is measured in what units?
   a. a number from 0 to 1
   b. A, B, or C
   c. pounds per square inch (psi)
   d. percentage

4. What type of disc brake pad should *not* be used to replace semi-metallic original equipment pads?
   a. semi-mets
   b. NAO
   c. organic (asbestos)
   d. NAS

5. The front brakes of a front-wheel-drive vehicle handle about what percentage of the braking power?
   a. 50%
   b. 60%
   c. 70%
   d. 80%

6. The letters "EF" on the edge of a brake lining are
   a. wear resistance codes.
   b. relative noise level codes.
   c. coefficient of friction codes.
   d. brake lining recipes.

7. What type of brake fluid is recommended by almost every vehicle manufacturer?
   a. DOT 2
   b. DOT 3
   c. DOT 4
   d. DOT 5

8. An owner of a vehicle equipped with ABS brakes complained that whenever he tried to stop on icy or slippery roads, the brake pedal would pulse up and down rapidly. Technician A says that this is normal for many ABS units. Technician B says that the ABS unit is malfunctioning. Which technician is correct?
   a. A only
   b. B only
   c. Both A and B
   d. Neither A nor B

◄ Chapter 3 ►

MASTER CYLINDER AND HYDRAULIC SYSTEM DIAGNOSIS AND SERVICE

OBJECTIVES

After studying Chapter 3, the reader will be able to:

1. Describe the function, purpose, and operation of the master cylinder.
2. Explain how hydraulic force can be used to supply high pressures to each individual wheel brake.
3. Describe the process of troubleshooting master cylinders and related brake hydraulic components.
4. List the purposes and functions of the metering valve and proportional valve.
5. Discuss the methods that can be used to bleed the hydraulic braking system.

The master cylinder is the heart of the entire braking system. No braking occurs until the driver depresses the brake pedal. The brake pedal linkage is used to apply the force of the driver's foot into a closed hydraulic system.

HYDRAULIC BRAKE SYSTEM

All braking systems require that a driver's force be transmitted to the drum or rotor attached to each wheel (see Figure 3–1). The force that can be exerted on the brake pedal varies due to the strength and size of the driver. Engineers design braking systems to require less than 150 lb (68 N) of force from the driver, yet provide the force necessary to stop a heavy vehicle from high speed.

◄ TECH TIP ►

TOO MUCH IS BAD

Some vehicle owners or inexperienced service people may fill the master cylinder to the top. Master cylinders should only be filled to the "maximum" level line or about 1/4 in. (6 mm) from the top to allow room for expansion when the brake fluid gets hot during normal operation. If the master cylinder is filled to the top, the expanding brake fluid has no place to expand and the pressure increases. This increased pressure can cause the brakes to "self-apply," shortening brake friction material life and increasing fuel consumption. Overheated brakes can result and the brake fluid may boil, causing a total loss of braking.

FIGURE 3–1 Hydraulic brake lines transfer the brake effort to each brake assembly attached to all four wheels. (Courtesy of General Motors)

PASCAL'S LAW

The hydraulic principles that permit a brake system to function were discovered by a French physicist, Blaise Pascal (1632–1662). He discovered that "when pressure is applied to a liquid confined in a container or an enclosure, the force is transmitted equal and undiminished in every direction." To help understand this principle, assume that a force of 10 lb is exerted on a piston with a surface area of 1 square inch (sq. in.). Since this *force* measured in pounds is applied to a piston with an area measured in square inches, the *pressure* is the force times the area, or 10 pounds per square inch (psi). It is this pressure that is transmitted, without loss, throughout the entire hydraulic system (see Figure 3–2).

Pascal's law can be stated mathematically as follows:

$$F = P \times A \quad \text{or} \quad P = F/A \quad \text{or} \quad A = F/P$$

where F = force, lb [newtons (n)]

P = pressure, in lbs per sq. in. (N/cm^2)

A = area, in sq. in. (cm^2)

A practical example involves a master cylinder with a piston area of 1 sq. in., one wheel cylinder, also with an area of 1 sq. in., and one wheel cylinder with a piston area of two square inches (see Figure 3–3).

On a typical vehicle, a driver input force of 150 lb (667 N) is boosted both mechanically (through the brake pedal linkage) and by the power booster to a fluid pressure of about 1700 psi (11,700 kPa). During a typical brake application only about 1 teaspoon (5 mL or cc) of brake fluid is actually moved from the master cylinder and into the hydraulic system to cause the pressure buildup to occur. With a drum brake, the wheel cylinder

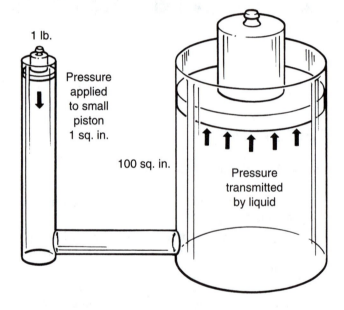

FIGURE 3–2 A 1-lb force exerted on a small piston in a sealed system transfers the pressure to each square inch throughout the system. In this example, the 1-lb force is able to lift a 100-lb weight because it is supported by a piston that is 100 times larger in area than the small piston.

expands and pushes the brake shoes against a brake drum. *The distance the shoes move is only about 0.005 to 0.012 in. (5 to 12 thousandths of an inch) (0.015 to 0.30 mm) (see Figure 3–4).*

With a disc brake, brake fluid pressure pushes on the piston in the caliper a small amount and causes a clamping of the disc brake pads against both sides of a rotor (disc) (see Figure 3–5). *The typical distance the pads move is only about 0.001 to 0.003 in. (1 to 3 thousandths of an inch) (0.025 to 0.076 mm).*

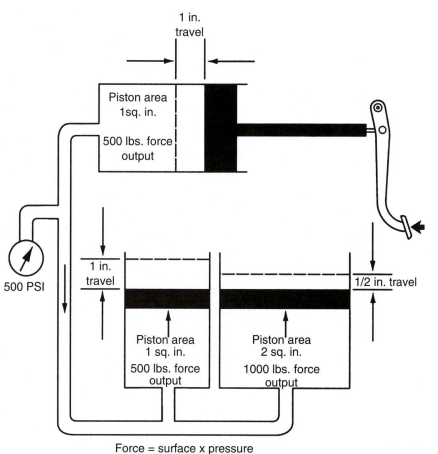

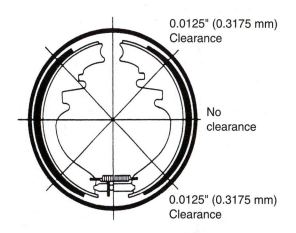Force = surface x pressure

FIGURE 3–3 The amount of force is the result of pressure multiplied by the surface area. The larger the surface area of the piston, the greater the force exerted. (Courtesy of General Motors)

MASTER CYLINDER RESERVOIRS

Most vehicles built since the early 1980s are equipped with see-through master cylinder reservoirs which permit owners and service technicians to check the brake fluid level without having to remove the top of the reservoir. Some countries have laws that require this type of reservoir (see Figure 3–6). The reservoir capacity is great enough to allow for the brakes to become completely worn out and still have enough reserve for safe operation.

MASTER CYLINDER OPERATION

The master cylinder is the heart of any hydraulic braking system. Brake pedal movement and force are transferred to the brake fluid and directed to wheel cylinders or calipers (see Figure 3–7). The master cylinder is also separated into two separate pressure building chambers (or circuits) to provide braking force to one-half of

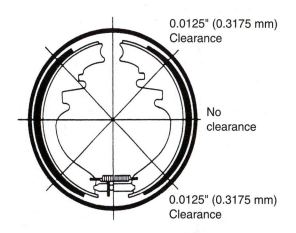

FIGURE 3–4 Drum brake illustrating the typical clearance between the brake shoes (friction material) and the rotating brake drum represented as the outermost black circle. (Courtesy of Cooper Industries)

the brake in the event of a leak or damage to one circuit (see Figures 3–8 and 3–9).

Both pressure-building sections of the master cylinder contain two holes from the reservoir. The Society of Automotive Engineers (SAE) term for the forward

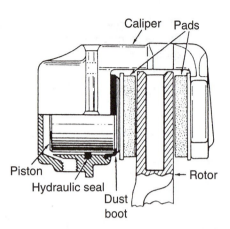

FIGURE 3–5 The brake pad (friction material) is pressed on both sides of the rotating rotor by the hydraulic pressure of the caliper. (Courtesy of EIS Brake Parts)

(tapered) hole is the **vent port** and the rearward straight-drilled hole is called the **replenishing port** (see Figure 3–10). Various vehicle and brake component manufacturers call these ports by various names. For example, the vent port is the high-pressure port. This tapered forward hole is also called the **compensating port** or **bypass** (a GM term). (See Figure 3–11 for an example of the terms used by General Motors). The replenishing port is the low-pressure rearward larger-diameter hole. The inlet port is also called the **bypass port**, **filler port**, **breather port**, or **compensating port** (a GM term).

HINT: Terms vary by vehicle and brake component manufacturer. Whatever name your boss calls it is the name you should use.

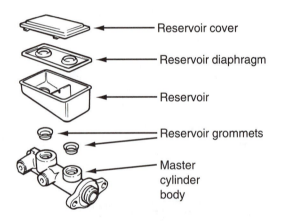

FIGURE 3–6 Typical master cylinder showing the reservoir and associated parts. The reservoir diaphragm lays directly on top of the brake fluid, which helps keep air from the surface of the brake fluid because brake fluid easily absorbs moisture from the air. (Courtesy of Allied Signal Automotive Aftermarket)

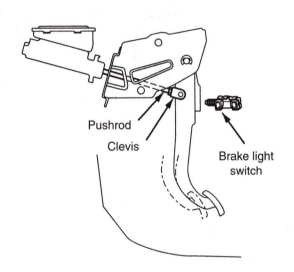

FIGURE 3–7 The typical brake pedal is supported by a mount and attached to the push rod by a U-shaped bracket. The pin used to retain the clevis to the brake pedal is usually called a *clevis pin.* (Courtesy of General Motors)

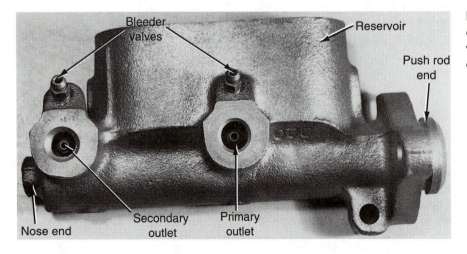

FIGURE 3–8 This cast iron master cylinder is equipped with bleeder valves for both the primary and secondary pressure building chambers.

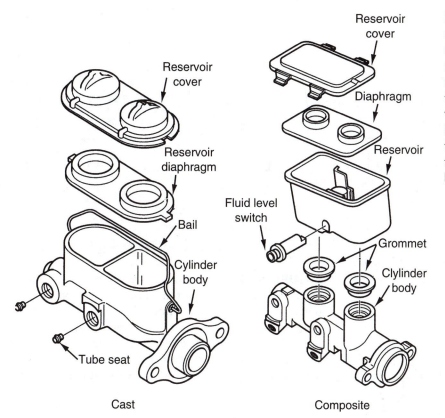

Cast

Composite

FIGURE 3–9 Note that the cast iron master cylinder (left) uses a wire reservoir cover lamp called a *bail*. The composite master cylinder is made from two different materials: aluminum for the body and plastic materials for the reservoir and reservoir cover. This type of reservoir feeds both primary and secondary chambers and therefore uses a fluid level switch that activates the red dash warning lamp if the brake fluid level drops.

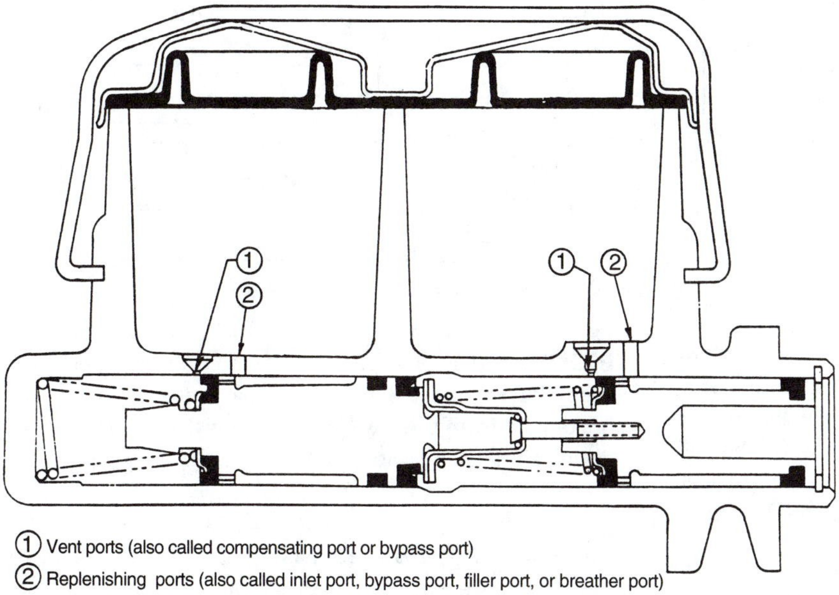

FIGURE 3–10 Note the various names for the vent port (front port) and the replenishing port (rear port). Names vary by vehicle and brake component manufacturers.

① Vent ports (also called compensating port or bypass port)

② Replenishing ports (also called inlet port, bypass port, filler port, or breather port)

The function of the master cylinder can be explained from the at-rest, applied, and released positions.

At-Rest Position. The primary sealing cups are between the compensating port hole and the inlet port hole. In this position the brake fluid is free to expand and move from the calipers, wheel cylinders, and brake lines up into the reservoir through the vent port (compensation port) if the temperature rises and the fluid expands. If the fluid was trapped, the pressure of the brake fluid would increase with temperature, causing the brakes to **self-apply** (see Figure 3–12).

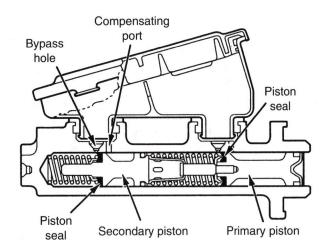

FIGURE 3–11 Typical General Motors (Delco Chassis) master cylinder showing the terms used by General Motors. (Courtesy of General Motors)

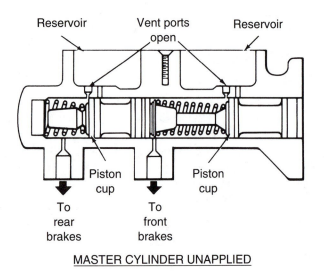

MASTER CYLINDER UNAPPLIED

FIGURE 3–12 The vent ports must remain open to allow brake fluid to expand when heated by the friction material and transferred to the caliper and/or wheel cylinder. As the brake fluid increases in temperature, it expands. The heated brake fluid can expand and flow back into the reservoir through vent ports. (Courtesy of Chrysler Corporation)

Applied Position. When the brake pedal is depressed, the pedal linkage forces the push rod and primary piston down the bore of the master cylinder (see Figure 3–13). As the piston moves forward, the primary sealing cup covers and blocks off the vent port. Hydraulic pressure builds in front of the primary seal as the push rod moves forward. The back of the piston is kept filled through the replenishing port (see Figure 3–14).

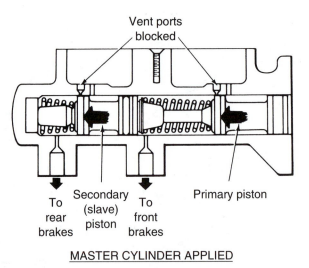

MASTER CYLINDER APPLIED

FIGURE 3–13 As the brake pedal is depressed, the push rod moves the primary piston forward, closing off the vent port. As soon as the port is blocked, pressure builds in front of the primary sealing cup, which pushes on the secondary piston. The secondary piston also moves forward, blocking the secondary vent port, and builds pressure in front of the sealing cup. (Courtesy of Chrysler Corporation)

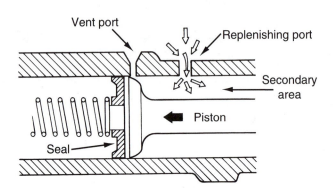

FIGURE 3–14 The purpose of the replenishing port is to keep the volume behind the primary position filled with brake fluid from the reservoir as the piston moves forward during a brake application.

Released Position. Releasing the brake pedal removes the pressure on the push rod and master cylinder pistons. A spring on the brake pedal linkage returns the brake pedal to its normal at-rest (up) position. The spring in front of the master cylinder piston expands, pushing the pistons rearward. At the same time, pressure is released from the entire braking system and the released brake fluid pressure is exerted on the master cylinder pistons, forcing them rearward. As the piston is pushed back, the lips of the seal fold forward, allowing fluid to move quickly past the piston, as shown in Figure

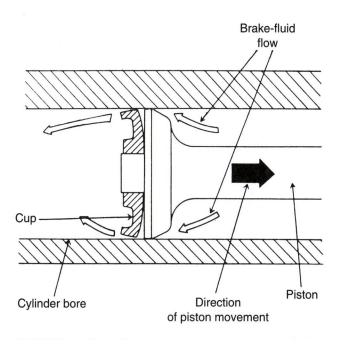

Brake-fluid
flow

Cup

Cylinder bore

Direction
of piston movement

Piston

FIGURE 3–15 When the brake pedal is released, the master cylinder piston moves rearward. Some of the brake fluid is pushed back up through the replenishing port, but most of the fluid flows past the sealing cup. Therefore, when the driver pumps the brake pedal, the additional fluid in front of the pressure-building sealing cup is available quickly.

3–15. Some pistons have small holes that allow the fluid to move more quickly. Once the primary seal passes the vent port, the remaining hydraulic pressure forces any excess fluid into the reservoir.

RESIDUAL CHECK VALVE

A residual check valve has been used on some drum brake systems to keep a slight amount of pressure on the entire hydraulic system for drum brakes (5 to 12 psi). This residual check valve is located in the master cylinder at the outlet for the drum brakes.

DUAL-SPLIT MASTER CYLINDER

Dual-split master cylinders use two separate pressure-building sections. One section operates the front brakes and the other section operates the rear brakes on vehicles equipped with a front/rear split system (see Figure 3–16). If there is a failure of the front-section hydraulic system, the primary piston (push rod end) operates normally and exerts pressure on the secondary piston. The secondary piston, however, will not be able to build pressure because of the leak in the system.

◀ **TECH TIP** ▶

ALWAYS CHECK FOR VENTING

Whenever diagnosing any braking problem, start the diagnosis at the master cylinder—the heart of any braking system. Remove the reservoir cover and observe the brake fluid for spurting while an assistant depresses the brake pedal.

1. *Normal operation (movement of fluid observed in the reservoir).* There should be a squirt or movement of brake fluid out of the vent port of both the primary and secondary chambers. This indicates that the vent port is open and that the sealing cup is capable of moving fluid upward through the port before the cup seals off the port as it moves forward to pressurize the fluid.
2. *No movement of fluid observed in the reservoir in the primary piston.* This indicates that brake fluid is not being moved as the brake pedal is depressed. This can be caused by:
   a. *Incorrect brake pedal height.* Brake pedal or push rod adjustment could be allowing the primary piston to be too far forward, causing the seal cup to be forward of the vent port. Adjust the brake pedal height to a higher level and check if the push rod is too long.
   b. *A defective or swollen rubber sealing cup on the primary piston.* This could cause the cup itself to block the vent port.

NOTE: If the vent port is blocked for any reason, the brakes of the vehicle may *self-apply* when the brake fluid heats up during normal braking. Since the vent port is blocked, the expanded hotter brake fluid has no place to expand and instead, increases the pressure in the brake lines. The increase in pressure causes the brakes to apply. Loosen the bleeder valves and release the builtup pressure to check if this is what is happening. Then check the master cylinder to see if it is "venting."

DIAGONAL-SPLIT MASTER CYLINDERS

With front-wheel-drive vehicles, the weight of the entire power train is on the front wheels, and 80 to 90 percent of the braking force is achieved by the front brakes.

This means that only 10 to 20 percent of the braking force is being handled by the rear brakes. If the front brakes fail, the rear brakes alone would not provide adequate braking force. The solution was the diagonal-split system (see Figures 3–17 and 3–18).

In a **diagonal-split braking system,** the left front brake and the right rear brake are on one circuit and the right front brake with the left rear brake is another circuit of the master cylinder. If one circuit fails, the remaining circuit can still stop the vehicle in a reasonable fashion because each circuit has one front brake. To prevent this one front brake from causing the vehicle to pull toward one side during braking, the front suspension is designed with negative scrub radius geometry. This effectively eliminates any handling problem in the event of a brake circuit failure.

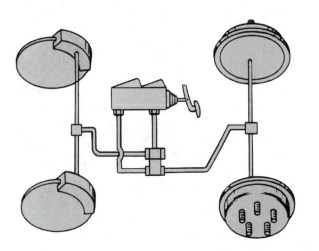

FIGURE 3–16 Rear-wheel-drive vehicles use a dual-split master cylinder, where one portion of the master cylinder applies the front brakes and the other section applies the rear brakes. The front brakes are not always activated by the front section of the master cylinder. The rear section of this master cylinder is used for the front brake. (Courtesy of Ford Motor Company)

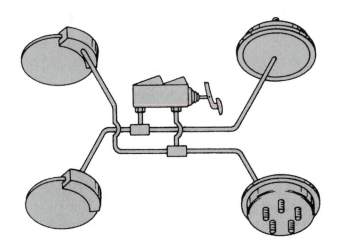

FIGURE 3–17 Front-wheel-drive vehicles use a diagonal-split master cylinder. In this design, one section of the master cylinder operates the right front and left rear brake, and the other section operates the left front and right rear. In the event of a failure in one section, at least one front brake will still function. (Courtesy of Ford Motor Company)

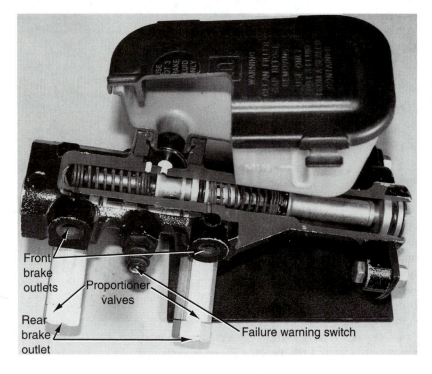

FIGURE 3–18 Typical General Motors diagonal-split master cylinder. Notice the two aluminum proportioner valves. These valves limit and control brake fluid pressure to the rear brakes to help eliminate rear wheel lockup during a rapid stop.

QUICK-TAKEUP MASTER CYLINDERS

Many newer vehicles use low-drag disc brake calipers to increase fuel economy. However, due to the larger distance between the rotor and the friction pads, excessive brake pedal travel would be required before the pads touched the rotor. The solution to this problem is a master cylinder design that can take up this extra clearance. The design of a **quick-takeup master cylinder** includes a larger-diameter primary piston (low-pressure chamber) and a quick-takeup valve. This type of master cylinder is also called a dual-diameter-bore, step-bore or fast-fill master cylinder.

A spring-loaded check ball valve holds pressure on the brake fluid in the large-diameter rear chamber of the primary piston. When the brakes are first applied, the movement of the larger, rear piston forces this larger volume of brake fluid forward past the primary piston seal and into the primary high-pressure chamber. This extra volume of brake fluid "takes up" the extra clearance of the front disc brake calipers without increasing the brake pedal travel distance (see Figure 3–19).

At 70 to 100 psi, the check ball valve in the quick-takeup valve allows fluid to return to the brake fluid reservoir.

◀ TECH TIP ▶

USING THE MASTER CYLINDER TO SHUT OFF BRAKE FLUID

The normal purpose and function of the master cylinder is to move brake fluid to the brake calipers and/or wheel cylinders and to transfer the braking force from the driver's foot to the wheel friction surfaces. The master cylinder can also be used to *block* the flow of brake fluid. Whenever any hydraulic brake component is removed, brake fluid tends to leak out because the master cylinder is usually higher than most other hydraulic components such as wheel cylinders and calipers (see Figure 3–20).

To prevent brake fluid loss, which can easily empty the master cylinder reservoir, simply *depress* the brake pedal slightly or prop up a stick or other pedal depressor to keep the brake pedal down. When the brake pedal is depressed, the piston sealing cups move forward, blocking off the reservoir from the rest of the braking system. The master cylinder stays full and the brake fluid stops dripping out of brake lines that have been disconnected.

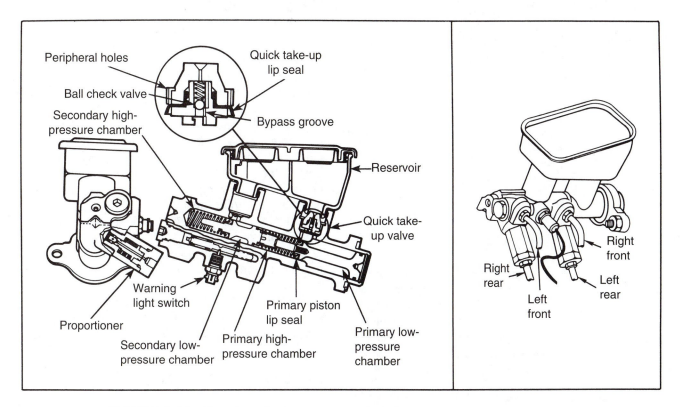

FIGURE 3–19 Typical quick-takeup master cylinder showing the ball check valve and spring. (Courtesy of Allied Signal Automotive Aftermarket)

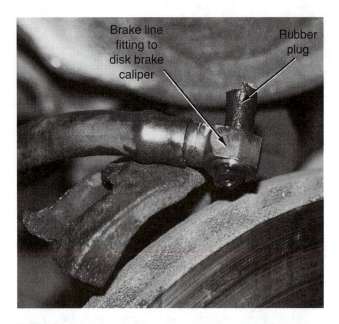

FIGURE 3–20 A rubber plug being used to block brake fluid from leaking out of a flexible brake line fitting.

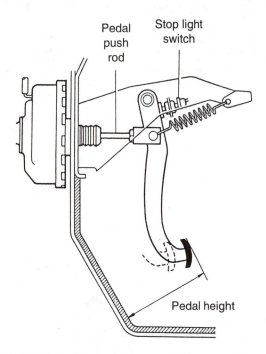

FIGURE 3–21 Pedal height is usually measured from the floor to the top of the brake pedal. Some vehicle manufacturers recommend removing the carpet and measure from the asphalt matting on the floor for an accurate measurement. Always follow the manufacturer's recommended procedures and measurements. (Courtesy of Toyota Motor Sales, U.S.A., Inc.)

Because the quick-takeup "works" until 100 psi is reached, a metering valve is not required to hold back the fluid pressure to the front brakes.

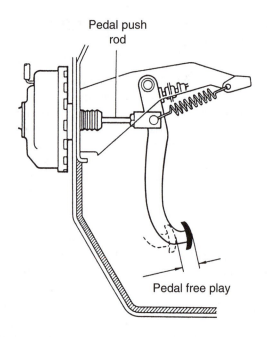

FIGURE 3–22 Brake pedal free-play is the distance between the brake pedal fully released and the position of the brake pedal when braking resistance is felt. (Courtesy of Toyota Motor Sales, U.S.A., Inc.)

DIAGNOSING AND TROUBLESHOOTING MASTER CYLINDERS

After a thorough visual inspection, check for proper operation of **pedal height, pedal free play,** and **pedal reserve distance** (see Figures 3–21 through 3–23). Proper brake pedal height is important for proper operation of the stop (brake) light switch. If the pedal is not correct, the push rod may be too far forward, causing the master cylinder cups from uncovering the vent port. If the pedal is too high, the free play will be excessive. Pedal reserve height is easily checked by depressing the brake pedal with the right foot and attempting to slide your left foot under the brake pedal (see Figure 3–22).

Free play is the distance the brake pedal travels before the primary piston in the master cylinder moves. *Most vehicles require brake pedal free play of between 1/8 and 1½ in. (3 to 38 mm).* Too little or too much free play can cause braking problems that can mistakenly be attributed to a defective master cylinder.

MASTER CYLINDER SERVICE

Many master cylinders can be disassembled, cleaned, and restored to service.

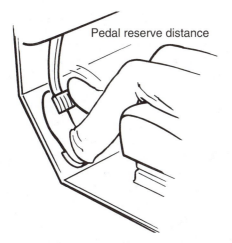

FIGURE 3–23 Brake pedal reserve is usually specified as the measurement from the floor to the top of the brake pedal with the brakes applied. A quick and easy test of pedal reserve is to try to place your left toe underneath the brake pedal while the brake pedal is depressed with your right foot. If your toe will not fit, the pedal reserve *may* not be enough.

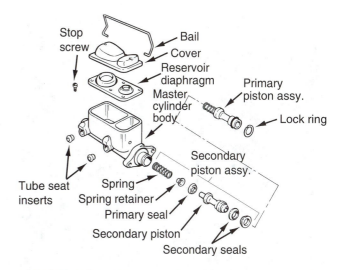

FIGURE 3–24 Whenever disassembling a master cylinder, note the exact order of parts as they are removed. Master cylinder overhaul kits (when available) often include entire piston assemblies rather than the individual seals. (Courtesy of Allied Signal Automotive Aftermarket)

NOTE: Check the vehicle manufacturer's recommendation before attempting to overhaul or service a master cylinder. Many manufacturers recommend replacing the master cylinder as an assembly.

CAUTION: If holding the master cylinder in a vise, use the flange area; never clamp the body of the master cylinder.

To disassemble a master cylinder, remove the snap ring and slowly release the pressure on the depressing tool. Spring pressure should push the primary piston out of the cylinder bore (see Figure 3–24). Remove the master cylinder from the vise and tap the open end of the bore against the top of a workbench to force the secondary piston out of the bore. If necessary, use compressed air into the outlet to force the piston out after removing the retaining bolt for the secondary piston, if present.

CAUTION: Use *extreme care* when using compressed air. The piston can be shot out of the master cylinder with a great force.

Inspect the master cylinder bore for pitting, corrosion, or wear. Most cast iron master cylinders cannot be honed because of the special bearingized surface finish

FIGURE 3–25 Photo of a nylon brush used to clean the bore of aluminum master cylinders by a national remanufacturer. The soft nylon will not harm the anodized surface coating.

that is applied to the bore during manufacturing. Slight corrosion or surface flaws can usually be removed with a hone or crocus cloth; otherwise, the master cylinder should be replaced as an assembly. Aluminum master cylinders cannot be honed. Aluminum master cylinders have an **anodized** surface coating applied that is hard and wear resistant. Honing would remove this protective coating (see Figure 3–25).

Thoroughly clean the master cylinder and any other parts to be reused (except for rubber components) in clean denatured alcohol. If the bore is okay, replacement **piston assemblies** can be installed into the master cylinder after dipping them into clean brake fluid.

FIGURE 3–26   Red brake warning lamp.

NOTE: While most master cylinder overhaul kits include the entire piston assemblies, some kits just contain the sealing cups and/or 0-rings. Always follow the installation instructions that accompany the kit and always use the installation tool that is included, to prevent damage to the replacement seals.

PRESSURE-DIFFERENTIAL SWITCH (BRAKE WARNING SWITCH)

A pressure-differential switch is used on all vehicles after 1967 with dual master cylinders to warn the driver of a loss of pressure in one of the two separate systems by lighting the dashboard red brake warning indicator lamp (see Figures 3–26). The brake lines from both the front and rear sections of the master cylinder are sent to this switch, which lights the brake warning indicator lamp in the event of a "difference in pressure" between the two sections (see Figure 3–27).

BRAKE FLUID LEVEL SENSOR SWITCH

Many master cylinders, especially systems that are a diagonal split, usually use a brake fluid level sensor switch in the master cylinder reservoir. This sensor will

CHECK FOR BYPASSING

If a master cylinder is leaking internally, brake fluid can be pumped from the rear chamber into the front chamber of the master cylinder. This internal leakage is called **bypassing**. When the fluid bypasses, the front chamber can overflow while emptying the rear chamber. Therefore, whenever checking the level of brake fluid, do not think that a low rear reservoir is always due to an external leak. Also, a master cylinder that is bypassing (leaking internally) will usually cause a lower-than-normal brake pedal.

---

light the red brake warning lamp on the dash if low brake fluid level is detected. A float-type sensor or a magnetic reed switch is commonly used and provides a complete electrical circuit when the brake fluid level is low.

DIAGNOSING A RED BRAKE DASH WARNING LAMP

The red brake dash warning lamp can be on for any one of several reasons, including:

1. **Parking brake on.** The same dash warning lamp is used to warn the driver that the parking brake is on.
2. **Low brake fluid.** This lights the red dash warning lamp on vehicles equipped with a master cylinder reservoir brake fluid level switch.
3. **Unequal brake pressure.** The pressure differential switch is used on most vehicles with a front/rear brake split system to warn the driver whenever there is low brake pressure to either the front or rear brakes.

NOTE: Brake systems use *either* a pressure differential switch *or* a low brake fluid switch to light the red brake dash lamp, but not both.

The most likely cause of the red brake warning lamp being on is low brake fluid, caused by a leaking brake line, wheel cylinder, or caliper. Therefore, the first step in diagnosis is to determine the cause of the lamp being on, then to repair the problem.

BRAKE MALFUNCTION

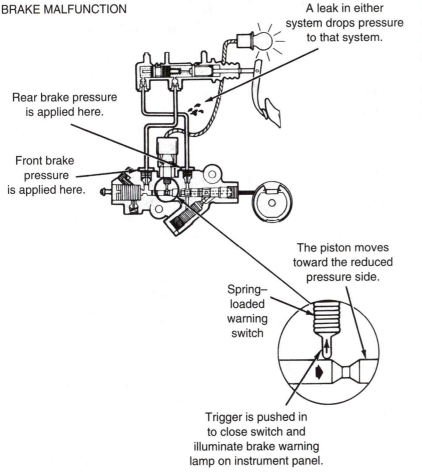

A leak in either system drops pressure to that system.

Rear brake pressure is applied here.

Front brake pressure is applied here.

The piston moves toward the reduced pressure side.

Spring–loaded warning switch

Trigger is pushed in to close switch and illuminate brake warning lamp on instrument panel.

FIGURE 3–27 A leak in the hydraulic system causes unequal pressures between the two different brake circuits. This difference in pressures causes the plunger inside the pressure differential switch to move, which completes the electrical circuit for the red brake warning lamp. (Courtesy of Ford Motor Company)

◄ **TECH TIP** ►

MASTER CYLINDER ONE-DRIP-PER-SECOND TEST

Excessive brake wear is often caused by brake cylinder linkage or brake light switches keeping the brake pedal from fully releasing. If the brake pedal is not fully released, the primary piston sealing cup blocks the vent port from the brake fluid reservoir. To test if this is the problem, loosen both lines from the master cylinder. Brake fluid should drip out of both lines about one drip per second. This is why this test is also called the *master cylinder drip test*. If the master cylinder does not drip, the brake pedal may not be allowing the master cylinder to release fully. Have an assistant pull up on the brake pedal. If dripping starts, the problem is due to a misadjusted brake light or speed (cruise) control switch or pedal stop. If the master cylinder still does not drip, loosen the master cylinder from the power booster. If the master cylinder now starts to drip, the push rod adjustment is too long.

If the master cylinder still does not drip, the problem is in the master cylinder itself. Check for brake fluid contamination. If mineral oil such as engine oil, power steering fluid, or automatic transmission fluid (ATF) has been used in the system, the rubber sealing cups swell and can block off the vent port. If contamination is discovered, *every* brake component that contains rubber *must* be replaced.

PROPORTIONING VALVE OPERATION

A proportioning valve may or may not be used on all types of braking systems. When used, it prevents rear wheel lockup by limiting the amount of pressure sent to the rear wheels. A proportioning valve reduces the pressure increase to the rear brakes. A proportioning valve is called a **pressure control valve** by some vehicle manufacturers.

A proportioning valve permits full master cylinder pressure to be sent to the rear brakes up to a certain point, called the **split point** or **changeover pressure**. Above the split point, the pressure is reduced to a certain ratio, called the **slope**, of the front brake pressure. The split point is usually 200 to 300 psi, and the ratio (slope) could vary from 0.25 to 0.50 (see Figure 3–28). With light brake pedal applications, approximately the same brake pressure is sent to both front and rear brakes. However, when higher pressures are required to stop a vehicle, the pressure is controlled by this proportioning valve to limit the pressure sent to the rear brakes to less than one half of the pressure being applied to the front brakes beyond the split point. This prevents rear wheel lockup due to weight transfer forward reducing the weight on the rear tires, plus allowing for the self-energizing forces of some rear drum brakes. In the event of a front brake system failure, a bypass opens, allowing full rear brake hydraulic pressure (see Figure 3–29).

DIAGONAL-SPLIT PROPORTIONING VALVES

Vehicles that are diagonal split use a proportioning valve for each rear brake. This can be two separate units, as shown in Figure 3–30, or one component part.

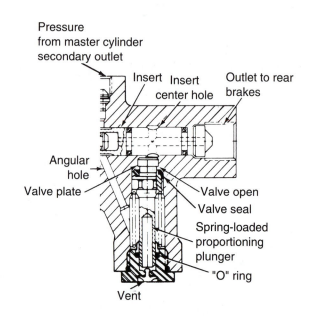

FIGURE 3–29   Cutaway view of a proportioning (proportioner) valve. (Courtesy of Chrysler Corporation)

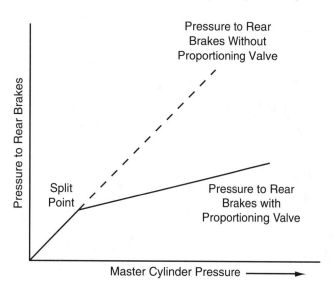

FIGURE 3–28   This graph shows a typical proportioner valve pressure relationship. Note that at low pressures, the pressure is the same to the rear brakes as is applied to the front brakes. After the split point, only a percentage (called the *slope*) of the master cylinder pressure is applied to the rear brakes.

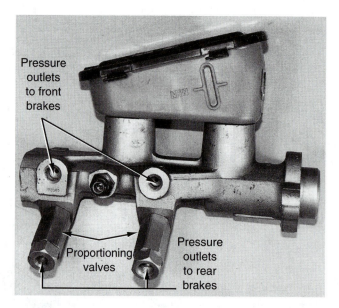

FIGURE 3–30   This diagonal split braking system used on a front-wheel-drive vehicle uses two proportioning valves, one for each rear wheel.

◄ TECH TIP ►

ALWAYS INSPECT BOTH FRONT AND REAR BRAKES

If the rear brakes of a vehicle tend to lock up during a stop, many technicians may try to repair the problem by replacing the proportioning valve or servicing the rear brakes. Proportioning valves are simple spring-loaded devices that are usually trouble-free. If the rear brakes lock up during braking, inspect the rear brakes carefully, looking for contaminated linings or other problems that can cause the rear brakes to grab. Do not stop there: Always inspect the front brakes, too. If the front brakes are rusted or corroded, they cannot operate efficiently and greater force must be exerted by the driver to stop the vehicle. Even if the proportioning valve is functioning correctly, the higher brake pedal pressure by the driver could easily cause the rear brakes to lock up.

A locked wheel has less traction with the road than does a rotating wheel. As a result, if the rear wheels become locked, the rear of the vehicle often "comes around" or "fishtails," which may cause the vehicle to skid. Careful inspection of the *entire* braking system is required to be assured of a safe vehicle.

HEIGHT-SENSING PROPORTIONING VALVES

Many vehicles use a proportioning valve that varies the amount of pressure that can be sent to the rear brakes depending on the height of the rear suspension. A lightly loaded vehicle required less braking force to stop than a heavily loaded vehicle.

When the vehicle is loaded, the rear suspension is forced downward. The lever on the proportioning valve moves and allows greater pressure to be sent to the rear brakes (see Figure 3–31). This greater pressure allows the rear brakes to achieve more braking force, helping to slow a heavier vehicle. When a vehicle is heavily loaded in the rear, the chances of rear wheel lockup are reduced.

CAUTION: Some vehicle manufacturers warn that service technicians should never install replacement air lift shock absorbers or springs that may result in a vehicle height that differs from that specified by the vehicle manufacturer.

METERING VALVE (HOLD-OFF) OPERATION

The metering valve is used on all front disc, rear drum brake–equipped vehicles. The metering valve prevents full operation of (holds off) the disc brakes until between 75 and 125 psi is sent to the rear drum brakes to overcome rear brake return spring pressure. This allows the front and rear brakes to apply at the same time, for even stopping. Most metering valves also allow for the pressure to the front brakes to be blended up gradually to the metering valve pressure to prevent front brake locking under light pedal pressures on icy surfaces. The metering valve remains open at pressures below 3 to 30 psi (20 to 200 kPa) to allow the pressure to be equalized when the brakes are not applied (see Figure 3–32).

NOTE: Braking systems that are diagonal split, such as those found on most front-wheel-drive vehicles, do *not* use a metering valve. A metering valve is used only on front/rear split braking systems such as those found on most rear-wheel-drive vehicles.

BRAKE LINES

High-pressure double-walled steel brake lines or high-strength flexible lines are used to connect the master cylinder to each wheel. The steel brake lines are also called **brake pipes** or **brake tubing** and are coated to help prevent corrosion. The steel brake lines leaving the master cylinder are usually coiled to allow for movement between the master cylinder and the mounting of

◄ TECH TIP ►

NO VALVES CAN CAUSE A PULL

When diagnosing a pull to one side during braking, some technicians tend to blame the metering valve, proportional valve, the pressure differential switch, or the master cylinder itself. Remember that if a vehicle pulls during braking, the problem *has* to be due to an individual wheel brake or brake line. The master cylinder and all the valves control front or rear brakes together or diagonal brakes and cannot cause a pull if not functioning correctly.

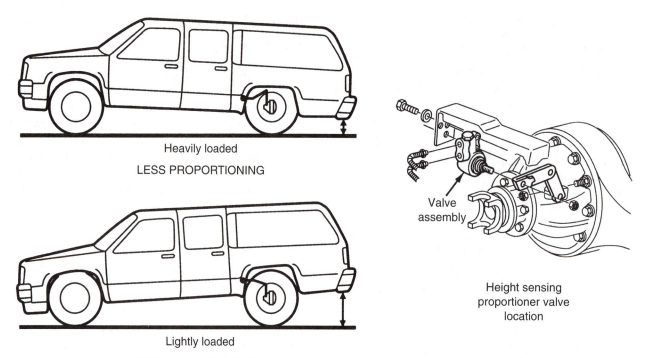

Heavily loaded
LESS PROPORTIONING

Lightly loaded

MORE PROPORTIONING

Valve
assembly

Height sensing
proportioner valve
location

FIGURE 3–31 A height-sensing proportioning valve allows a higher pressure to be applied to the rear brakes when the vehicle is heavily loaded and less pressure when the vehicle is lightly loaded. (Courtesy of General Motors)

METERING (HOLD-OFF)

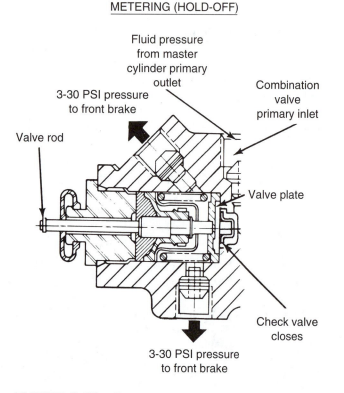

Fluid pressure
from master
cylinder primary
outlet

3-30 PSI pressure
to front brake

Combination
valve
primary inlet

Valve rod

Valve plate

Check valve
closes

3-30 PSI pressure
to front brake

FIGURE 3–32 Cutaway view of a metering (hold-off) valve. (Courtesy of Chrysler Corporation)

◀ **TECH TIP** ▶

PUSH-IN OR PULL-OUT METERING VALVE?

Whenever bleeding the air out of the hydraulic brake system, the metering valve should be bypassed. The metering valve stops the passage of brake fluid to the front wheels until pressure exceeds about 125 psi (860 kPa). It is important to not push the brake pedal down with a great force so as to keep from dispersing any trapped air into small and hard to bleed bubbles. To bypass the metering valve, the service technician has to push or pull a small button located on the metering valve. An easy way to remember whether to push in or to pull out is to inspect the button itself. *If the button is rubber coated, push in. If the button is made of steel, pull out.* See Figure 3–33 which shows a combination valve that includes a metering valve.

Special tools allow the metering valve to be held in the bypass position. Failure to remove the tool after bleeding the brakes can result in premature application of the front brakes before the rear drum brakes have enough pressure to operate.

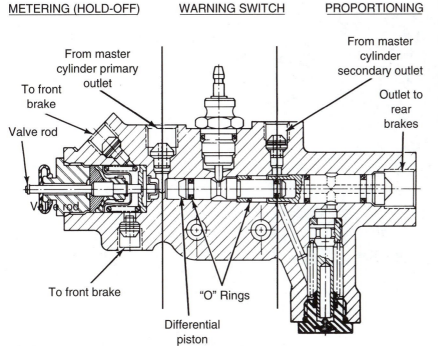

METERING (HOLD-OFF) WARNING SWITCH PROPORTIONING

From master cylinder primary outlet

To front brake

Valve rod

Valve rod

To front brake

"O" Rings

Differential piston

From master cylinder secondary outlet

Outlet to rear brakes

FIGURE 3–33 Combination valve containing metering, pressure differential (warning switch), and proportioning valves all in one unit. This style is often called a "pistol grip" design because the proportioning valve section resembles the grip section of a hand gun. (Courtesy of Chrysler Corporation)

the brake line to the frame, which could cause fatigue and brake line failure.

CAUTION: Copper tubing should *never* be used for brake lines. Copper tends to burst at a lower pressure than steel.

The ends of all steel brake lines should have a **double lap flare** or **ISO flare** at each end to ensure that the connection will have the necessary strength. **ISO** stands for **International Standards Organization**. ISO flare may also be called a **ball flare** or **bubble flare** (see Figure 3–34). Whenever replacing steel brake line, new steel tubing can be used and a double lap flare or an ISO flare completed at each end using a special flaring tool. Brake line can also be purchased in selected lengths already correctly flared. They are available in different diameters, the most commonly used being 3/16 in. (4.8 mm), 1/4 in. (6.4 mm), and 5/16 in. (7.9 mm) outside diameter (OD).

CAUTION: According to vehicle manufacturers' recommended procedures, compression fittings should never be used to join two pieces of steel brake line. Only use double-flared ends and connections, if necessary, when replacing damaged steel brake lines (see Figures 3–35 and 3–36).

See Figures 3–37 and 3–38 for brake line flaring procedures.

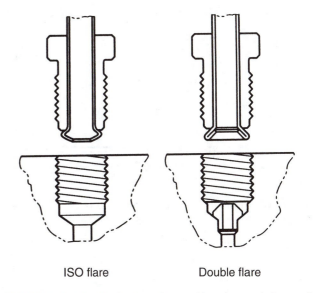

ISO flare Double flare

FIGURE 3–34 Brake line flares. (Courtesy of General Motors)

ARMORED BRAKE LINE

In many areas of the brake system, the steel brake line is covered with a wire coil wrap, as shown in Figure 3–39. This armor is designed to prevent damage from stones and other debris that could dent or damage the brake line. If a section of armored brake line is to be replaced, armored replacement line should be installed. Braided flexible brake line is also used on some vehicles

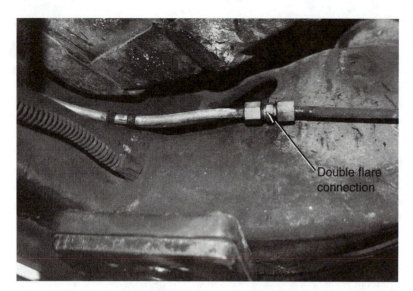

FIGURE 3–35 Note that the replacement line was reinstalled into the factory clips to keep it in the same location as the original factory brake line.

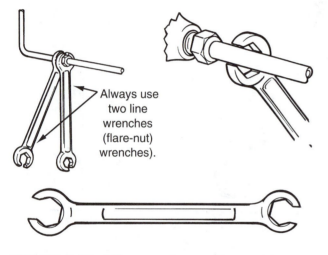

FIGURE 3–36 Whenever disconnecting or tightening a brake line, always use the correct-size flare nut wrench. A flare nut wrench is also called a *tube nut wrench* or a *line wrench*.

to allow flexibility and protection against stone damage, as shown in Figure 3–40.

FLEXIBLE BRAKE HOSE

Flexible brake hoses are used on each front wheel to allow for steering and suspension movement and at the rear to allow for rear suspension travel (see Figure 3–42). These rubber high-strength hoses can crack, blister, or leak and should be inspected at least every six months. Typical flexible brake hose is constructed of rubber hose surrounded by layers of woven fabric, as

◄ TECH TIP ►

BEND IT RIGHT THE FIRST TIME

Replacing rusted or damaged brake line can be a difficult job. It is important that the replacement brake line be located in the same location as the original to prevent possible damage from road debris or heat from the exhaust. Often, this means bending the brake line with many angles and turns. To make the job a lot easier, use a stiff length of wire and bend the wire into the exact shape necessary. Then use the wire as a pattern to bend the brake line. Always use a tubing bender to avoid kinking the brake line. A kink not only restricts the flow of brake fluid but also weakens the line. To bend brake line without a tubing bender tool, use an old V-belt pulley. Clamp the pulley in a vise and lay the tubing in the groove and bend the tubing smoothly. Different-diameter pulleys will create various radius bends (see Figure 3–41).

NOTE: Always use a tubing cutter instead of a hacksaw when cutting brake line. A hacksaw will leave a rough and uneven end that will not flare properly.

---

shown in Figure 3–43. An outside jacket is made from rubber and protects the reinforcement fabric from moisture and abrasion. The outside covering is also

(a)

(b)

(c)

(d)

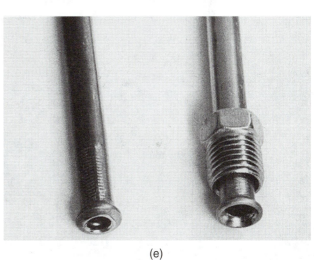

(e)

FIGURE 3–37 Double flaring the end of a brake line. (a) Clamp the line at the correct height above the surface of the clamping tool using the shoulder of the insert as a gauge. (b) The insert is pressed into the end of the tubing. This creates the first bend. (c) Remove the insert and use the pointed tool to complete the overlap double flare. (d) The completed operation as it appears while still in the clamp. (e) The end of the line as it appears after the first operation on the left and the completed double flare on the right.

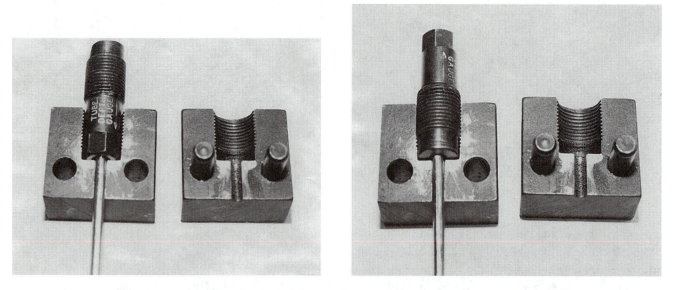

FIGURE 3–38 Making an ISO flare requires this special tool. Position the brake line into the two-part tool at the correct height using the gauge end of the tool. Assemble the two blocks of the tool together and clamp in a vise. Turn the tool around and thread it into the tool block. The end of the threaded part of the tool forms the "bubble" or ISO flare.

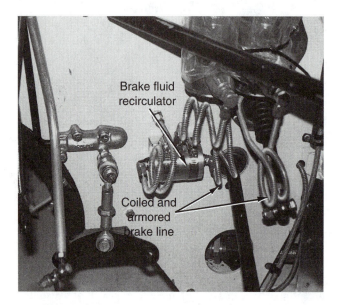

FIGURE 3–39 Brake lines on a race car showing the coiled brake line to allow for movement (vibration) without breaking and armored to help protect the lines from damage. The brake fluid recirculator is used in racing to circulate brake fluid from the calipers back to the master cylinder, then back to the calipers. The system uses the bleeder valve openings in the caliper to circulate brake fluid to keep it cooler, to help prevent it from boiling. Boiling brake fluid causes bubbles in the hydraulic system and a total loss of braking.

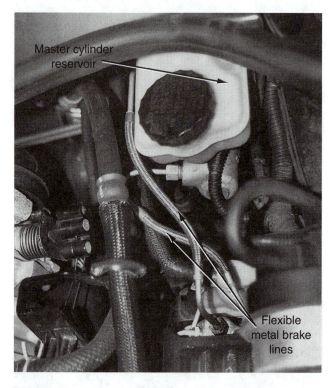

FIGURE 3–40 Some vehicles such as this front-wheel-drive Chrysler vehicle use flexible braided metal brake line from the master cylinder.

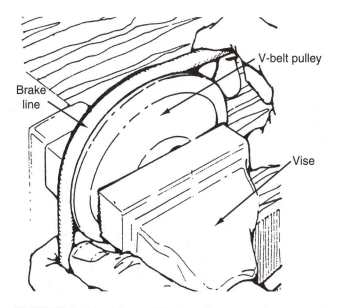

FIGURE 3–41   Using a V-belt pulley in a vise to bend brake line.

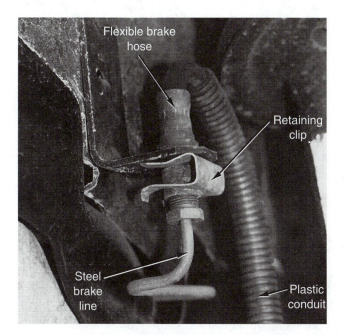

FIGURE 3–42   Flexible brake hoses are used between the flame or body of the vehicle and the wheel brakes. Because of suspension and/or steering movement, these flexible brake lines must be strong enough to handle high brake fluid pressures, yet remain flexible. Note that this flexible brake hose is further protected against road debris with a plastic conduit covering.

ribbed so that a technician can see if the hose is twisted. It is not unusual for flexible brake lines to become turned around and twisted when the disc brake caliper is removed and then replaced during a brake pad change (see Figure 3–44).

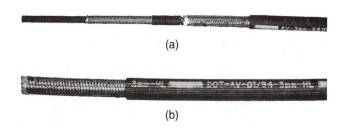

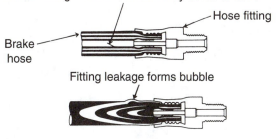

FIGURE 3–43   (a) Typical flexible brake hose showing the multiple layers of rubber and fabric. (b) The inside diameter (ID) is printed on the hose (3 mm) and the date it was manufactured (01-94).

FIGURE 3–44   Typical flexible brake hose faults. Many faults cannot be seen, yet can cause the brakes to remain applied after the brake pedal is released.

◀   **TECH TIP**   ▶

DON'T FILL THE MASTER CYLINDER WITHOUT SEEING ME!

The boss explained to the beginning technician that there are two reasons why the customer should be told not to fill the master cylinder reservoir when the brake fluid is down to the "minimum" mark, as shown in Figure 3–45.

1. If the master cylinder reservoir is low, there may be a leak that should be repaired.
2. As the brakes wear, the disc brake piston moves outward to maintain the same distance between friction materials and the rotor. Therefore, as the disc brake pads wear, the brake fluid level goes down to compensate.

If the customer notices that the brake fluid is low in the master cylinder reservoir, the vehicle should be serviced—either for new brakes or to repair a leak.

◀ TECH TIP ▶

BLEEDER VALVE LOOSENING TIPS

Attempting to loosen a bleeder valve often results in breaking (shearing off) of the bleeder valve. Several of these service procedures can be tried that help prevent the *possibility* of breaking a bleeder valve. Bleeder valves are tapered and become wedged in the caliper on the wheel cylinder housing (see Figures 3–46 and 3–47). All of these methods use shock to "break the taper" and to loosen the stuck valve.

- **Air-impact method.** Use a six-point socket for the bleeder valve and use the necessary adapters to fit an air-impact wrench to the socket. Apply some penetrating oil to the bleeder valve and allow to flow around the threads. Turn the pressure down on the impact wrench to limit the force. The hammering effect of the impact wrench loosens the bleeder valve without breaking it off.

- **Hit-and-tap method.**

  Step 1. Tap on the end of the bleeder valve with a steel hammer. This shock often "breaks the taper" at the base of the bleeder valve. The shock also breaks loose any rust or corrosion on the threads.

  Step 2. Using a six-point wrench or socket, *tap* the bleeder valve in the clockwise direction (tighten).

  Step 3. Using the same six-point socket or wrench, *tap* the bleeder valve counterclockwise to loosen and remove the bleeder valve.

NOTE: It is the *shock* of the tap on the wrench that breaks loose the bleeder valve. Simply pulling on the wrench often results in breaking off the bleeder.

If the valve is still stuck (frozen), repeat steps 1 through 3.

- **Air Punch Method** Use an air punch near the bleeder valve while attempting to loosen the bleeder valve at the same time (see Figure 3–48). The air punch creates a shock motion that often loosens the taper and threads of the bleeder valve from the caliper or wheel cylinder. It is also helpful, first, to attempt to turn the bleeder valve in the clockwise (tightening) direction, then turn the bleeder in counterclockwise direction to loosen and remove the bleeder valve.

- **Heat-and-tap method.** Heat the area around the bleeder valve with a torch. The heat expands the size of the hole and usually allows the bleeder to be loosened and removed.

CAUTION: The heat from a torch will damage the rubber seals inside the caliper or wheel cylinder. Using heat to free a stuck bleeder valve will *require* that all internal rubber parts be replaced.

FIGURE 3–45 Master cylinder with brake fluid level at the"min"(minimum) line.

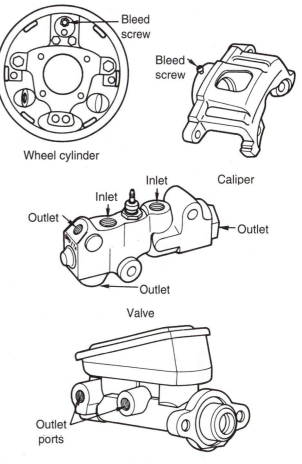

FIGURE 3–47 Typical bleeder locations. Note that the combination valve and master cylinder shown do not have bleeder valves; therefore, bleeding is accomplished by loosening the brake line at the outlet ports. (Courtesy of Allied Signal Automotive Aftermarket)

FIGURE 3–46 Typical bleeder valve from a disc brake caliper. The arrows point to the taper section that does the actual sealing. It is this taper that requires a shock to loosen. If the bleeder is simply turned with a wrench, the bleeder usually breaks off because the tapered part at the bottom remains adhered to the caliper or wheel cylinder. Once loosened, brake fluid flows around the taper and out through the hole in the side of the bleeder valve.

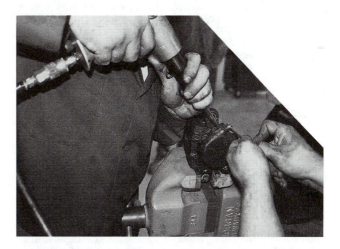

FIGURE 3–48 Using an air punch next to the bleeder valve to help"break the taper"on the bleeder valve.

BLEEDING THE MASTER CYLINDER

Whenever the master cylinder is replaced or the hydraulic system has been left opened for several hours, the air may have to be bled from the master cylinder. Bleed the master cylinder "on the bench" before installing it on the vehicle. If bleeding the master cylinder after working on the hydraulic system, follow these steps:

Step 1. Fill the master cylinder with clean brake fluid from a sealed container up to the recommended "full" level.

Step 2. Have an assistant slowly depress the brake pedal as you "crack open" the master cylinder bleed screw, starting with the section closest to the bulkhead. It is very important that the primary section of the master cylinder be bled before attempting to bleed the air out of the secondary section of the master cylinder. Before the brake pedal reaches the floor, close the bleeder valve.

HINT: A proper manual bleeding of the hydraulic system requires that accurate communications occur between the person depressing the brake pedal and the person opening and closing the bleeder valve(s). The bleeder valve (also called a *bleed valve*) should be open only when the brake pedal is being depressed. The valve *must* be closed when the brake pedal is released to prevent air from being drawn into the system.

Step 3. Repeat the procedure several times until a solid flow of brake fluid is observed leaving the bleeder valve. If the master cylinder is not equipped with bleeder valves, the outlet tube nuts can be loosened instead.

BLEEDING SEQUENCE

After bleeding the master cylinder, the combination valve should be bled if equipped. Follow the same procedure as when bleeding the master cylinder, being careful not to allow the master cylinder to run dry.

NOTE: The master cylinder is located in the highest section of the hydraulic braking system. All master cylinders can be bled using the same procedure used for bleeding calipers and wheel cylinders. If master cylinder is not equipped with bleeder valves, it can be bled by loosening the brake line fittings at the master cylinder.

Check the level in the master cylinder frequently and keep it filled with clean brake fluid throughout the brake bleeding procedure.

For most rear-wheel-drive vehicles equipped with a front/rear split system, start the bleeding with the wheel farthest from the master cylinder and working toward the closest (see Figure 3–49). For most vehicles, this sequence is:

1. Right rear
2. Left rear
3. Right front
4. Left front

NOTE: Before bleeding the front brakes, attach a holding tool to the stem or the metering valve to allow the brake fluid to flow through the valve unrestricted (see Figure 3–50).

For vehicles equipped with a diagonal split section, follow the recommended brake bleeding procedure as specified in the service manual or service information for the vehicle you are servicing.

NOTE: If the vehicle has two wheel cylinders on one brake, bleed the upper cylinder first.

MANUAL BLEEDING

Manual bleeding is the process of applying pressure (with the help from an assistant) to the brake pedal (ser-

FIGURE 3–49 Most vehicle manufacturers recommend starting the brake bleeding process at the rear wheel farthest from the master cylinder.

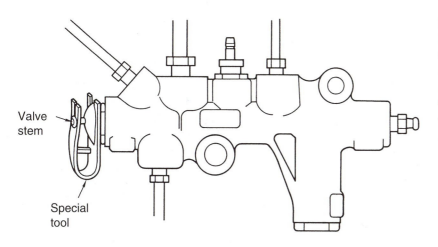

Valve stem

Special tool

FIGURE 3–50 Special tool being used to hold out the valve stem on the metering valve. This allows brake fluid to flow unrestricted to the front brakes. If this valve was not released, a lot of brake pedal pressure would have to be used to overcome the metering valve.

vice brakes) while the bleeder valve is opened slightly. The brake pedal applies pressure on the fluid, which is forced out through the bleeder valve along with any trapped air.

NOTE: See the service manual for the exact brake bleeding procedure for the vehicle you are servicing.

For best results do not allow the brake pedal to go to the floor. All vehicle manufacturers and brake experts agree that the technician should *wait at least 15 seconds* before repeating the bleeding procedure. This wait time is important to allow time for any air bubbles present to re-form into larger bubbles that can be bled out of the system more easily. See the Tech Tip "Tiny Bubbles" for details.

NOTE: Make certain that all the brake components such as calipers and wheel cylinders are installed correctly with the bleeder valve located on the highest section of the part. Some wheel cylinders and calipers (such as many Ford calipers) can be installed upside down! This usually occurs whenever both front calipers are off the vehicle and they accidentally get reversed left to right. If this occurs, the air will never be bled from the caliper completely.

PRESSURE BLEEDING

Pressure bleeding is a term used to describe the use of pressure on top of the master cylinder to rid the hydraulic brake system of trapped air. Pressure bleeders must use the correct master cylinder adapter (see Figures 3–51 and 3–52). Whenever using a power bleeder, do not exceed 20

◀ **TECH TIP** ▶

TINY BUBBLES

Do not use excessive brake pedal force while bleeding and never normally bleed the brakes with the engine running! The extra assist from the power brake unit greatly increases the force exerted on the brake fluid in the master cylinder. The trapped air bubbles may be dispersed into tiny bubbles that often cling to the inside surface of the brake lines. These tiny air bubbles may not be able to be bled from the hydraulic system until enough time has been allowed for the bubbles to re-form. To help prevent excessive force, do *not* start the engine. Without power assistance, the brake pedal force can be kept from becoming excessive. If the dispersal of the air into tiny bubbles is suspected, try tapping the calipers or wheel cylinders with a plastic hammer. After this tapping, simply waiting for a period of time will cause the bubbles to re-form into larger and easier- to-bleed air pockets. Most brake experts recommend waiting 15 seconds or longer between attempts to bleed each wheel. This waiting period is critical and allows time for the air bubbles to form.

HINT: To help prevent depressing the brake pedal too far down, some experts recommend placing a piece of 2"by 4"lumber under the brake pedal. This helps prevent the seals inside the master cylinder from traveling over unused sections inside the bore that may be corroded or rusty.

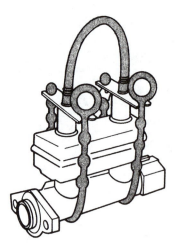

FIGURE 3–52   Brake fluid under pressure from the power bleeder is applied to the top of the master cylinder. It is very important that the proper adapter be used for the master cylinder. Failure to use the correct adapter or failure to release the pressure on the brake fluid before removing the adapter can cause brake fluid to escape under pressure. (Courtesy of Allied Signal Automotive Aftermarket)

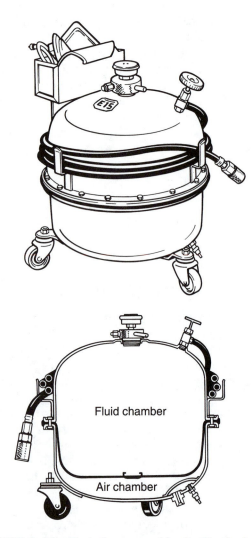

FIGURE 3–51   Typical pressure bleeder. The brake fluid inside is pressurized with air pressure in the air chamber. This air pressure is applied to the brake fluid in the upper section. A rubber diaphragm separates the air from the brake fluid. (Courtesy of EIS Brake Parts)

psi (140 kPa) of pressure. Since the pressure is low, the button on the metering valve (if equipped) must be either pushed in or pulled out, depending on the design and manufacturer. If the metering valve is not released, no fluid pressure will be applied to the front disc brakes. A major disadvantage to using the pressure bleeding method is use of the correct master cylinder reservoir adapter fitting. Using the wrong adapter or not installing it correctly can cause brake fluid to leak out under pressure, causing damage to the vehicle.

VACUUM BLEEDING

Another popular bleeder uses a vacuum pump (either manual or electric) to draw the brake fluid from the bleeder valve and into a container. This type of bleeder can also be used to bleed the hydraulic brake system without the need of an assistant (see Figure 3–53). Loosen the bleeder valve at least three-fourths of a turn and operate the vacuum bleeder. Continue bleeding until clean brake fluid flows into the bleeder container. Some air bubbles may be seen in the clear plastic tubing leading from the bleeder valve to the vacuum bleeder unit. Often, these bubbles are the result of air being drawn in around the threads of the bleeder valve and do not indicate that there is air in the brake system (see Figure 3–54).

GRAVITY BLEEDING

Gravity bleeding is a slow, but effective method that will work on many vehicles to rid the hydraulic system of air. The procedure involves simply opening the bleeder valve and waiting until brake fluid flows from the open valve. Any air trapped in the part being bled will rise and escape from the port when the valve is opened. It may take several minutes before brake fluid escapes. If no brake fluid comes out, remove the bleeder valve entirely—it may be clogged. Remember, nothing but air and brake fluid will be *slowly* coming out of the wheel cylinder or caliper when the bleeder valve is removed. *Do not press on the brake pedal with the bleeder valve out while gravity bleeding.*

◄ TECH TIP ►

QUICK AND EASY TEST FOR AIR IN THE LINES

If air is in the brake lines, the brake pedal will be low and will usually feel "spongy" or "mushy." To confirm that trapped air in the hydraulic system is the cause, perform this simple and fast test:

Step 1. Remove the cover from the master cylinder. Have an assistant pump the brake pedal several times and then hold the pedal down.

Step 2. Observe the squirts of brake fluid from the master cylinder when the brake pedal is released quickly. (This is best performed by allowing the foot to slip off the end of the brake pedal.)

Results: If the brake fluid squirts higher than 3 in. (8 cm) from the surface, air is trapped in the hydraulic system (see Figure 3–55).

CAUTION: Always use a fender cover whenever performing this test. Brake fluid will remove paint if it gets onto the unprotected fender. See Figure 3–56 for an example of how to keep fender covers from slipping off.

Explanation: Air can be compressed; liquid cannot be compressed. When pumping the brake pedal, the assistant is compressing any trapped air. When the pedal is released quickly, the compressed air expands and takes up more volume, forcing the brake fluid upward through the vent ports into the reservoir. Some upward movement is normal because of the return spring pressure on the valves in the master cylinder and springs in the wheel cylinders. If, however, the spurt is higher than normal, this is a sure sign of air being trapped in the system. This test is also called the *air entrapment test.*

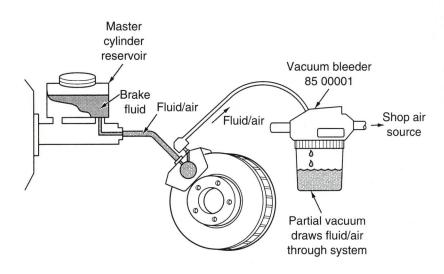

Master cylinder reservoir

Brake fluid

Fluid/air

Fluid/air

Vacuum bleeder 85 00001

Shop air source

Partial vacuum draws fluid/air through system

FIGURE 3–53 Vacuum bleeding is an easy method to use and does not require special adapters or an assistant. The vacuum bleeder shown uses compressed air flowing through a restriction to create a vacuum. A hand-operated vacuum pump is also available and does an excellent job at a moderate cost. (Courtesy of General Motors)

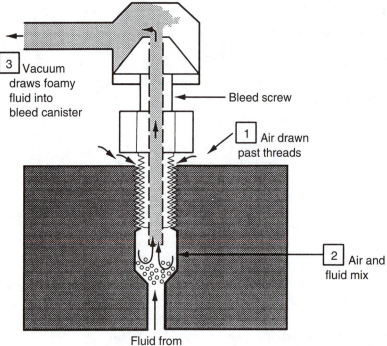

FIGURE 3–54 Air is often drawn past the screw threads and may cause foam in the fluid being bled out of the component. (Courtesy of General Motors)

3 Vacuum draws foamy fluid into bleed canister

Bleed screw

1 Air drawn past threads

2 Air and fluid mix

Fluid from wheel circuit

FIGURE 3–55 Note the large drops being squirted from the top of the master cylinder due to trapped air when the brakes were released.

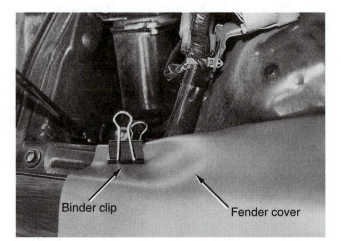

Binder clip

Fender cover

FIGURE 3–56 It is very important to use fender covers to protect the paint of the vehicle from being splashed with brake fluid. Use a binder clip available at local office supply stores to clip the fender cover to the lip of the fender, preventing the fender cover from slipping.

SUMMARY

1. The brake pedal mechanism increases the force of the driver's foot and applies it to the master cylinder.

2. During a typical brake application, only about 1 teaspoon (5 mL or cc) of brake fluid is actually moved from the master cylinder into the hydraulic system.

3. Master cylinder reservoirs are large enough for the brakes to be worn completely down and still have a small reserve.

4. The front port of the master cylinder is called the vent port and the rear port is called the replenishing port.

5. Brake system diagnosis should always start with checking for venting.

6. Dual-split master cylinders that separate the front brakes from the rear brakes are used on rear-wheel-drive vehicles.

7. Diagonal-split master cylinders that separate right front and left rear from the left front and right rear brakes are used on front-wheel-drive vehicles.

8. A pressure differential or a brake fluid level sensor will light the red brake warning lamp if there is a leak in the hydraulic system.

9. A proportioning valve is used in braking systems to limit and control the pressure sent to the rear brakes to help prevent rear wheel lockup during heavy braking.

10. A metering valve (hold-off valve) delays the operation of the front disc brakes until enough pressure has been sent to the rear drum brakes to overcome return spring pressure. This valve helps prevent front wheel lockup during light braking on slippery surfaces.

11. Air trapped in the hydraulic system is removed by bleeding the brakes. Bleeder valves are used to bleed the air from disc brake calipers and drum brake wheel cylinders.

REVIEW QUESTIONS

1. Explain Pascal's law.

2. Describe how a master cylinder works.

3. Discuss the difference between a dual-split and a diagonal-split master cylinder.

4. Explain the operation of the pressure differential switch.

5. Describe the purpose of the metering and proportioning valves.

6. List the procedure for bleeding air from the hydraulic brake system.

MULTIPLE-CHOICE QUESTIONS

1. Two technicians are discussing master cylinders. Technician A says that it is normal to see fluid movement in the reservoir when the brake pedal is depressed. Technician B says that a defective master cylinder can cause the brake pedal to sink slowly to the floor when depressed. Which technician is correct?
   a. A only
   b. B only
   c. Both A and B
   d. Neither A nor B

2. Technician A says that a pull to the right during braking could be caused by a defective metering valve. Technician B says that a pull to the left could be caused by a defective proportioning valve. Which technician is correct?
   a. A only
   b. B only
   c. Both A and B
   d. Neither A nor B

3. If the brake pedal linkage is not adjusted correctly, brake fluid may not be able to expand back into the reservoir through the _____ port of the master cylinder when the brakes get hot.
   a. vent port (forward hole)
   b. replenishing port (rearward hole)

4. The primary brake circuit fails due to a leak in the lines leaving the rear section of a dual-split master cylinder. Technician A says that the driver will notice a lower-than-normal brake pedal and some reduced braking power. Technician B says that the brake pedal will "grab" higher than normal. Which technician is correct?
   a. A only
   b. B only
   c. Both A and B
   d. Neither A nor B

5. Two technicians are discussing a problem where the brake pedal travels too far before the vehicle starts to

slow. Technician A says that the brakes may be out of adjustment. Technician B says that one circuit from the master cylinder may be leaking or defective. Which technician is correct?

a. A only

b. B only

c. Both A and B

d. Neither A nor B

6. The rear brakes lock up during a regular brake application. Technician A says that the metering valve could be the cause. Technician B says that stuck front disc brake calipers could be the cause. Which technician is correct?

a. A only

b. B only

c. Both A and B

d. Neither A nor B

7. A spongy brake pedal is being diagnosed. Technician A says that air in the hydraulic system could be the cause. Technician B says that a defective pressure differential switch could be the cause. Which technician is correct?

a. A only

b. B only

c. Both A and B

d. Neither A nor B

8. The button on the _____ valve should be held when bleeding the brakes.

a. metering

b. proportioning

c. pressure differential

d. residual check

9. A double lap flare and an ISO flare are interchangeable.

a. True

b. False

10. The brake bleeding procedure usually specified for a rear-wheel-drive vehicle with a dual-split master cylinder is

a. RR, LR, RF, LF.

b. LF, RF, LR, RR.

c. RF, LR, LF, RR.

d. LR, RR, LF, RF.

◀ Chapter 4 ▶

WHEEL BEARINGS AND SERVICE

OBJECTIVES

After studying Chapter 4, the reader will be able to:

1. Discuss the various types, designs, and parts of automotive antifriction wheel bearings.
2. Describe the symptoms of defective wheel bearings.
3. Explain wheel bearing inspection procedures and causes of spalling and brinelling.
4. List the installation and adjustment procedures for wheel bearings.
5. Explain how to inspect, service, and replace wheel bearings and seals.

Bearings allow the wheels of a vehicle to rotate and still support the weight of the entire vehicle.

ANTIFRICTION BEARINGS

Antifriction bearings use rolling parts inside the bearing to reduce friction. Four styles of rolling contact bearings are ball, roller, needle, and tapered roller bearings, shown in Figure 4–1. All four styles convert sliding friction into rolling motion. All of the weight of a vehicle or load on the bearing is transferred through the rolling part.

In a ball bearing, all of the load is concentrated into small spots where the ball contacts the inner and outer race (rings) (see Figure 4–2). While ball bearings cannot support the same weight as roller bearings, there is less friction in ball bearings and they generally operate at higher speeds.

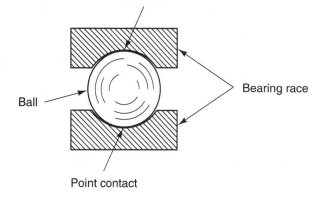

FIGURE 4-2 Ball bearing point contact.

FIGURE 4-1 Rolling contact bearings include (left to right) ball, roller, needle, and tapered roller.

A roller bearing, having a greater (longer) contact area than a ball bearing, can support heavier loads (see Figure 4–3).

A needle bearing is a type of roller bearing that uses smaller rollers called **needle rollers**.

The clearance between the diameter of the ball or straight roller is manufactured into the bearing to provide the proper **radial clearance,** and is **not adjustable**.

TAPERED ROLLER BEARINGS

The most commonly used automotive wheel bearing is the tapered roller bearing. Not only is the bearing itself tapered, but the rollers are tapered. By design, this type of bearing can withstand **radial** (up and down) as well as **axial** (thrust) loads in one direction (see Figure 4–4).

Most non-drive-wheel bearings are tapered rollers rotating between races and assembled on an angle. The taper allows more weight to be handled by the friction-reducing bearings because the weight is directed over the entire length of each roller rather than concentrated on a small spot, as with ball bearings. The rollers are held in place by a **cage** between the **inner race** (also called the **inner ring,** or **cone**) and the **outer race** (also called the **outer ring,** or **cup**). Tapered roller bearings must be loose in the cage to allow for heat expansion. Tapered roller bearings should always be adjusted with a certain amount of free play to allow for heat expansion. On non-drive-axle vehicle wheels, the cup is tightly fitted to the wheel hub and the cone is loosely fitted to the wheel spindle. New bearings come packaged with rollers, cage, and inner race assembled together, wrapped with moisture, resistant paper.

INNER AND OUTER WHEEL BEARINGS

Most rear-wheel-drive vehicles use an inner and an outer wheel bearing on the front wheels. The inner wheel bearing is always larger because it is designed to carry most of the vehicle weight and transmit this weight to the suspension through to the spindle (see Figure 4–5). Between the inner wheel bearing and the spindle, there is a grease seal that prevents grease from getting onto the braking surface.

STANDARD BEARING SIZES

Bearings use standard dimensions for inside diameter, widths, and outside diameters. The standardization of bearing sizes helps interchangeability. The dimensions that are standardized include bearing bore size (inside diameter), bearing series (light to heavy usage), and external dimensions. When replacing a wheel bearing, note the original bearing brand name and number. Replacement bearing catalogs usually have crossover charts from one brand to another. The bearing number is usually the same because of the

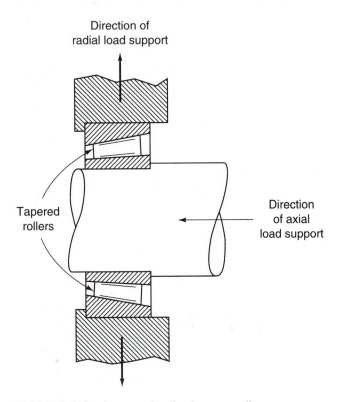

FIGURE 4-4 A tapered roller bearing will support a radial load and an axial load in only one direction.

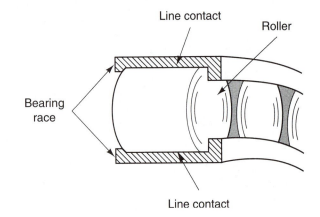

FIGURE 4-3   Roller bearing line contact.

FIGURE 4-5 Non-drive-wheel hub with inner and outer tapered roller bearings. By angling the inner and outer in opposite directions, axial (thrust) loads are supported in both directions.

interchangeability and standardization within the wheel bearing industry.

SEALED FRONT-WHEEL-DRIVE BEARINGS

Most front-wheel-drive vehicles use a sealed nonadjustable front wheel bearing. This type of bearing can include either two preloaded tapered roller bearings or a double-row ball bearing. This type of sealed bearing is also used on the rear of many front-wheel-drive vehicles (see Figure 4–6).

Double-row ball bearings are often used because of their reduced friction and greater seize resistance. Figures 4–7 and 4–8 show sealed bearings.

BEARING GREASES

Vehicle manufacturers specify the type and consistency of grease for each application. The service technician should know what these specifications mean. Grease is an oil with a thickening agent to allow it to be installed in places where a liquid lubricant would not stay. Greases are named for their thickening agents, such as aluminum, barium, calcium, lithium, or sodium.

The **American Society for Testing and Materials (ASTM)** specifies the consistency using a penetration test.

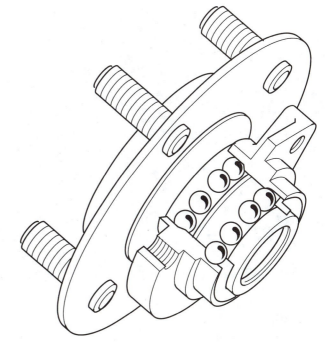

FIGURE 4-6 This front wheel bearing assembly is a double-row ball bearing design. It is a prelubricated sealed bearing. The bearing assembly is a loose fit in the steering knuckle. The drive axle shaft is a splined fit through the bearing.

FIGURE 4-7 Hub and bearing assembly pulled from the knuckle and drive axle shaft.

The **National Lubricating Grease Institute (NLGI)** specifies grease by designation as to its use:

- "GC" designation is acceptable for wheel bearings.
- "LB" designation is acceptable for chassis lubrication.

Many greases are labeled with both GC and LB and are therefore acceptable for both wheel bearings and

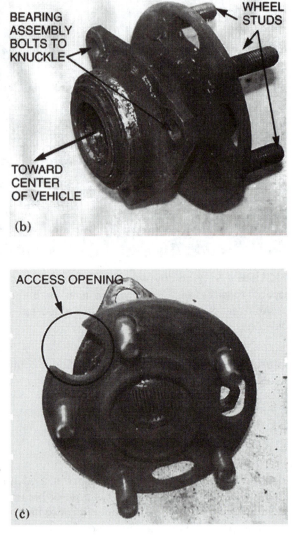

FIGURE 4-8 (a) Front-wheel-drive bearing shown after removal of the drive axle shaft. (b) Bearing assembly as removed from a front-wheel-drive General Motors vehicle. The entire assembly as shown is replaced as a unit. (c) The wheel mounting flange is notched to provide access to the bearing-to-knuckle bolts.

chassis use, such as in lubricating ball joints, tie-rods, etc. NLGI also uses the penetration test as a guide to assign the grease a number. Low numbers are very fluid and higher numbers are more firm. For example, a typical grease used for wheel bearings is labeled NLGI #2 GC. See the chart.

Timken OK Load is from a test that determines the maximum load the lubricant will carry. The *OK Load* is the maximum weight that can be applied without scoring the test block.

More rolling bearings are destroyed by overlubrication than by underlubrication because the heat generated in the bearings cannot be transferred easily to the air through the excessive grease. Bearings should never be filled beyond one-third to one-half of their grease capacity by volume. Molybdenum disulfide is added to grease in amounts up to 10 percent for use as a multipurpose lubrication on automotive equipment parts, such as chassis joints, steering joints, U-joints, and kingpins.

National Lubricating Grease Institute (NLGI) Numbers

NLGI number	Relative consistency
000	Very fluid
00	Fluid
0	Semifluid
1	Very soft
2	Soft (typically used for wheel bearings)
3	Semifirm
4	Firm
5	Very firm
6	Hard

SEALS

Seals are used in all vehicles to keep lubricant, such as grease, from leaking out and to prevent dirt, dust, or water from getting into the bearing or lubricant. Two general applications of seals include static and dynamic. **Static seals** are used between two surfaces that do not move. **Dynamic seals** are used to seal between two surfaces that move. Wheel bearing seals are dynamic seals that must seal between rotating axle hubs and the stationary spindles or axle housing. Most dynamic seals use a synthetic rubber lip seal encased in metal. The lip is often held in contact with the moving part with the aid of a **garter spring**, as seen in Figure 4–9.

The sealing lip should be installed toward the grease or fluid being contained (see Figure 4–10).

SYMPTOMS AND DIAGNOSIS OF DEFECTIVE BEARINGS

Wheel bearings control the positioning and reduce the rolling resistance of vehicle wheels. Whenever a bearing fails, the wheel may not be kept in position and noise is usually heard. Symptoms of defective wheel bearings include:

1. A hum, rumbling, or growling noise which increases with vehicle speed.
2. Roughness felt in the steering wheel which changes with vehicle speed or cornering.

3. Looseness or excessive play in the steering wheel, especially while driving over rough road surfaces.
4. A loud grinding noise produced in severe cases by a defective front wheel bearing.
5. Pulling to one side during braking.
6. Bearing roughness—with the vehicle off the ground, rotate the wheel by hand, listening and feeling carefully.
7. Bearing looseness—grasp the wheel at the top and bottom and wiggle it back and forth.

NOTE: Excessive looseness in the wheel bearings can cause a low brake pedal.

If any of these symptoms are present, carefully clean and inspect the bearings.

NON-DRIVE-WHEEL BEARING INSPECTION AND SERVICE

1. Hoist the vehicle safely.
2. Remove the wheel.
3. Remove the brake caliper assembly and support it with a coat hanger or other suitable hook to avoid allowing the caliper to hang by the brake hose.

FIGURE 4-9 Typical lip seal with a garter spring.

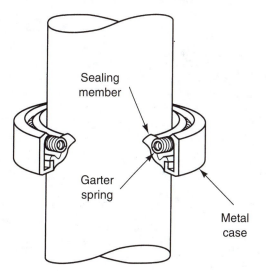

FIGURE 4-10 A garter spring helps hold the sharp lip edge of the seal tight against the shaft. (Courtesy of Dana Corporation)

4. Remove the grease cap (dust cap). See Figure 4–11.
5. Remove the old cotter key and discard.

NOTE: The term *cotter* as in *cotter key* or *cotter pin* is derived from the old English verb meaning to "close or fasten."

---

6. Remove the spindle nut (castle nut).
7. Remove the washer and the outer wheel bearing (see Figure 4–12).
8. Remove the bearing hub from the spindle. The inner bearing will remain in the hub and may be removed (simply lifted out) after the grease seal is pried out (see Figure 4–13).
9. Most vehicle and bearing manufacturers recommend cleaning the bearing thoroughly in solvent or acetone. If there is no acetone, clean the solvent off the bearings with denatured alcohol to make certain that the thin solvent layer is completely washed off and dry. **All solvent must be removed or allowed to dry from the bearing**

◄ **TECH TIP** ►

BEARING OVERLOAD

It is not uncommon for vehicles to be overloaded. This is particularly common with pickup trucks and vans. Whenever there is a heavy load, the axle bearings must support the entire weight of the vehicle, including its cargo. If a bump is hit while driving with a heavy load, the balls of a ball bearing or the rollers of a roller bearing can dent the race of the bearing. This dent or imprint is called **brinelling,** named after Johann A. Brinell, a Swedish engineer who developed a process of testing for surface hardness by pressing a hard ball with a standard force into a sample material to be tested.

Once this imprint is made, the bearing will make noise whenever the roller or ball rolls over the indentation. Continued use causes wear to occur on all of the balls or rollers, with eventual failure. While this may take months, the *cause* of the bearing failure is often overloading of the vehicle. Avoid shock loads and overloading for safety and for longer vehicle life.

because the new grease will not stick to a layer of solvent.

10. Carefully inspect the bearings and the races for the following:
    a. The outer race for lines, scratches, or pits.
    b. The cage should be round; if the cage has straight sections, this is an indication of an overtightened adjustment or the cage has been dropped.

If either of these are observed, the bearing must be replaced, including the outer race. Failure to replace the outer race (which is included when

FIGURE 4-11 Removing the grease cap with grease cap pliers.

FIGURE 4-12 After wiggling the brake rotor slightly, the washer and outer bearing can easily be lifted out of the wheel hub.

FIGURE 4-13 Some technicians remove the inner wheel bearing and the grease seal at the same time by jerking the rotor off the spindle after reinstalling the spindle nut. Although this is an quick-and-easy method, sometimes the bearing is damaged (deformed) from being jerked out of the hub.

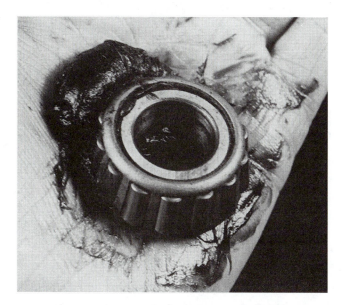

FIGURE 4-14 When packing grease into a cleaned bearing, force grease around each roller as shown.

FIGURE 4-15 Installing a grease seal with a special tool after installing the inner bearing.

NOTE: Some vehicle manufacturers do *not* recommend that stringy-type wheel bearing grease be used. Centrifugal force can cause the grease to be thrown outward from the bearing. Because of the stringy texture, the grease may not flow back into the bearing after it has been thrown outward. The result is lack of lubrication and eventual bearing failure.

---

you purchase a bearing) could lead to rapid failure of the new bearing.

11. Pack the cleaned or new bearing thoroughly with clean, new, approved wheel bearing grease using hand packing or a wheel bearing packer. Always clean out all of the old grease before applying the recommended type of new grease. *Because of compatibility problems, it is not recommended that greases be mixed* (see Figure 4–14).

12. Place a thin layer of grease on the outer race.
13. Apply a thin layer of grease to the spindle, being sure to cover the outer bearing seat, inner bearing seat, and shoulder at the grease seal seat.
14. Install a new **grease seal** (also called a **grease retainer**) flush with the hub (see Figure 4–15).
15. Place approximately 3 tablespoons of grease into the grease cavity of the wheel hub. Excessive grease could cause the inner grease seal to fail, with the possibility of grease getting on the brakes. Place the rotor with the inner

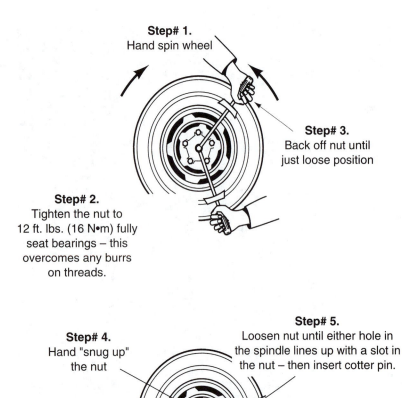

Step# 1.
Hand spin wheel

Step# 3.
Back off nut until
just loose position

Step# 2.
Tighten the nut to
12 ft. lbs. (16 N•m) fully
seat bearings – this
overcomes any burrs
on threads.

Step# 4.
Hand "snug up"
the nut

Step# 5.
Loosen nut until either hole in
the spindle lines up with a slot in
the nut – then insert cotter pin.

NOTE: When the bearing is properly
adjusted there will be from .001-.005
inches (.03-.13mm) end-play (looseness).

FIGURE 4-16 The wheel bearing adjustment procedure as specified for rear-wheel-drive General Motors vehicles. (Courtesy of Oldsmobile Division)

bearing and seal in place over the spindle until the grease seal rests on the grease seal shoulder.

16. Install the outer bearing and the bearing washer.

17. Install the spindle nut and, while rotating the tire assembly, tighten to about 12 to 30 lb-ft with a wrench to seat the bearing correctly in the race (cup) and on the spindle (see Figure 4–16).

18. While rotating the tire assembly, loosen the nut approximately one-half turn and then *hand tighten only* (about 5 in.-lb).

NOTE: If the wheel bearing is properly adjusted, the wheel will still have about 0.001 to 0.005 in. (0.03 to 0.13 mm) end play. This looseness is necessary to allow the tapered roller bearing to expand when hot and not bind or cause the wheel to lock up.

---

19. Install a new cotter key. (An old cotter key could break a part off where it was bent and lodge in the bearing, causing major damage.)

HINT: Most vehicles use a cotter key that is ⅛ in. in diameter by 1½ in. long.

---

20. If the cotter key does not line up with the hole in the spindle, loosen slightly (no more than 1/16 in. of a turn) until the hole lines up. **Never tighten more than hand tight.**

21. Bend the cotter key ends up and around the nut, not over the end of the spindle where the end of the cotter key could rub on the grease cap, causing noise (see Figure 4–17).

22. Install the grease cap (dust cap) and the wheel cover.

FIGURE 4-17 Properly installed cotter key.

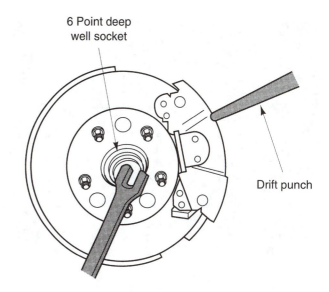

FIGURE 4-18 Removing the drive axle shaft hub nut. This nut is usually very tight; the drift (tapered) punch wedged into the cooling fins of the brake rotor keeps the hub from revolving when the nut is loosened. (Courtesy of Oldsmobile Division)

CAUTION: Clean grease off disc brake rotors or drums after servicing the wheel bearings. Use a brake cleaner and a rag. Even a slight amount of grease on the friction surfaces of the brakes can harm the friction lining and/or cause brake noise.

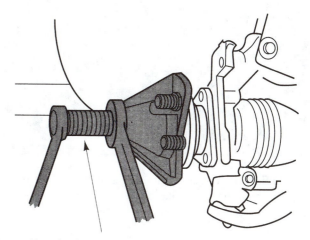

Turn forcing screw until axle splines are just loose

FIGURE 4-19 A special puller makes the job of removing the hub bearing from the knuckle easy without damaging any component. (Courtesy of Oldsmobile)

FRONT-WHEEL-DRIVE SEALED BEARING REPLACEMENT

Most front-wheel-drive vehicles use a sealed bearing assembly that is bolted to the steering knuckle and supports the drive axle. Many designs incorporate the splined drive hub, which transfers power from the drive axle to the wheels, which are bolted to the hub (see Figures 4–18, 4–19, and 4–20).

Many front-wheel-drive vehicles use a bearing that must be pressed off the steering knuckle. Special aftermarket tools are also available to remove many of the bearings without removing the knuckle from the vehicle. Check the factory service manual and tool manufacturers for exact procedures for the vehicle being serviced.

Diagnosing a defective front bearing on a front-wheel-drive vehicle is sometimes confusing. A defective wheel bearing is usually noisy while driving straight. The noise increases with vehicle speed (wheel speed). A drive axle shaft U-joint (CV joint) can also be the cause of noise on a front-wheel-drive vehicle but usually makes *more noise* while turning and accelerating.

OUTER
CV JOINT

DRIVE AXLE
SHAFT

WHEEL SPEED
SENSOR "TONE
RING"

STEERING
KNUCKLE

FIGURE 4-20 Be careful not to nick or damage the wheel speed sensor used for input information for traction control and antilock braking functions.

RETAINER PLATE

ACCESS HOLE

BACKING
PLATE

AXLE FLANGE

FIGURE 4-21 Retainer-plate rear axles can be removed by first removing the brake drum. The fasteners can be reached through a hole in the axle flange.

REAR-AXLE BEARING AND SEAL REPLACEMENT

The rear bearings used on rear-wheel-drive vehicles are constructed and serviced differently from other types of wheel bearings. Rear-axle bearings are either sealed or lubricated by the rear end lubricant. The rear axle must be removed from the vehicle to replace the rear-axle bearing. There are two basic types of axle retaining methods: **retainer-plate type** and the **C-lock.**

A **retainer plate** rear axle uses four fasteners that retain the axle to the axle housing. To remove the drive shaft and the rear-axle bearing and seal, the retainer bolts or nuts must be removed.

HINT: If the axle flange has an access hole, a retainer-plate axle is used (see Figure 4–21).

The hole or holes in the wheel flange permit access to the fasteners using a socket wrench. After the fasteners have been removed, the axle shaft must be removed from the rear-axle housing. With the retainer-plate rear axle, the bearing is press-fit onto the axle, and the bearing cup (outer race) is press-fit into the axle housing tube (see Figures 4–22, 4–23, 4–24, and 4–25 for ways to remove the axle shaft).

Vehicles that use **C-clips** use a straight roller bearing supporting a semifloating axle shaft inside the axle housing. The straight rollers do not have an inner race; the rollers ride on the axle itself. If a bearing fails, both the axle and the bearing usually need to be replaced. The outer bearing race holding the rollers is pressed into the rear-axle housing. The axle bearing is usually lubricated by the rear end lubricant and a grease seal is located on the outside of the bearing.

NOTE: Some replacement bearings are available that are designed to ride on a fresh, unworn section of the old axle. These bearings allow the use of the original axle, saving the cost of a replacement axle.

The C-clip rear-axle retaining method requires that the differential cover plate be removed (see Figure 4–26). After removal of the cover, the differential pinion shaft has to be removed before the C-clip that retains the axle can be removed (see Figure 4–27).

FIGURE 4-22 (a) To remove the axle from this vehicle equipped with a retainer-plate rear axle, the brake drum was placed back onto the axle studs backward so that the drum itself can be used as a slide hammer to pull the axle out of the axle housing. (b) A couple of pulls and the rear axle is pulled out of the axle housing.

FIGURE 4-23 A slide-hammer axle puller can also be used.

NOTE: When removing the differential cover, rear-axle lubricant will flow from between the housing and the cover. Be sure to dispose of the old rear-axle lubricant in the environmentally approved way, and refill with the proper type and viscosity (thickness) of rear end lubricant. Check the vehicle specifications for the grade recommended.

Once the C-clip has been removed, the axle simply is pulled out of the axle tube. Axle bearings with inner races are pressed onto the axle shaft and must be pressed off using a hydraulic press. A bearing retaining collar should be chiseled or drilled into to expand the collar to allow it to be removed (see Figure 4–28).

Always follow the manufacturer's recommended bearing removal and replacement procedures. Always replace the rear-axle seal whenever replacing a rear-axle

FIGURE 4-24   The brake backing plate came off with the rear axle and bearing on this vehicle. The backing plate made it more difficult to press off the old bearing.

"C" washer

FIGURE 4-26   The C-clip (C washer) can be seen after removing the differential cover plate. The C-clip fits into a groove in the axle.

FIGURE 4-25   The ball bearings fell out onto the ground when this axle was pulled out of the axle housing. Diagnosing the cause of the noise and vibration was easy on this vehicle.

◀ **TECH TIP** ▶

WHEEL BEARING LOOSENESS TEST

Looseness in a front-wheel bearing can allow the rotor to move whenever the front wheel hits a bump, forcing the caliper piston in, causing the brake pedal to kickback. This causes the feeling that the brakes are locking up.

Loose wheel bearings are easily diagnosed by removing the cover of the master cylinder reservoir and watching the brake fluid as the front wheels are turned left and right with the steering wheel. If the brake fluid moves while the front wheels are being turned, caliper piston(s) are moving in and out caused by loose wheel bearing(s). If everything is OK, the brake fluid should not move.

bearing. See Figure 4–29 for an example of seal removal. See Figure 4–30 for an example of a rear-axle bearing with a broken outer race. When refilling the differential, check for a tag or lettering as to the correct lubricant, as shown in Figure 4–31. Always check the differential vent to make sure that it is clear (see Figure 4–32).

A clogged vent can cause excessive pressure to build up inside the differential and cause the rear axle seals to leak. If rear-end lubricant gets on the brake

linings, the brakes will not have the proper friction and the linings themselves will be ruined and must be replaced.

BEARING FAILURE ANALYSIS

Whenever a bearing is replaced, the old bearing must be inspected and the cause of the failure eliminated. See

(a)

(b)

(c)

FIGURE 4-27 (a) Removing the pinion shaft lock bolt. (b) After the lock bolt has been removed, the pinion shaft can be removed. (c) The axle can be pushed inward slightly to allow the C-clip to be removed. After the C-clip has been removed, the axle can be easily pulled out of the axle housing.

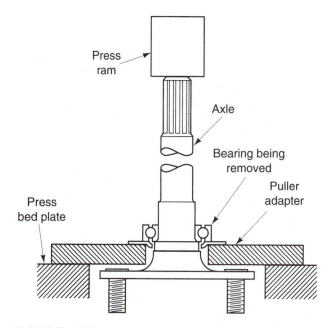

FIGURE 4-28 Using an hydraulic press to press an axle bearing from the axle. When pressing a new bearing back onto the axle, pressure should only be on the inner bearing race, to prevent damaging the bearing.

FIGURE 4-29 Removing an axle seal using the axle shaft as the tool.

FIGURE 4-30 This axle bearing came from a high-mileage vehicle. The noise was first noticed when turning because weight transfer increased the load on the bearing. Later, the rumbling sound occurred all the time, increasing in noise level as the vehicle speed increased.

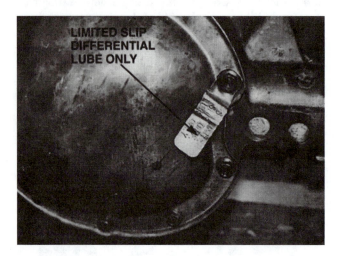

FIGURE 4-31

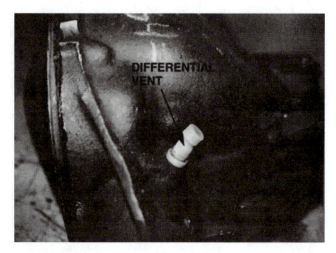

FIGURE 4-32

FIGURE 4-33 This is a normally worn bearing. If it does not have too much play, it can be reused. (Courtesy of SKF, USA, Inc.)

Figures 4–33 through 4–39 for examples of normal and abnormal bearing wear. A wheel bearing may fail for several reasons including:

Metal Fatigue. Long vehicle usage, even under normal driving conditions, causes metal to fatigue. Cracks often appear. Eventually, these cracks expand downward into the metal from the surface. The metal between the cracks can break out into small chips, slabs, or scales of metal. This process of breaking up is called **spalling** (see Figure 4–40).

Shock Loading. Causes dents to be formed in the race of a bearing, which eventually leads to bearing

failure. See the Tech Tip "Bearing Overload" and Figure 4–41.

Electrical Arcing. Electrical current flowing through a bearing can cause arcing and damage to the bearing, as shown in Figure 4–42. Electrical current can result from electrical welding on the vehicle without a proper ground connection. Without a proper return path, the electrical flow often travels throughout the vehicle, attempting to find a ground path. Always place the ground cable as close as possible to the area being welded. Another very common cause of bearing failure due to electrical arcing is due to a poor body ground wire connection between the body of the vehicle and the engine. All electrical current for accessories, lights, sound systems, etc., must return to the negative (-) terminal of the battery. If this ground wire connection becomes loose or corroded, the electrical current takes alternative paths to ground. The engine

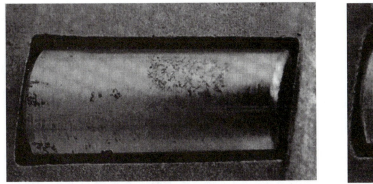
(a)

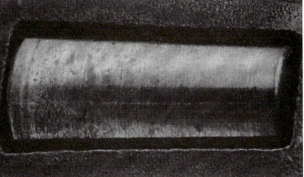
(b)

FIGURE 4-34 (a) When corrosion etches into the surface of a roller or race, the bearing should be discarded. (b) If light corrosion stains can be removed with an oil-soaked cloth, the bearing can be reused. (Courtesy of SKF, USA, Inc.)

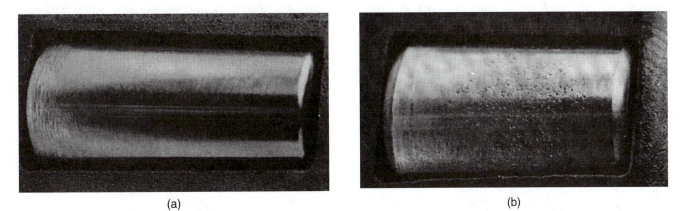

(a)   (b)

FIGURE 4-35 (a) When just the end of a roller is scored, it is because of excessive preload. Discard the bearing. (b) This is a more advanced case of pitting. Under load, it will rapidly lead to spalling. (Courtesy of SKF, USA, Inc.)

(a)   (b)

FIGURE 4-36 (a) Always check for faint grooves in the race. This bearing should not be reused. (b) Grooves like this are often matched by grooves in the race (above). Discard the bearing. (Courtesy of SKF, USA, Inc.)

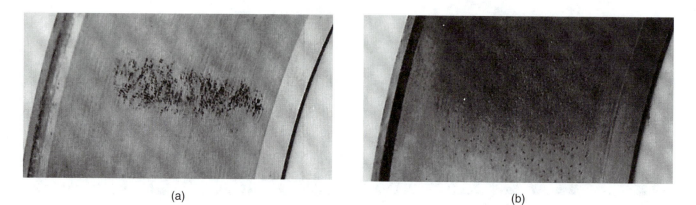

(a)   (b)

FIGURE 4-37 (a) Regular patterns of etching in the race are from corrosion. This bearing should be replaced. (b) Light pitting comes from contaminants being pressed into the race. Discard the bearing. (Courtesy of SKF, USA, Inc.)

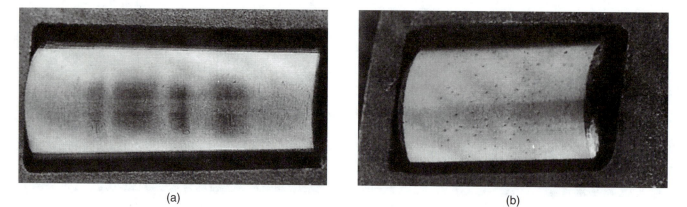

(a)   (b)

FIGURE 4-38 (a) This bearing is worn unevenly. Notice the stripes. It should not be reused. (b) Any damage that causes low spots in the metal renders the bearing useless. (Courtesy of SKF, USA, Inc.)

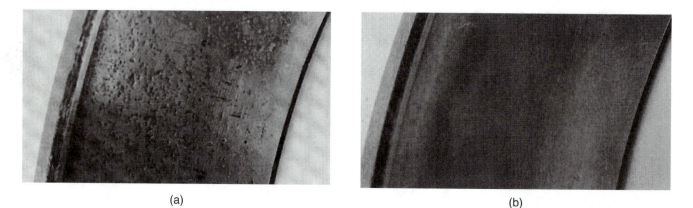

(a)   (b)

FIGURE 4-39 (a) In this more advanced case of pitting, you can see how the race has been damaged. (b) Discoloration is a result of overheating. Even a lightly burned bearing should be replaced. (Courtesy of SKF, USA, Inc.)

(a)   (b)

FIGURE 4-40   (a) Pitting eventually leads to spalling, a condition where the metal falls away in large chunks. (b) In this spalled roller, the metal has actually begun to flake away from the surface. (Courtesy of SKF, USA, Inc.)

and entire drivetrain are electrically insulated by rubber mounts. Even the exhaust system is electrically insulated from the body or from the vehicle with rubber insulating **hangers**. Suspension and wheel bearings are also insulated electrically by rubber control arm and shock absorber **bushings**. However, dirt is a conductor of electricity, especially when wet. Dirt acts as an electrical conductor and allows electrical current to flow through the suspension system *through the bearing* to the engine block, where the main starter ground cable connects to the battery negative (-) terminal.

FIGURE 4-41   These dents result from the rollers "hammering" against the race, a condition called brinelling. (Courtesy of SKF, USA, Inc.)

◄ TECH TIP ►

WHAT'S THAT SOUND?

Defective wheel bearings usually make noise. The noise most defective wheel bearings make sounds like noisy snow tires. Wheel bearing noise will remain constant while driving over different types of road surfaces, whereas tire tread noise usually changes with different road surfaces. In fact, many defective bearings have been ignored by vehicle owners and technicians because it was thought that the source of the noise was the aggressive tread design of the mud and snow tires. Always suspect defective wheel bearings whenever you hear what seems to be extreme or unusually loud tire noise.

FIGURE 4-42   This condition results from an improperly grounded arc welder. Replace the bearing. (Courtesy of SKF, USA, Inc.)

Therefore, whenever any bearing is replaced in the chassis or driveline systems, such as wheel bearings, U-joints, or drive axle shaft bearings, always check to see that the ground wires between the body of the vehicle and the engine block or negative (-) terminal of the battery are OK and the connections at both ends are clean and tight. If there is any doubt as to whether or not the body ground wires are OK, additional wires can always be added between the body and the engine without causing any harm.

SUMMARY

1. Wheel bearings support the entire weight of a vehicle and are used to reduce rolling friction. Ball and straight roller-type bearings are non adjustable, while **tapered** roller-type bearings must be adjusted for proper clearance.

2. Most wheel bearings are standardized sizes.

3. Most front-wheel-drive vehicles use sealed bearings, either two preloaded tapered roller bearings or double-row ball bearings.

4. Bearing grease is an oil with a thickener. The higher the NLGI number of the grease, the thicker or harder its consistency.

5. Defective wheel bearings usually make more noise while turning because more weight is applied to the bearing as the vehicle turns.

6. A defective bearing can be caused by metal fatigue that leads to **spalling** or shock loads that cause **brinelling,** bearing damage from electrical arcing due to poor body ground wires or improper electrical welding on the vehicle.

7. Tapered wheel bearings must be adjusted by hand-tightening the spindle nut after properly seating the bearings. A new cotter key must always be used.

8. All bearings must be serviced, replaced, and/or adjusted using the vehicle manufacturer's recommended procedures as stated in the service manual.

REVIEW QUESTIONS

1. List three common types of automotive antifriction bearings.

2. Explain the adjustment procedure for a typical tapered roller wheel bearing.

3. List four symptoms of a defective wheel bearing.

4. Describe how the rear axle is removed from a C-clip axle.

MULTIPLE-CHOICE QUESTIONS

1. Which type of automotive bearing can withstand radial and thrust loads, yet must be adjusted for proper clearance?
   a. Roller bearing
   b. Tapered roller bearing
   c. Ball bearings
   c. Needle roller bearing

2. Most sealed bearings used on the front wheels of front-wheel-drive vehicles are of which type?
   a. Roller bearing
   b. Single tapered roller bearing
   c. Double-row ball bearing
   d. Needle roller bearing

3. On a bearing that has been shock loaded, the race (cup) of the bearing can be dented. This type of bearing failure is called
   a. spalling.
   b. arcing.
   c. brinelling.
   d. fluting.

4. The bearing grease most often specified is rated NLGI
   a. No. 00.
   b. No. 0.
   c. No. 1.
   d. No. 2.

5. A non-drive-wheel bearing adjustment procedure includes a final spindle nut tightening torque of
   a. finger tight.
   b. 5 in.-lb.
   c. 12 to 30 lb-ft.
   d. 10 to 15 lb-ft plus 1/16 turn.

6. After a non-drive-wheel bearing has been adjusted properly, the wheel should have how much end play?

    a. zero

    b. 0.001 to 0.005 in.

    c. 0.10 to 0.30 in.

    d. $\frac{1}{16}$ to $\frac{3}{32}$ in.

7. The differential cover must be removed before removing the rear axle on which type of axle?

    a. Retainer plate

    b. C-clip

    c. Press fit

    d. Welded tube

8. What part *must* be replaced when servicing a wheel bearing on a nondrive wheel?

    a. The bearing cup

    b. The grease seal

    c. The cotter key

    d. The retainer washer

◄ Chapter 5 ►

DRUM BRAKE OPERATION, DIAGNOSIS, AND SERVICE

OBJECTIVES

After studying Chapter 5, the reader will be able to:

1. Identify drum brake component parts.
2. Describe the operation of dual servo and leading-trailing brakes.
3. Discuss the procedure recommended for brake drum removal.
4. Discuss the inspection and lubrication points of the backing plate.
5. Explain the importance of proper drum brake hardware.
6. Disassemble and reassemble a drum brake assembly.

Drum brakes were the first type of brakes used on motor vehicles. Even today, over 100 years after the first

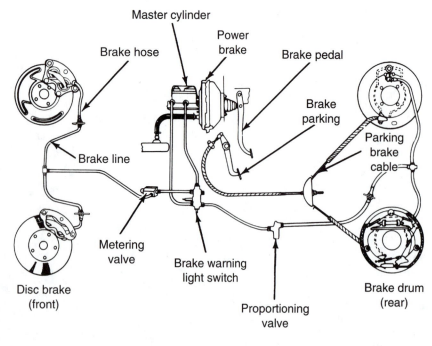

FIGURE 5–1 Typical brake system components showing disc brakes on the front and drum brakes on the rear. (Courtesy of Chrysler Corporation)

Master cylinder
Power brake
Brake hose
Brake pedal
Brake parking
Parking brake cable
Brake line
Metering valve
Brake warning light switch
Proportioning valve
Disc brake (front)
Brake drum (rear)

"horseless carriages," drum brakes are still used on the rear of most vehicles, as shown in Figure 5–1.

DRUM BRAKE PARTS

Drum brakes use two **brake shoes** mounted on a stationary **backing (back) plate** also called a **support plate** (see Figures 5–2 and 5–3). Hydraulic force from the master cylinder to the **wheel cylinder** pushes brake shoes against the rotating drum as shown in Figure 5–4.

DUAL-SERVO (DUO-SERVO) DRUM BRAKES

Drum brakes use outward-expanding brake shoes that contact the rotating brake drum when the driver de-

presses the brake pedal. Since the wheels of the vehicle are attached to the drums, the wheels also slow and stop when the brakes are applied. Because of the curved surface of the brake lining and the rotating brake drum, a wedging action occurs whenever the brakes are applied. This **self-energizing** action increases the amount of force applied to the drums beyond that provided by hydraulic pressure alone. During braking, the primary lining wedges into the drum and tends to pivot in the direction of rotation. This movement is transmitted through a lower connecting link to the secondary lining, which is forced into the drum with even greater pressure. This type of braking action is called **servo-self-energizing, dual-servo** or **Duo-Servo** (Duo-Servo is a trade name of the Bendix Corporation) (see Figures 5–5 through 5–7).

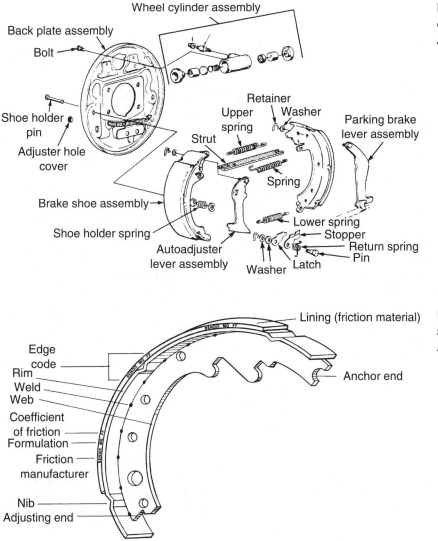

FIGURE 5–2 Exploded view of a typical drum brake showing all parts. (Courtesy of Allied Signal Automotive Aftermarket)

FIGURE 5–3 Typical drum brake shoe and the names of the parts. (Courtesy of Allied Signal Automotive Aftermarket)

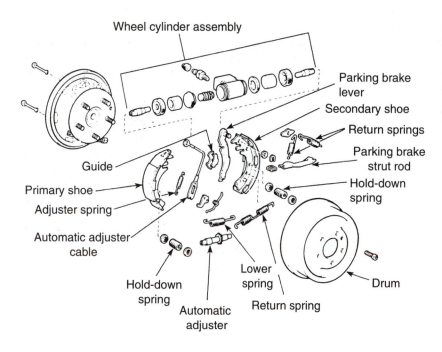

FIGURE 5–4 The drum rotates with the wheels. The brake shoes are attached to the vehicle. When the brake shoes expand outward, friction between the brake shoes and the rotating brake drum slows and stops the wheels. (Courtesy of Allied Signal Automotive Aftermarket)

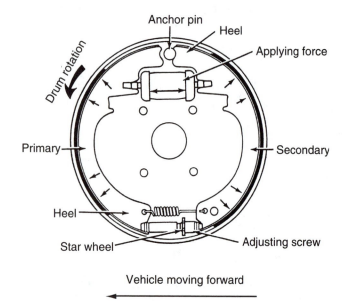

FIGURE 5–5 The rotation of the brake drum causes the curved brake shoe to wedge tighter into the drum. (Courtesy of Chrysler Corporation)

The self-energizing effect depends on free movement of the linings against the backing plate. This is why the rear (secondary) lining is usually longer and made from a different mix of materials, to be able to handle the greater forces and heat (see Figure 5–8).

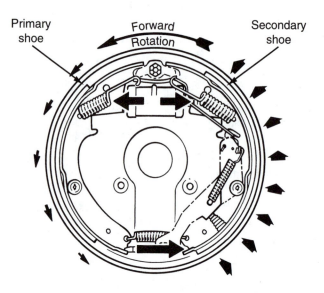

FIGURE 5–6 The primary shoe on the left exerts a force on the secondary shoe on the right.

LEADING-TRAILING DRUM BRAKES

In a leading-trailing drum brake system, the forward shoe is held stationary at the bottom and pushed against the drum by the wheel cylinder at the top. As the drum rotates, this leading shoe is pulled tighter into the drum

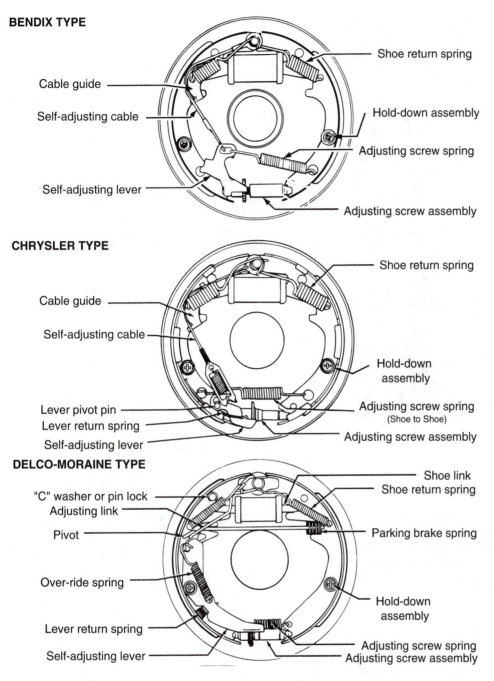

BENDIX TYPE

Cable guide

Self-adjusting cable

Self-adjusting lever

Shoe return spring

Hold-down assembly

Adjusting screw spring

Adjusting screw assembly

CHRYSLER TYPE

Cable guide

Self-adjusting cable

Lever pivot pin
Lever return spring
Self-adjusting lever

Shoe return spring

Hold-down assembly

Adjusting screw spring
(Shoe to Shoe)

Adjusting screw assembly

DELCO-MORAINE TYPE

"C" washer or pin lock
Adjusting link

Pivot

Over-ride spring

Lever return spring

Self-adjusting lever

Shoe link
Shoe return spring

Parking brake spring

Hold-down assembly

Adjusting screw spring
Adjusting screw assembly

FIGURE 5–7 Typical drum brake assemblies. (Courtesy of Gibson Products, a Division of Rolero, Inc.)

FIGURE 5–8 Brake lining on the assembly line at Delco Chassis. The darker-colored linings are the secondary shoes.

and tends to rotate with the drum. But because the shoe is anchored at the bottom, the shoe simply wedges itself into the drum. This **leading shoe** is also called the **forward shoe** or **energized shoe** (see Figure 5–9). The **trailing shoe** is also called the **deenergized shoe** or **reverse shoe**. Leading-trailing drum brakes are also called **nonservo brakes**.

BRAKE DRUM REMOVAL

The drum has to be removed before inspection or repair of the drum brake can start. There are two basic types of drums, and the removal procedure depends on which type is being serviced. With either type, it is usually recommended that the drums be marked with an "L" for left or an "R" for right so that they can be replaced in the same location.

CAUTION: Precautions should be taken to prevent any asbestos that may be present in the brake system from becoming airborne. Removal of the brake drum should occur inside a sealed vacuum enclosure equipped with a HEPA filter or washed with water or solvent.

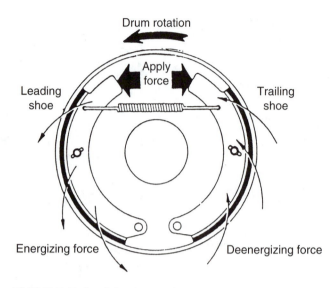

FIGURE 5–9 A leading-trailing design drum brake uses two brake shoes, but they are the same length and are not connected. The applying force is provided by the wheel cylinder.

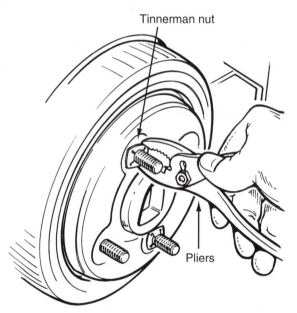

FIGURE 5–10 Tinnerman nuts are used at the vehicle assembly plant to prevent the brake drum from falling off until the wheels can be installed. These sheet metal retainers can be discarded after removal.

Hub or Fixed Drums. To remove the brake drum, remove the dust cap and the cotter key that is used to retain the spindle nut. Remove the spindle nut and washer, and then the brake drum can be carefully pulled off the spindle.

Hubless or Floating Drums. New vehicles have **tinnerman nuts (clips)**, also called **speed nuts**, on the

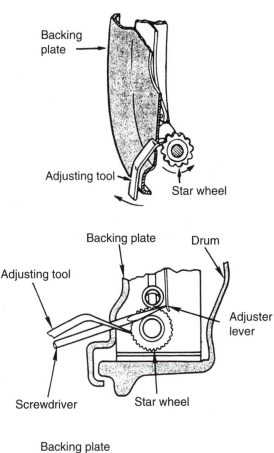

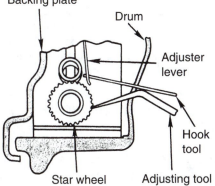

FIGURE 5–11 If the brake shoes have worn into the brake drum, the adjuster can be backed in after removing the access plug. Often, the plug is an outline on the brake drum that must be knocked out. After removing the plug, use a wire or screwdriver to move the adjusting lever away from the star wheel, then turn the star wheel with a brake adjusting tool, often called a brake "spoon." (Courtesy of Allied Signal Automotive Aftermarket)

stud when the vehicle is being assembled. These nuts can be discarded because they are not needed after the vehicle leaves the assembly plant (see Figure 5–10). After removing the wheels, the drum *should* move freely on the hub and slip off over the brake shoes. See

FIGURE 5–12 Using side-cut pliers to cut the heads off the hold-down spring pins (nails) from the backing plate to release the brake drum from the shoes.

Figure 5–11 for an example of how to adjust the brake shoes to allow for removal of the drum.

DRUM BRAKE DISASSEMBLY AND INSPECTION

After removal of the brake drum, the brake shoes and other brake hardware should be wetted down with a

◀ **TECH TIP** ▶

EASY SOLUTION TO A DRUM REMOVAL PROBLEM

Often, a brake drum cannot be removed because the linings have worn a groove into the drum. Attempting to adjust the brakes inward is often a frustrating and time-consuming operation. The easy solution is to use a pair of diagonal side-cut pliers and cut the heads off the hold-down pins (nails) at the backing plate. This releases the brake shoes from the backing plate and allows enough movement of the shoes to permit the removal of the brake drum without bending the backing plate.

The hold-down pins (nails) must obviously be replaced, but they are included in most drum brake hardware kits. Since most brake experts recommend replacing all drum brake hardware anyway, this solution does not cost more than normal, may save the backing plate from damage, and saves the service technician lots of time (see Figure 5–12).

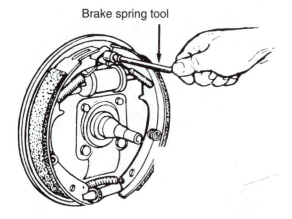

FIGURE 5–13 Using a brake spring tool to release a return (retracting) spring from the anchor pin.

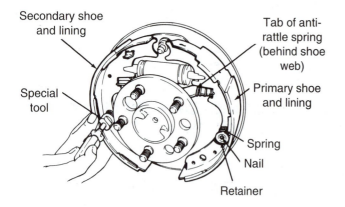

FIGURE 5–14 A special tool called a *hold-down spring tool* being used to depress and rotate the retainer. (Courtesy of Chrysler Corporation)

solvent or enclosed in an approved evacuation system to prevent possible asbestos release into the air. Usually, the first step in the disassembly of a drum brake system is removal of the return (retracting) springs (see Figure 5–13). After the return springs have been removed, the hold-down springs and the other brake hardware can be removed (see Figure 5–14).

NOTE: Generally "exact" disassembly or reassembly procedures are not specified by the manufacturer. The order in which the parts are disassembled or reinstalled is based on the experience and personal preference of the technician.

When brakes are serviced, the six raised contact surfaces called the **pads**, **ledges** or **shoe contact area** of the backing plate, should be inspected because they rub against the sides of the shoes (see Figure 5–15).

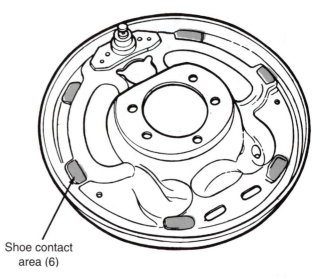

FIGURE 5–15 Most drum brakes use six shoe contact areas. (Courtesy of Chrysler Corporation)

Lithium high-temperature brake grease, synthetic brake grease, or **antiseize** should be used to lubricate drum brake parts. Both primary (front-facing) and secondary (rear-facing) lining material must be thicker than 0.060 in. (1.5 mm).

HINT: Most vehicle and brake lining manufacturers recommend replacing worn brake lining when the thickness of the riveted lining reaches 0.060 in. or less. An American nickel is about 0.060 in. thick, so simply remember that you must always have at least "a nickel's worth of lining."

The lining must be replaced if cracked, as shown in Figure 5–16.

Each lining has a return spring (retracting spring) which returns the brake shoes back from the drums whenever the brakes are released. The springs are called the **primary return spring** and **secondary return spring**. The primary return spring attaches to the primary brake shoe and the secondary return spring attaches to the secondary brake shoe. Some drum brakes use a spring that connects the primary and secondary shoes, commonly called a **shoe-to-shoe spring**. Return springs can get weak due to heat and age and can cause the linings to remain in contact with the drum. See the Tech Tip "Drop Test" for one testing method.

Hold-down springs (one on each shoe) are springs used with a retainer and a hold-down **spring pin** (or **nail**) to keep the linings on the backing plate (see Figure 5–18). Other types of hold-down springs include U-shape, flat spring-steel type, and the combination return-and-hold spring. Hold-down springs keep the

FIGURE 5–16   A cracked brake lining must be replaced.

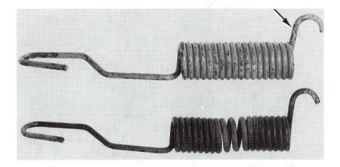

FIGURE 5–17   The top spring looks good because all coils of the spring are touching each other. The bottom spring is stretched and should be discarded. The arrow points to the back side of the spring that goes into a hole in the brake shoe. The open loop of the spring is not strong enough to keep from straightening out during use. Using the back side of the hook provides a strong, long-lasting hold in the brake shoe.

brake shoes against the shoe contact area of the backing plate. See Figure 5–19.

ANCHOR PIN

The anchor pin is located at the top of the backing plate and supports the top of both linings as well as providing an anchor for both return springs. Covering the linings over the anchor pin is a **shoe guide plate**, also called an **anchor pin support plate** or **butterfly** (see Figure 5–20).

SELF-ADJUSTING MECHANISMS

The operation of the **adjusting lever** on the **star wheel adjuster** involves applying the brakes while driving in reverse (see Figures 5–21 and 5–22). Because of the free movement built into drum brakes, the forces that cause the servo-self-energizing, cause the secondary lining to become the primary lining in reverse, which causes rotation of the linings. This action pulls on the adjusting lever and upon release of the brakes causes this lever to move one tooth of the star-wheel adjuster if there was enough movement (slack) to need adjustment.

WHEEL CYLINDERS

Hydraulic pressure is transferred from the master cylinder to each wheel cylinder through the brake fluid. The

◀ TECH TIP ▶

DROP TEST

Brake return (retracting) springs can be tested by dropping them to the floor. A good spring should "thud" when the spring hits the ground. This noise indicates that the spring has not been stretched and that all coils of the springs are touching each other. If the spring "rings" when dropped, the spring should be replaced because the coils are not touching each other (see Figure 5–17).

Although the drop test is often used, many experts recommend *replacing* all brake springs every time the brake linings are replaced. Heat generated by the brake system often weakens springs enough to affect their ability to retract brake shoes, especially when hot, yet not "ring" when dropped.

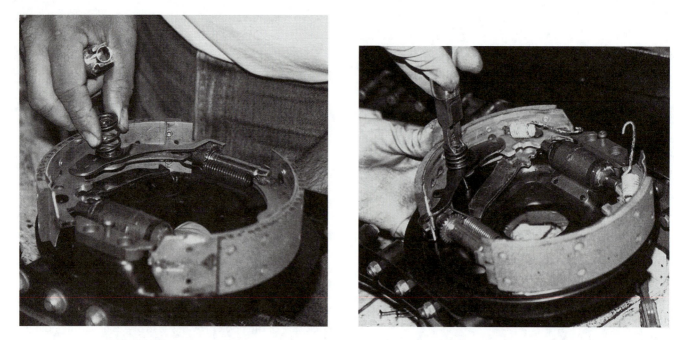

FIGURE 5–18 (a) Typical hold-down spring being installed on a drum brake on an assembly line at the factory. (b) The hold-down spring retainer is being pushed down and rotated using a special hold-down spring tool.

Shoe contact area

FIGURE 5–19 The brake shoes rest on the backing plate on raised pads called the shoe contact area or shoe ledge. As the shoes move, these raised areas can wear. Lack of lubrication in this area often causes a squeaking sound when the brake pedal is depressed or released.

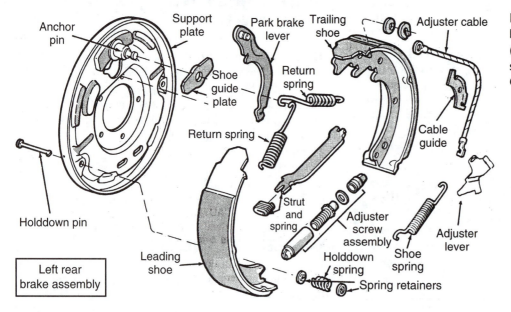

FIGURE 5–20 Typical drum brake showing support plate (backing plate), anchor pin, and shoe guide plate. (Courtesy of Chrysler Corporation)

Anchor pin
Support plate
Park brake lever
Trailing shoe
Adjuster cable
Shoe guide plate
Return spring
Return spring
Cable guide
Holddown pin
Strut and spring
Adjuster screw assembly
Adjuster lever
Leading shoe
Holddown spring
Shoe spring
Spring retainers

Left rear brake assembly

force exerted on the brake fluid by the driver forces the piston inside the wheel cylinder to move outward. Through **push rods** or **links**, this movement acts on the

brake shoes, forcing them outward against the brake drum (see Figure 5–23).

Drum brake wheel cylinders are cast iron with a bore (hole) drilled and finished to provide a smooth finish for the wheel cylinder seals and pistons. This special finish is called **bearingized**. It is this bearingized surface finish on the inside of the wheel cylinder that is often destroyed when a wheel cylinder is honed. Outside each wheel cylinder piston dust boots are installed to keep dirt out of the cylinder bore. Between both piston

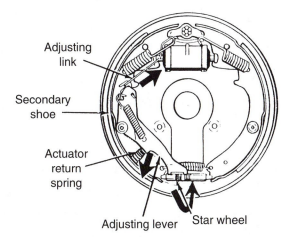

FIGURE 5–21 This self-adjusting mechanism operates only when the brake pedal is being depressed while traveling in reverse. If the secondary brake shoe travels far enough, the adjusting lever is raised up enough to cause the star wheel to move when the brake pedal is released. (Courtesy of EIS Brake Parts)

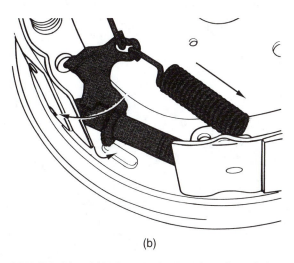

(b)

FIGURE 5-22 (b) When the brakes are released, the tension on the cable is released and the adjusting lever is returned by the spring. As the lever moves downward, the lever rotates the star wheel. (Courtesy of Ford Motor Company)

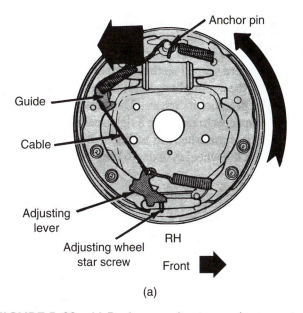

(a)

FIGURE 5–22 (a) Dual-servo adjusting mechanism using a cable between the anchor pin and the adjusting lever. A cable guide is used on the secondary brake shoe and is held in position by the secondary return spring. Note the arrows showing how the secondary shoe is moved off the anchor pin during a brake application when the vehicle is moving backward. As the shoe moves away from the anchor pin, the shoe guide pulls upward on the cable, which raises the adjusting lever. (Courtesy of Ford Motor Company)

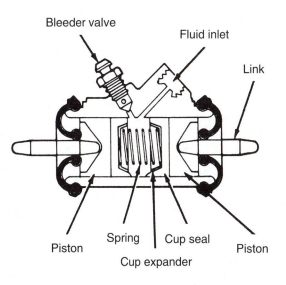

FIGURE 5–23 Cross section of a wheel cylinder that shows all of its internal parts. The brake line attaches to the fluid inlet. The cup expander prevents the cup seal lip from collapsing when the brakes are released.

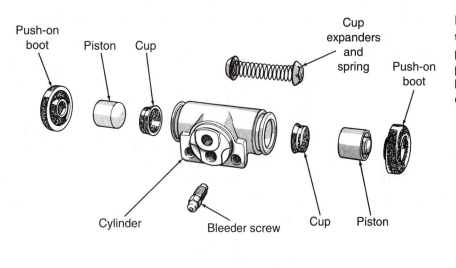

Push-on boot · Piston · Cup · Cup expanders and spring · Push-on boot · Cylinder · Bleeder screw · Cup · Piston

FIGURE 5–24 Exploded view of a typical wheel cylinder. Note how the flat part of the cups touch the flat part of the piston. The cup expander and spring go between the cups. (Courtesy of Chrysler Corporation)

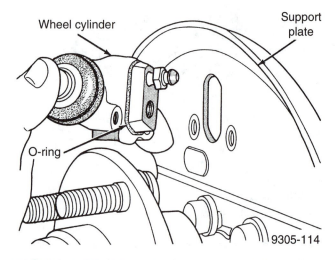

Wheel cylinder · Support plate · O-ring · 9305-114

FIGURE 5–25 Many wheel cylinders are bolted to the support plate (backing plate). The seal helps keep water and dirt out of the drum brake. (Courtesy of Chrysler Corporation)

seals there is a spring with piston **seal expanders** to keep the seals from collapsing toward each other and to keep pressure exerted on the lips of both seals to ensure proper sealing (see Figure 5–24).

On the back of each wheel cylinder there is a threaded hole for the brake line and a bleeder valve that can be loosened to remove (bleed) air from the hydraulic system. The wheel cylinder is bolted or clipped to the backing plate (see Figures 5–25 and 5–26).

The wheel cylinders should be checked regularly for possible brake fluid leakage past the piston seals. With a dull tool, pry the wheel cylinder dust boots aside slightly and check for wetness. Remember, the dust boots are not designed to hold hydraulic pressure, but they do act as a small reservoir for seeping fluid.

NOTE: When inspecting a drum brake wheel cylinder, look for brake fluid under the dust boot. According to vehicle manufacturers, a slight amount of brake fluid behind the boot is normal and serves to lubricate the pistons. However, if there is enough fluid to run or spill out when the boot is pried back, excessive leakage is indicated and the wheel cylinder must be rebuilt or replaced.

OVERHAUL OF THE WHEEL CYLINDERS

Wheel cylinders can be overhauled if defective or leaking, if recommended by the vehicle manufacturer. The following steps and procedures should be followed:

Step 1. Loosen the bleeder valve.

Step 2. To remove the wheel cylinder, the brake line must first be removed from the wheel cylinder. Unbolt or remove the wheel cylinder retainer clip.

Step 3. Remove all internal parts of the wheel cylinder if the bleeder valve can be removed.

NOTE: With some vehicles, to remove the seals and piston the wheel cylinder must be unbolted from the backing plate.

Step 4. Clean and/or hone a wheel cylinder as specified by the manufacturer to remove any rust and corrosion (see Figure 5–27).

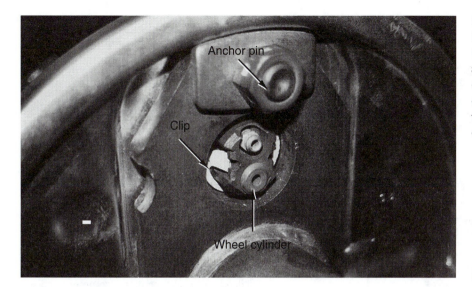

FIGURE 5–26   (a) Some wheel cylinders are simply clipped to the backing plate. (b) This special tool makes it a lot easier to remove the wheel cylinder clip. A socket (1 ⅛ in., 12 point) can be used to push the clip back onto the wheel cylinder. (Courtesy of General Motors)

(a)

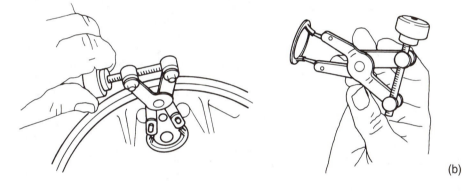

(b)

◀ TECH TIP ▶

TIME—NOT MILEAGE—IS IMPORTANT

Many brake experts recommend rebuilding or replacing wheel cylinders with every *other* brake job. Some experts recommend that the wheel cylinders be overhauled or replaced every time the brake linings are replaced. If the wheel cylinders are found to be leaking, they *must* be replaced or overhauled. The most important factor is *time*, not mileage, when determining when to repair or replace hydraulic components.

The longer the elapsed time, the more moisture is absorbed by the brake fluid. The greater the amount of moisture absorbed by the brake fluid, the greater the corrosion to metal hydraulic components. For example, the brakes in a vehicle that is used all day every day will probably wear out much sooner than those in a vehicle driven only a short distance every week. In this example, the high-mileage vehicle may need replacement brake linings every year, whereas the short-distance vehicle will require several years before replacement brakes are needed. The service technician should try to determine the amount of time the brake fluid has been in the vehicle. The longer the brake fluid has been in the system, the greater the chances that the wheel cylinders need to be replaced or overhauled.

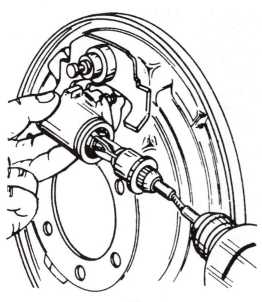

(a)

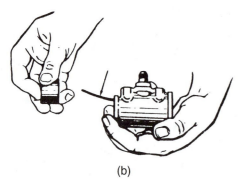

(b)

FIGURE 5–27 (a) A special wheel cylinder hone being turned with an electric drill. Use brake fluid to lubricate the hone stones during the honing operation. Thoroughly clean the wheel cylinder using a soft cloth and clean brake fluid. (b) Use a narrow [1/4 in. (6 mm)] thickness (feeler) gauge 0.005 in. (0.13 mm) thick held into the wheel cylinder. If the cylinder piston fits, the wheel cylinder bore is too large and the wheel cylinder must be replaced. Typical piston-to-wall clearance should range from 0.001 in. (0.03 mm) to 0.003 in. (0.08 mm).

NOTE: Some vehicle manufacturers do not recommend honing wheel cylinders because honing would destroy the special bore surface.

Step 5. Install the pistons (usually *not* included in a wheel cylinder overhaul kit), seals, spring, and dust boots. Then bleed the system.

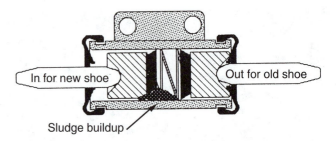

Sludge buildup

FIGURE 5–28 When new, thicker brake shoes are installed, the pistons and cups are forced back into the wheel cylinder and pushed through the sludge that is present in every cylinder.

NOTE: Even though the wheel cylinder is not leaking, many brake experts recommend replacing or rebuilding the wheel cylinder every time replacement linings are installed. Any sludge buildup in the wheel cylinder can cause the wheel cylinder to start to leak shortly after a brake job. When the new thicker replacement linings are installed, the wheel cylinder piston may be pushed inward enough to cause the cup seals to ride on a pitted or corroded section of the wheel cylinder. As the cup seal moves over this rough area, the seal can lose its ability to maintain brake fluid pressure and an external brake fluid leak can occur (see Figure 5–28).

REASSEMBLING THE DRUM BRAKE

Carefully clean the backing plate. Check the anchor pin for looseness. Lubricate the shoe contact surfaces (shoe pads) and self-adjuster with antiseize, brake grease, or synthetic grease (see Figure 5–29). Reassemble the primary and secondary shoes and brake strut along with all springs.

HINT: Many technicians preassemble the primary and secondary shoes with the connecting (lower retracting) spring as a unit before installing them onto the backing plate (see Figure 5–30).

Finish assembling the drum brake, being careful to note the correct location of all springs and parts. Most self-adjusters operate off the rear (secondary) shoe and should therefore be assembled toward the rear of the vehicle.

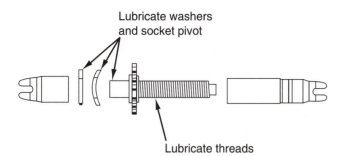

Lubricate washers
and socket pivot

Lubricate threads

FIGURE 5–29 Star-wheel adjusters are designed with left- and right-handed threads. They are not interchangeable from side to side! Using an adjuster on the wrong side of the vehicle would cause the brake to self-adjust inward (farther from the brake drum) instead of closer to the brake drum. The wavy washer acts on the flat thrust washer to help prevent noise. The threads and end caps should be cleaned and lubricated before reuse.

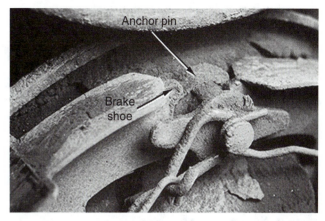

Anchor pin

Brake shoe

FIGURE 5–31 Notice that the brake shoe is not contacting the anchor pin. This often occurs when the parking brake cable is stuck or not adjusted properly.

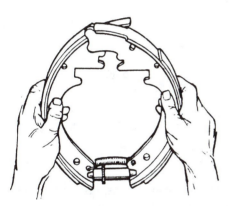

FIGURE 5–30 Sometimes it is necessary to cross the shoes when preassembling the star-wheel adjuster and connecting spring. (Courtesy of Allied Signal Automotive Aftermarket)

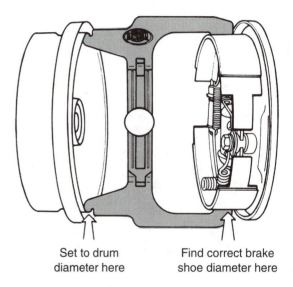

Set to drum
diameter here

Find correct brake
shoe diameter here

FIGURE 5–32 A brake shoe clearance gauge is first placed in the drum and the knob tightened to maintain the measured distance. The gauge is then turned around and the outside arm of the tool can be used as a gauge to adjust the shoes. When the shoes touch the gauge, the brakes are correctly matched to the brake drum. (Courtesy of Ford Motor Company)

ADJUSTING DRUM BRAKES

Most drum brakes are adjusted by rotating a star wheel or rotary adjuster. As the adjuster is moved, the brake shoes move toward the drum. If the brakes have been assembled correctly with the parking brake fully released, both brake shoes should make contact with the anchor pin at the top. See Figure 5–31 for an example where one shoe does not make contact with the anchor pin. Many technicians use a **brake shoe clearance gauge** to adjust the brake shoes before installing the drum (see Figure 5–32).

CAUTION: Before installing the brake drum, be sure to clean any grease off the brake lining. Do not use sandpaper on the lining to remove grease. The sandpaper may release asbestos fiber into the air. Cover the linings with masking tape prior to installation on the vehicle to help protect the lining. Grease on the linings can cause the brakes to grab.

DRUM BRAKE TROUBLESHOOTING GUIDE

Low Pedal or Pedal Goes to the Floor

Possible Causes:

1. Excessive clearance between linings and drum
2. Automatic adjusters not working
3. Leaking wheel cylinder
4. Air in the system

Springy, Spongy Pedal

Possible Causes:

1. Drums worn below specifications
2. Air in the system

Excessive Pedal Pressure Required to Stop Vehicle

Possible Causes:

1. Grease- or fluid-soaked linings
2. Frozen wheel cylinder pistons
3. Linings installed on wrong shoes

Light Pedal Pressure—Brakes Too Sensitive

Possible Causes:

1. Brake adjustment not correct
2. Loose backing plate
3. Lining loose on the shoe
4. Excessive dust and dirt in the drums
5. Scored, bellmouthed, or barrel-shaped drum
6. Improper lining contact pattern

Brake Pedal Travel Decreasing

Possible Causes:

1. Weak shoe retracting springs
2. Wheel cylinder pistons sticking

Pulsating Brake Pedal (Parking Brake Apply Also Pulsates)

Possible Causes:

1. Drums out of round

Brakes Fade (Temporary Loss of Brake Effectiveness When Hot)

Possible Causes:

1. Poor lining contact
2. Drums worn below the discard dimension
3. Charred or glazed linings

Shoe Click

Possible Causes:

1. Shoes that lift off the backing plate and snap back
2. Weak hold-down springs
3. Shoe bent
4. Grooves in the backing plate pads

Snapping Noise

Possible Causes:

1. Grooved backing plate pads
2. Loose backing plates

Thumping Noise When Brakes Are Applied

Possible Causes:

1. Cracked drum; hard spots in the drum
2. Retractor springs unequal—weak

Grinding Noise

Possible Causes:

1. Shoe hitting the drum
2. Bent shoe web
3. Brake improperly assembled

One Wheel Drags

Possible Causes:

1. Weak or broken shoe retracting springs
2. Brake shoe to drum clearance too tight—brake shoes not adjusted properly
3. Brake improperly assembled
4. Wheel cylinder piston cups swollen and distorted
5. Pistons sticking in the wheel cylinder

6. Drum out of round
7. Loose anchor pin/plate
8. Parking brake cable not free
9. Parking brake not adjusted properly

Vehicle Pulls to One Side

Possible Causes:

1. Brake adjustment not correct
2. Loose backing plate
3. Linings not of specified kind, primary and secondary shoes reversed or not replaced in pairs
4. Water, mud, or other material in brakes
5. Wheel cylinder sticking
6. Weak or broken shoe retracting springs
7. Drums out of round
8. Wheel cylinder size different on opposite sides
9. Scored drum

Brakes Squeak

Possible Causes:

1. Backing plate bent or shoes twisted
2. Shoes scraping on backing plate pads
3. Weak or broken hold-down springs
4. Loose backing plate, anchor, or wheel cylinder
5. Glazed linings
6. Dry shoe pads and hold-down pin surfaces

Brakes Chatter

Possible Causes:

1. Incorrect lining to drum clearance
2. Loose backing plate

3. Weak or broken retractor spring
4. Drums out of round
5. Tapered or barrel-shaped drums
6. Improper lining contact pattern

SUMMARY

1. Brake shoes are forced outward against a brake drum by hydraulic action working on the brake shoes by the piston of a wheel cylinder.
2. The curved arch of the brake shoe causes a wedging action between the brake shoe and the rotating drum. This wedging action increases the amount of force applied to the drum.
3. Dual-servo brakes use primary and secondary brake shoes that are connected at one end. The wedge action on the front (primary) shoe forces the secondary shoe into the drum with even greater force. This action is called servo self-energizing.
4. Leading-trailing brakes use two brake shoes that are *not* connected. Leading-trailing brakes operate on a more linear basis and are therefore better suited than dual-servo brakes for ABS.
5. Care should be exercised when removing a brake drum so as not to damage the drum, backing plate, or other vehicle components.
6. After disassembly of the drum brake component, the backing plate should be inspected and cleaned.
7. Most experts recommend replacing the wheel cylinder as well as all brake springs as part of a thorough drum brake overhaul.
8. When a drum brake is assembled properly, both shoes should contact the anchor pin.

REVIEW QUESTIONS

1. Describe the difference between a dual-servo and a leading-trailing drum brake system.
2. List all the parts of a typical drum brake.
3. List all items that should be lubricated on a drum brake.
4. Explain how a self-adjusting brake mechanism works.

MULTIPLE-CHOICE QUESTIONS

1. Technician A says that the tinnerman nuts used to hold the brake drum on should be reinstalled when the drum is replaced. Technician B says that a drum should be removed inside a sealed vacuum enclosure or washed with water or solvent to prevent possible asbestos dust released into the air. Which technician is correct?
   a. A only
   b. B only
   c. Both A and B
   d. Neither A nor B

2. The backing plate should be replaced if the shoe contact areas (pads or ledges) are worn more than
   a. 1/2 in. (13 mm).
   b. 1/4 in. (7 mm).
   c. 1/8 in. (4 mm).
   d. 1/16 in. (2 mm).

3. Technician A says that the hold-down spring can be cut to release a stuck brake drum. Technician B says that only synthetic brake grease, lithium brake grease, or antiseize compound should be used as a brake lubricant. Which technician is correct?
   a. A only
   b. B only
   c. Both A and B
   d. Neither A nor B

4. Most brake experts and vehicle manufacturers recommend replacing brake lining when the lining thickness is
   a. 0.030 in. (0.8 mm).
   b. 0.040 in. (1.0 mm).
   c. 0.050 in. (1.3 mm).
   d. 0.060 in. (1.5 mm).

5. Technician A says that star-wheel adjusters use different threads (left and right handed) for the left and right sides of the vehicle. Technician B says that the threads and end caps of the adjusters should be lubricated with brake grease before being installed. Which technician is correct?
   a. A only
   b. B only
   c. Both A and B
   d. Neither A nor B

6. Technician A says that many vehicle manufacturers recommend that wheel cylinders *not* be honed because of the special surface finish inside the bore. Technician B says that seal expanders are used to help prevent the lip of the cup seal from collapsing when the brakes are released. Which technician is correct?
   a. A only
   b. B only
   c. Both A and B
   d. Neither A nor B

7. Most manufacturers recommend that brake parts be cleaned with
   a. carburetor cleaner.
   b. denatured alcohol.
   c. Stoddard solvent.
   d. detergent and water.

8. Old brake shoes are often returned to the manufacturer, where new friction material is installed. These old shoes are usually called the
   a. core.
   b. web.
   c. rim.
   d. nib.

9. After assembling a drum brake, it is discovered that the brake drum will not fit over the new brake shoes. Technician A says that the parking brake cable may not have been fully released. Technician B says to check to see if both shoes are contacting the anchor pin. Which technician is correct?
   a. A only
   b. B only
   c. Both A and B
   d. Neither A nor B

10. Technician A says to use masking tape temporarily over lining material to help prevent getting grease on the lining. Technician B says that grease on the brake lining can cause the brakes to grab. Which technician is correct?
    a. A only
    b. B only
    c. Both A and B
    d. Neither A nor B

◀ Chapter 6 ▶

DISC BRAKE OPERATION, DIAGNOSIS, AND SERVICE

OBJECTIVES

After studying Chapter 6, the reader will be able to:

1. Describe how disc brakes function.
2. Name the parts of a typical disc brake system.
3. Identify the types of disc brake pads.
4. Explain how to disassemble and reassemble a disc brake caliper assembly.
5. Describe how to prevent disc brake noise.

Disc brakes use a piston(s) to squeeze friction material (pads) on both sides of a rotating disc (rotor). The rotor is attached to and stops the wheel. Disc brakes are capable of converting energy of a moving vehicle into heat (see Figure 6–1).

FIXED CALIPERS

The first disc brakes on American vehicles often used four pistons, two on each side of the rotors. The caliper (containing the four pistons with rubber piston seals on each) was bolted directly to the steering knuckle and did not move. This is called a **fixed caliper** (see Figures 6–2 and 6–3).

When the brake pedal is depressed, hydraulic brake fluid is forced into the caliper cylinder bores. This forces the pistons outward and against the pads as shown in Figure 6–4. Because of Pascal's law, all four pistons received the same pressure and because of the large surface area of the four pistons combined, the pressure was great enough to stop big, heavy American vehicles.

FIGURE 6–I Disc brakes can absorb and dissipate a great deal of heat. During this demonstration, the brakes were applied gently as the engine drove the front wheels until the rotor became cherry red. During normal braking, the rotor temperature can exceed 350°F (180°C) and about 1500°F (800°C) on a race vehicle.

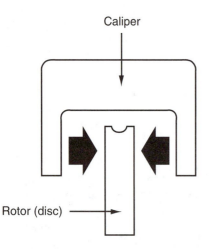

Caliper

Rotor (disc)

FIGURE 6–2 A fixed caliper disc brake uses two or four pistons, with one or two on each side of the rotating rotor (disc).

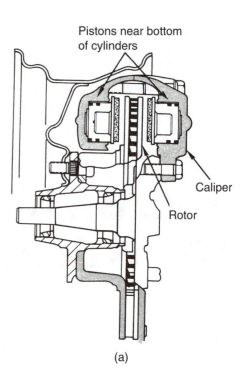

Pistons near bottom of cylinders

Caliper

Rotor

(a)

FIGURE 6–3 A four-piston fixed caliper assembly on a race vehicle.

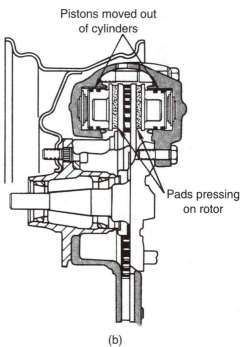

Pistons moved out of cylinders

Pads pressing on rotor

(b)

FIGURE 6–4 (a) Fixed caliper in the released position. (b) When the brake pedal is depressed, brake fluid from the master cylinder is applied behind the caliper piston, forcing the friction pads against the rotor. (Courtesy of Allied Signal Automotive Aftermarket)

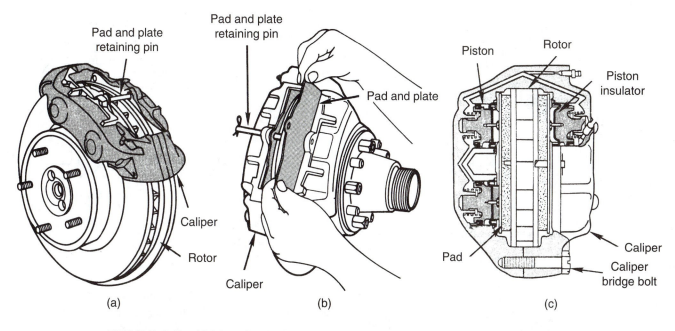

Pad and plate retaining pin

Pad and plate retaining pin

Pad and plate

Caliper

Rotor

Caliper

Caliper

Piston

Rotor

Piston insulator

Pad

Caliper

Caliper bridge bolt

(a)

(b)

(c)

FIGURE 6–5 (a) Many fixed caliper disc brakes use a simple retaining pin to hold the disc brake pads. (b) Removing the retainer pads allows the pads to be removed. (c) Notice the crossover hydraulic passage that connects both sides of the caliper. (Courtesy of Allied Signal Automotive Aftermarket)

When the brake pedal is released, a small amount of brake fluid returns to the master cylinder, lowering the hydraulic pressure on the piston. The clamping action of the brake pads against the rotor is released and the caliper pistons retract slightly back into the caliper bore (see Figure 6–5).

FLOATING CALIPERS

Most modern disc brake calipers use only one large piston in a caliper, which moves slightly, allowing it to squeeze the rotor between the two disc pads. This type of caliper is called a **single-piston floating caliper** (see Figure 6–6). When the brakes are applied, the hydraulic pressure in the caliper bore is exerted equally against the bottom of the piston and on the back of the caliper itself (remember Pascal's law; see Figures 6–7 and 6–8).

The pressure on the piston forces the inside pad against the rotor. At the same time the piston is pushing against the rotor, the caliper itself is being forced toward the center of the vehicle. Since the outboard pad is attached to this inward-moving caliper, it contacts the outer surface of the rotor with the same force that is being exerted on the inboard pad, as shown in Figure 6–9.

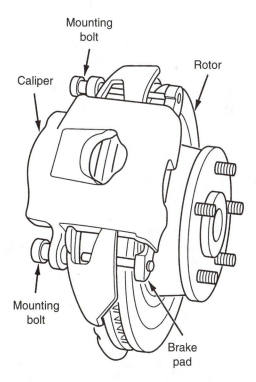

Mounting bolt

Caliper

Rotor

Mounting bolt

Brake pad

FIGURE 6–6 Typical single-piston floating caliper. In this type of design, the entire caliper moves when the single piston is pushed out of the caliper during a brake application. When the caliper moves, the outboard pad is pushed against the rotor. (Courtesy of General Motors)

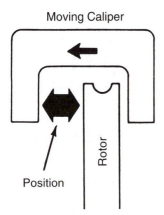

FIGURE 6–7 In a floating caliper, a piston on one side applies force to both the rotor and the caliper causing the caliper to move. (Courtesy of Ford Motor Company)

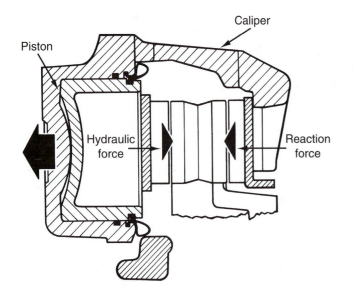

FIGURE 6–8 Hydraulic force on the piston (left) is applied to the inboard pad and the caliper housing itself. The reaction of the piston pushing against the rotor causes the entire caliper to move toward the inside of the vehicle (large arrow). Since the outboard pad is retained by the caliper, the reaction force of the moving caliper applies pressure of the outboard pad against the outboard surface of the rotor.

NOTE: If a floating caliper is stuck and cannot move easily when the brakes are applied, the inboard pad will wear more than the outboard pad. Uneven pad wear will also occur if the piston is stuck ("frozen") inside the caliper. Therefore, when uneven disc brake pad wear occurs, inspect the caliper piston(s) and caliper slides carefully.

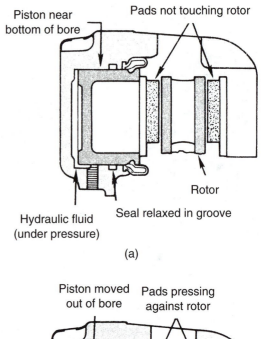

(a)

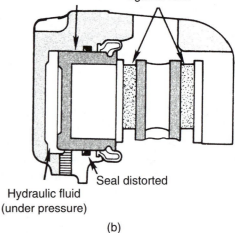

(b)

FIGURE 6–9 (a) Single-piston floating caliper brake in the released position. (b) When the brake pedal is depressed, hydraulic fluid, under pressure, moves the caliper piston outward, pressing the disc brake pads against the rotor, bringing the vehicle to a stop. (Courtesy of Allied Signal Automotive Aftermarket)

The square-cut O-ring piston seal allows the piston to move and also acts as a piston return. As the brake pad wears, the piston is forced to move through the caliper seal to make up for any pad wear.

DISC BRAKE PADS

Disc brake pads have friction material attached to a steel backing. Three methods are used to attach the

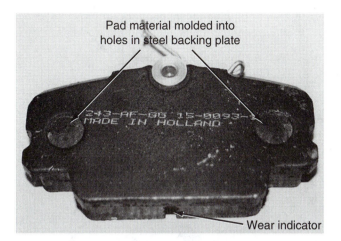

FIGURE 6–10 These replacement pads have both a wear indicator groove and are integrally molded.

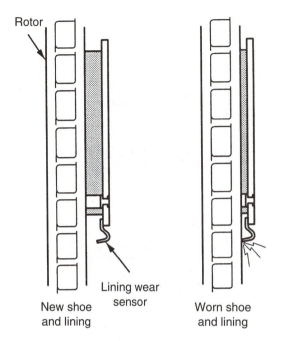

FIGURE 6–11 Typical pad wear sensor operation. (Courtesy of General Motors)

◀ **TECH TIP** ▶

BLEED AND SQUIRT TEST

If you suspect that a brake is not fully being released, simply loosen the bleeder valve. If brake fluid squirts out under pressure, the brake is being kept applied. Look for a defective flexible brake hose.

If the vehicle is off the ground, the wheels should be able to be rotated with the brakes off. If a wheel is difficult or hard to turn by hand and is easy to turn after opening the bleeder valve, there is a brake fluid restriction between the master cylinder and the brake.

metal tag are rubbing against the rotor (see Figure 6–11).

***

NOTE: With many vehicles, the wear indicator noise *stops* when the brakes are applied. The fact that the noise disappears when the brakes are applied has wrongly convinced many drivers (and some service technicians) that the problem is *not* due to the brakes. Some vehicles have indicators that make noise only while the vehicle is being driven in reverse. Other vehicles have sensors that tend to make the most noise during braking. Any noise while driving should be investigated.

***

friction material to the steel backing: *riveted, bonded, or integrally molded.* Integral molded linings have holes in the steel backing to allow the friction material to become a part of the steel backing.

WEAR SENSORS

Many disc brake pads are equipped with a groove or notch molded into the pad, as shown in Figure 6–10. When the pads are worn down to the depth of the notch, the pads should be replaced. Another type of wear indicator uses a soft metal tang that contacts the rotor when the pad is worn down to the thickness requiring replacement. These wear indicators make a "chirp, chirp" sound when the wheels are rotating and the ends of the

◀ **TECH TIP** ▶

BE SURE TO HAVE A NICKEL'S WORTH OF LININGS

While the *exact* thickness of allowable brake lining varies with the vehicle manufacturer, most experts agree that the lining should be thicker than the thickness of an American nickel, about ¹⁄₁₆ in. (0.060 in.) or 1.5 mm. Another rule of thumb that is easily used is to replace the brake pads if the thickness of the friction material is the same or less than the thickness of the steel backing of the disc brake pad.

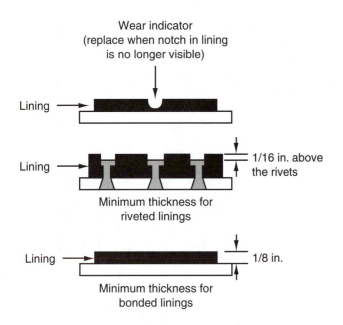

FIGURE 6–12   Minimum thickness for various types of disc brake pads. Disc brake pads, of course, can be replaced before they wear down to the factory recommended *minimum* thickness.

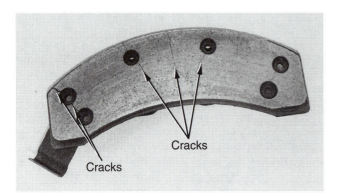

FIGURE 6–13   This cracked disc brake pad must be replaced even though it is thicker than the minimum allowable by the vehicle manufacturer.

VISUAL INSPECTION

Even with operating-wear-indicating sensors, a thorough visual inspection is very important (see Figure 6–12). A lining thickness check alone should not be the only inspection performed on a disc brake. *A thorough visual inspection can be accomplished only by removing the friction pads.* See Figure 6-13 for an example of a disc brake pad that shows usable lining

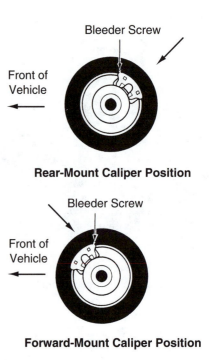

FIGURE 6–14   Both rear-mount and forward-mount calipers have the bleeder valve at the top. Some calipers *will* fit on the wrong side of the vehicle, yet not be able to be bled correctly because the bleeder valve would point down, allowing trapped air to remain inside the caliper bore. If both calipers are being removed at the same time, mark them left and right.

thickness but is severely cracked and *must* be replaced.

NOTE: Some disc brake pads use a heat barrier (thermo) layer between the steel backing plate and the friction material. The purpose of the heat barrier is to prevent heat from transferring into the caliper piston, where it may cause the brake fluid to boil. Do not confuse the thickness of the barrier as part of the thickness of the friction lining material. The barrier material is usually of a different color and thus can be distinguished from the lining material.

DISC BRAKE CALIPER

Removal.   Hoist the vehicle and remove the wheel(s). Before removing the caliper, note the caliper mount position as shown in Figure 6–14. It may be necessary to know whether the caliper is "rear mount" or "forward mount" when purchasing replacement calipers. Remove the caliper following the steps shown in Figures 6–15 through 6–21.

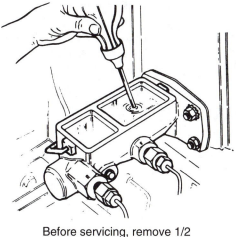

Before servicing, remove 1/2
fluid from reservoir

FIGURE 6-15 Many manufacturers recommend removing one-half of the brake fluid from the master cylinder before servicing disc brakes. Use a squeeze bulb and dispose of the used brake fluid correctly. (Wagner Division, Cooper Industries Inc.)

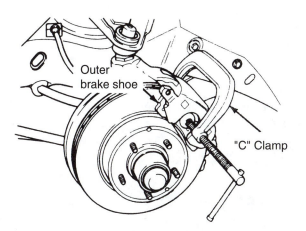

Outer
brake shoe

"C" Clamp

FIGURE 6-16 Use a C-clamp on the outboard pad and the caliper housing to squeeze the disc brake caliper piston back into its bore. As the piston is being forced into the caliper, brake fluid is being forced back into the master cylinder reservoir. (Wagner Division, Cooper Industries Inc.)

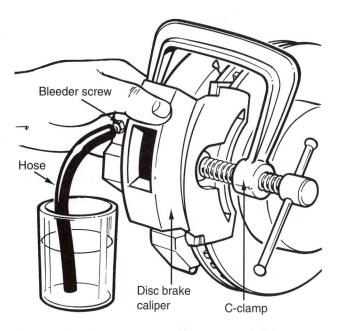

Bleeder screw

Hose

Disc brake
caliper

C-clamp

FIGURE 6-17 Some manufacturers recommend that the bleeder valve be opened and the brake fluid forced into a container rather than back into the master cylinder reservoir. This helps prevent contaminated brake fluid from being forced into the master cylinder, where the dirt and contamination could cause problems.

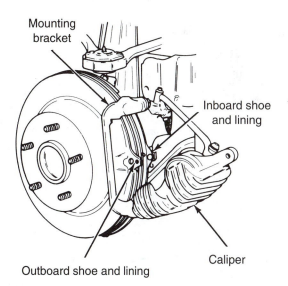

Mounting
bracket

Inboard shoe
and lining

Outboard shoe and lining

Caliper

FIGURE 6-18 Many disc brake calipers can simply be pivoted upward or downward on one of the mounting bolts to gain access to the disc brake pads. (Courtesy of General Motors)

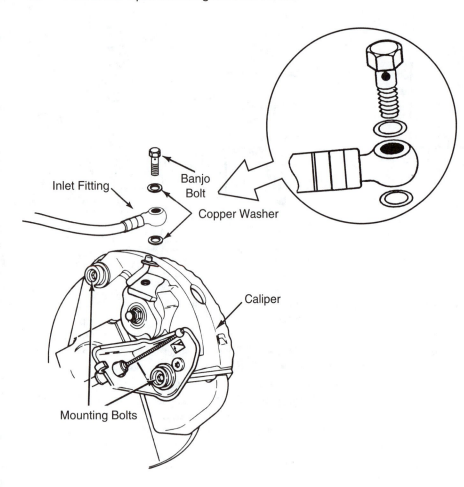

FIGURE 6–19 Many calipers use a hollow "banjo bolt" to retain the flexible brake line to the caliper housing. The fitting is usually round like a banjo musical instrument. The copper washers should always be replaced and not reused.

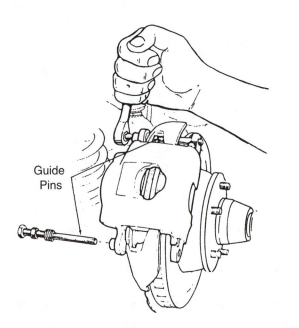

FIGURE 6–20 Caliper retaining bolts are often called guide pins. These pins are used to retain the caliper to the steering knuckle. These pins also slide through metal bushings and rubber O-rings. (Courtesy of EIS Brake Parts)

Inspection and Disassembly. Check for brake fluid in and around the piston boot area. If the boot is damaged or a fluid leak is visible, a repair or caliper assembly replacement is required (see Figures 6–22 and 6–23).

CALIPER PISTONS

Phenolic caliper pistons are made from a phenol-formaldehyde resin combined with various reinforcing fibers. Phenolic caliper pistons are natural thermal insulators and help keep heat generated by the disc brake pads from transferring through the caliper piston to the brake fluid. Phenolic brake caliper pistons are also lighter in weight than steel caliper pistons and are usually brown in color (see Figure 6–24).

Many manufacturers still use steel pistons. The stamped steel pistons are plated first with nickel, then chrome to achieve the desired surface finish (see Figure 6–25). Unlike phenolic caliper pistons, steel pistons can transfer heat from the brake pads to the brake fluid. The surface finish on a steel piston is critical. Steel can rust and corrode. Any surface pitting can cause the piston to stick.

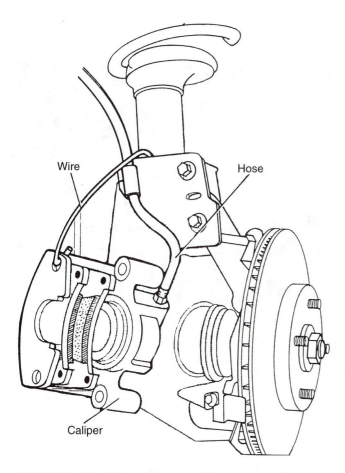

Wire

Hose

Caliper

FIGURE 6–21 If the caliper is not being removed, it must be supported properly so that the weight of the caliper is not pulling on the flexible rubber brake line. A suitable piece of wire such as a coat hanger may be used.

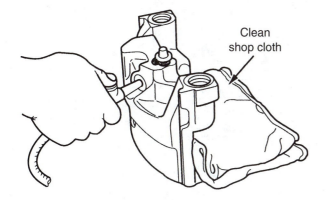

Clean shop cloth

FIGURE 6–22 With the caliper removed from the vehicle, the caliper piston can be removed by applying compressed air into the inlet port with the bleeder valve closed. The compressed air will force the piston out of the caliper. A cloth or other soft material should be used to prevent damage to the piston as it is being forced out. Obviously, hands and fingers should also be kept out of the path of the piston. (Courtesy of General Motors)

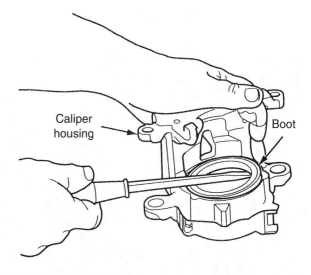

Caliper housing

Boot

FIGURE 6–23 After the piston is removed from the caliper housing, the dust boot can often be removed using a straight-blade screwdriver. (Courtesy of General Motors)

Cracks, chips, gouges may not enter dust boot groove

Cracks, chips, gouges may be 1/2" long and may go inward almost to piston seal groove.

If no cracks, chips, gouges, or any other surface damage on ground seal surface (piston o.d.), the pistons are acceptable.

FIGURE 6–24 Phenolic (plastic) pistons should be inspected carefully. (Courtesy of Allied Signal Automotive Aftermarket)

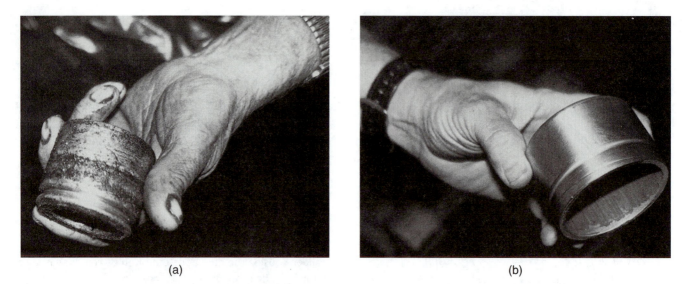

(a)   (b)

FIGURE 6–25 The outside surface of caliper pistons should be inspected carefully. The square-cut O-ring inside the caliper rides on this outside surface of the piston. Dirty pistons can sometimes be cleaned and reused. If there are any surface flaws such as rust pits on the piston, it should be replaced.

NOTE: Care should be taken when cleaning steel pistons. Use crocus cloth to remove any surface staining. Do not use sandpaper or emery cloth or any other substance that may remove or damage the chrome surface finish.

REASSEMBLING DISC BRAKE CALIPERS

After disassembly, the caliper should be cleaned thoroughly in denatured alcohol and closely examined. If the caliper bore is rusted or pitted, some manufacturers recommend that a special hone be used, as shown in Figure 6–26. Some manufacturers do *not* recommend honing the caliper bore because the actual sealing surface in the caliper is between the piston seal and the piston itself. This is the reason the surface condition of the piston is so important.

Carefully clean the caliper bore with clean brake fluid from a sealed container. Coat a new piston seal with clean brake fluid and install it in the groove inside the caliper bore as shown in Figure 6–27. Check the piston-to-caliper bore clearance. Typical piston-to-caliper bore clearance is:

Steel piston: 0.002 to 0.005 in.(0.05 to 0.13 mm)

Phenolic piston: 0.005 to 0.010 in.(0.13 to 0.25 mm)

Coat a new piston boot with brake fluid and install the piston into the caliper piston, as shown in Figures 6–28 and 6–29. Some caliper boots require a special boot seating tool as shown in Figure 6-30.

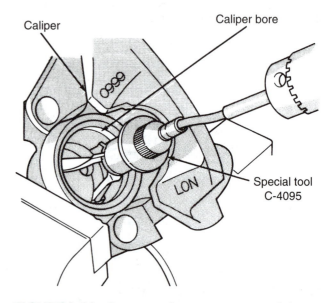

FIGURE 6–26 Some manufacturers recommend cleaning the inside of the caliper bore using a honing tool as shown. Even though the caliper piston does not contact the inside of this bore, removing any surface rust or corrosion is important to prevent future problems. If the honing process cannot remove any pits or scored areas, the caliper should be replaced. (Courtesy of Chrysler Corporation)

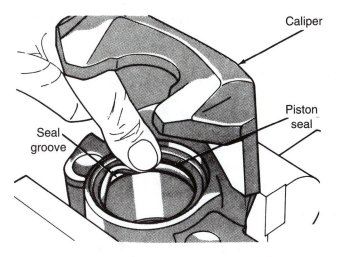

FIGURE 6–27   Installing a new piston seal. Never reuse old rubber parts. (Courtesy of Chrysler Corporation)

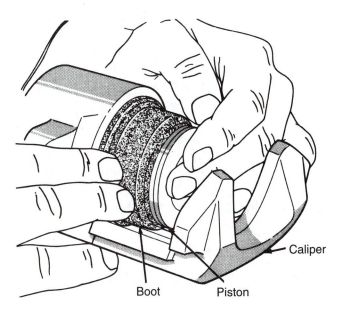

FIGURE 6–28   Installing the caliper piston. Many calipers require that the dust boot be installed in the groove of the piston and/or caliper before installing the piston. (Courtesy of Chrysler Corporation)

Always lubricate caliper bushings, shims, and other brake hardware as instructed by the manufacturer (see Figure 6–31). The pads should also be attached securely to the caliper, as shown in Figures 6–32 and 6–33.

CAUTION: Installing disc brake pads on the wrong side of the vehicle (left versus right) will often prevent the sensor from making noise when the pads are worn down.

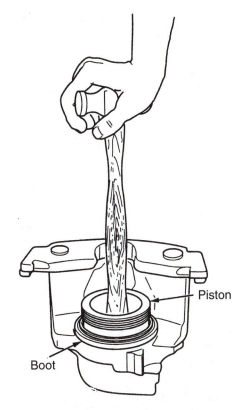

FIGURE 6–29   Installing a piston into a caliper. Sometimes a C-clamp is needed to install the piston. Both the piston and the piston seal should be coated in clean brake fluid before assembly. (Courtesy of General Motors)

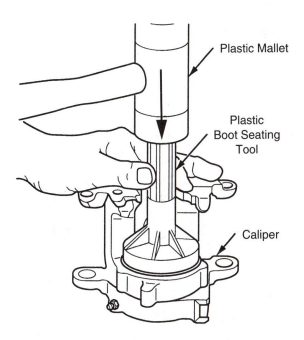

FIGURE 6–30   Seating the dust boot into the caliper housing using a special plastic seating tool.

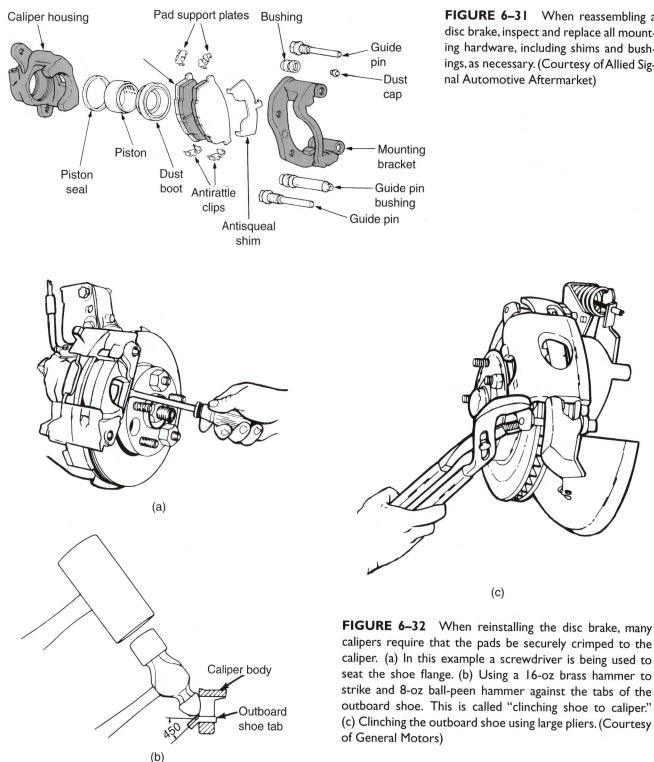

FIGURE 6–31 When reassembling a disc brake, inspect and replace all mounting hardware, including shims and bushings, as necessary. (Courtesy of Allied Signal Automotive Aftermarket)

FIGURE 6–32 When reinstalling the disc brake, many calipers require that the pads be securely crimped to the caliper. (a) In this example a screwdriver is being used to seat the shoe flange. (b) Using a 16-oz brass hammer to strike and 8-oz ball-peen hammer against the tabs of the outboard shoe. This is called "clinching shoe to caliper." (c) Clinching the outboard shoe using large pliers. (Courtesy of General Motors)

CALIPER MOUNTS

When the hydraulic force from the master cylinder applies pressure to the disc brake pads, the entire caliper tends to be forced in the direction of rotation of the rotor.

All calipers are mounted to the steering knuckle or axle housing (see Figure 6–34). All braking force is transferred through the caliper to the mount. Where the caliper contacts the caliper mount are called the **abutments, reaction pads,** or **ways.** The sliding surfaces of the caliper

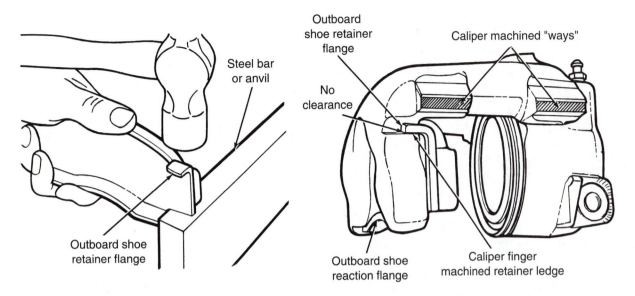

FIGURE 6-33 Often, a hammer is necessary to bend the retainer flange to make certain that the pads fit tightly to the caliper. If the pads are loose, a click may be heard every time the brakes are depressed. The click occurs when the pad(s) move and hit the caliper or caliper mount. If the pads are loose, a clicking noise may be heard while driving over rough road surfaces. (Courtesy of Chrysler Corporation)

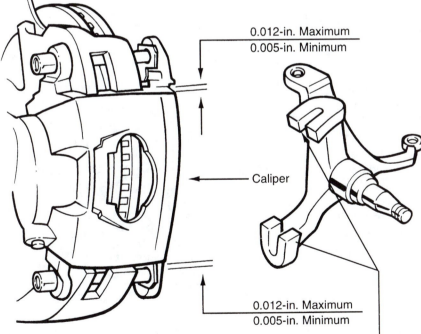

FIGURE 6-34 Floating calipers must be able to slide during normal operation. Therefore, there must be clearance between the caliper and the caliper mounting pads (abutments). Too little clearance will prevent the caliper from sliding, and too much clearance will cause the caliper to make a clunking noise when the brakes are applied.

LOADED CALIPERS

Many technicians find that disassembly, cleaning, and rebuilding calipers can take a lot of time. Often, the bleeder valve breaks off or the caliper piston is too corroded to reuse. This means that the technician has to get a replacement piston, caliper overhaul kit (piston seal and boot), and replacement friction pads and hardware kit.

To save time (and sometimes money), many technicians are simply replacing the old used calipers with *loaded calipers*, which are remanufactured calipers that include (come loaded with) the correct replacement friction pads and all the necessary hardware. Therefore, only one part number is needed for each side of the vehicle for a complete disc brake overhaul.

support should be cleaned with a wire brush and coated with a synthetic grease or antiseize compound according to the manufacturer's recommendations.

TEST DRIVE AFTER BRAKE REPLACEMENT

After installing replacement disc brake pads or any other brake work, depress the brake pedal several times before driving the vehicle. This is a very important step! New brake pads are installed with the caliper piston pushed all the way into the caliper. The first few brake pedal applications usually result in the brake pedal going all the way to the floor. The brake pedal must be depressed ("pumped") several times before enough brake fluid can be moved from the master cylinder into the calipers to move the piston tight against the pads and the pads against the rotors.

CAUTION: Never allow a customer to be the first to test drive a vehicle after brake work has been performed.

REAR DISC BRAKES

Rear disc brakes are used on the rear of many vehicles. A parking brake is more difficult to use with disc brakes; therefore, rear disc brakes are commonly found on high-performance or more expensive vehicles. Most vehicles equipped with rear disc brakes use one of two different styles of parking brake.

1. An integral parking brake built into the piston assembly (see Figures 6–35 through 6–37).
2. A more conventional disc brake that uses a small drum brake for the parking brake function (see Figure 6–38).

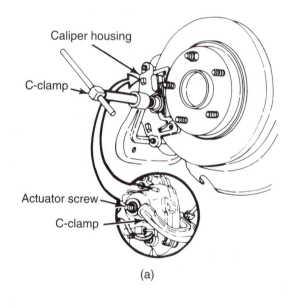

(a)

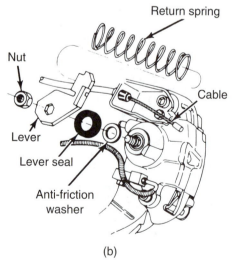

(b)

FIGURE 6–35 (a) Care should be taken not to damage the actuator screw when using a C-clamp on a rear disc brake. (b) Carefully note the location of the seal and washer when disassembling a rear disc brake. (Courtesy of General Motors)

1. Nut
2. Lever
3. Return spring
4. Bolt
5. Bracket
6. Sleeve
7. Bushing
8. Bolt
9. Copper washer
10. Fitting
11. Bushing
12. Caliper housing
13. Shaft seal
14. Thrust washer
15. Balance spring
16. Actuator screw
17. Piston seal
18. Piston assembly
19. Two-way check valve
20. Bleeder valve
21. Anti-friction washer
22. Lever seal
23. Mounting bolt
24. Boot
25. Inboard shoe & lining
26. Wear sensor
27. Outboard shoe & lining
28. Shoe dampening spring

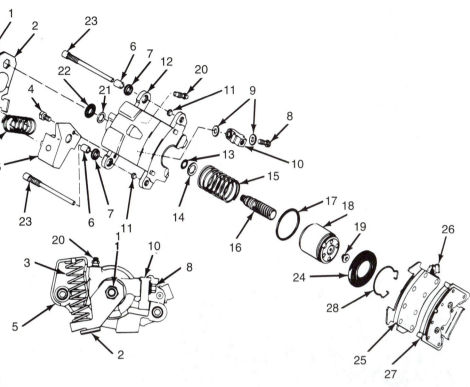

FIGURE 6–36 An exploded view of a typical rear disc brake with an integral parking brake. The parking brake lever mechanically pushes the caliper piston against the rotor. (Courtesy of General Motors)

CORRECTING DISC BRAKE SQUEAL

Brake squeal can best be *prevented* by careful attention to details whenever servicing any disc brake. Some of these precautions include:

1. **Keep the disc brake pads clean.** Grease on brake lining material causes the friction surface to be uneven. When the brakes are applied, this uneven brake surface causes the brake components to move.

2. **Use factory clips and antisqueal shims.** The vehicle manufacturer has designed the braking system to be as quiet as possible. To assure that the brakes are restored to like-new performance, all of the original hardware should be used. Many original equipment brake pads use **constrained layer shims** (CLS) on the back of the brake pads. These shims are constructed with dampening material between two layers of steel (see Figure 6–39).

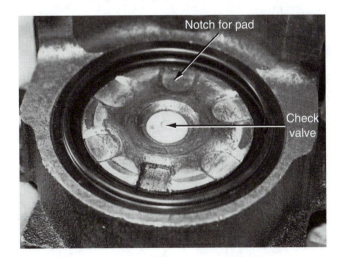

FIGURE 6–37 Many rear disc brakes have a notched piston as shown. The disc brake pad must fit into the notch to keep the piston from rotating inside the caliper bore when the parking brake is applied. The two-way check valve should be pried out. If there is brake fluid under the valve, the caliper should be overhauled or replaced because the inside caliper seal has failed. Slight dampness should be considered normal.

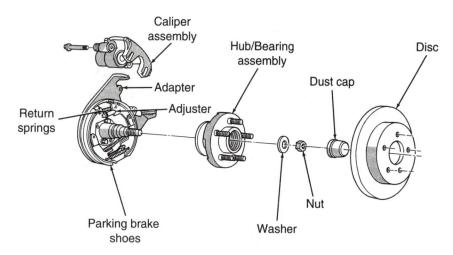

FIGURE 6–38 A typical rear disc brake that uses a small drum brake as the parking brake. The drum brake shoes move outward and contact the inner surface (hat) section of the rotor. (Courtesy of Chrysler Corporation)

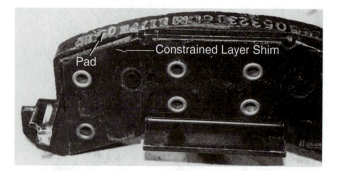

FIGURE 6–39 Many original equipment disc brake pads are equipped with noise-reducing constrained layer (dampening) shims.

NOTE: Many aftermarket disc brake pads do *not* include replacement hardware, which usually includes noise-reducing shims and clips. One of the advantages of purchasing original-equipment (OE) disc brake pads is that they usually come equipped with all necessary shims and often with special grease that is recommended to be used on metal shims.

3. **Lubricate all caliper slide points according to the manufacturer's recommendation.** Lubrication of moving or sliding components prevents noise being generated as the parts move over each other. Many vehicle manufacturers recommend one or more of the following greases:

a. Lithium-based brake grease

b. Silicone grease

c. Molybdenum disulfide (MOS_2) grease ("Molykote")

◀ **TECH TIP** ▶

INCREASING PAD LIFE

Many vehicles seem to wear out front disc brakes more often than normal. Stop-and-go city-type driving is often the cause. Driving style, such as rapid stops, also causes a lot of wear to occur.

The service technician can take some actions to increase brake pad life that is easier than having to cure the driver's habits. These steps include:

1. Make sure that the rear brakes are adjusted properly and working correctly. If the rear brakes are not functioning, all of the braking is accomplished by the front brakes alone.

HINT: Remind the driver to apply the parking brake regularly to help maintain proper rear brake clearance on the rear brakes.

2. Use factory brake pads or premium brake pads from a known manufacturer. Tests performed by vehicle manufacturers show that many aftermarket replacement brake pads fail to deliver original equipment brake pad life.

d. Synthetic grease [usually polyalphaolefin (PAO)] - sometimes mixed with graphite, Teflon, and/or MOS_2

e. Antiseize compound

◄ TECH TIP ►

SCREWDRIVER TRICK

A low brake pedal on GM vehicles equipped with rear disc brakes is a common customer complaint. Often, the reason is a lack of self-adjustment, which *should* occur whenever the brake pedal (or parking brake) is released. During brake release, the pressure is removed from the caliper piston and the spring inside the caliper piston is free to adjust. Often this self-adjustment does not occur and a low brake pedal results.

A common trick that is used on the vehicle assembly line is to use a screwdriver to hold the piston against the rotor while an assistant releases the brake pedal (see Figure 6–40). As the brake pedal is released, the adjusting screw inside the caliper piston is free to move. It may sometimes be necessary to tap on the caliper itself with a dead-blow hammer to free the adjusting screw. Repeat the process as necessary until the proper brake pedal height returns. If this method does not work, replace the caliper assembly.

In summary:

Step 1. Have an assistant depress the brake pedal.

Step 2. Using a screwdriver through the hole in the top of the caliper, hold the piston against the rotor.

NOTE: Be careful not to damage the dust boot.

Step 3. While holding the piston against the rotor, have the assistant release the brake pedal. The adjusting screw adjusts when the brake pedal is *released* and a slight vibration or sound will be noticed as the brake is released. This vibration or sound is created by the self-adjusting mechanism inside the caliper piston taking up the excess clearance.

Step 4. Repeat as necessary until normal brake pedal height is achieved.

The grease should be applied on both sides of shims used between the pad and the caliper piston.

CAUTION: Grease should be applied only to the nonfriction steel side of the disc brake pads.

ROTATING PISTONS BACK INTO THE CALIPER

Many disc brake calipers used on the rear wheels require that the piston be rotated to reseat the pistons. When the parking brake is applied, the actuating screw moves the piston outward, forcing the pads against the disc brake rotor. The piston is kept from rotating because of an antirotation device or notch on the inboard pad and piston.

When the disc brake pads are being replaced, use a special tool to rotate the piston back into the brake calipers. Insert the tip of the tool in the holes or slots in the piston. Exert inward pressure while turning the piston. Make sure that the piston is retracting into the caliper and continue to turn the piston until it bottoms out.

NOTE: Some pistons are activated with left-handed threads.

After replacing the pads in the caliper, check that the clearance does not exceed ¹⁄₁₆ in. (1.5 mm) from the rotor. Clearance greater than ¹⁄₁₆ in. may allow the adjuster to be pulled out of the piston when the service brake is applied. If the clearance is greater than 1/16 in., readjust by rotating the piston outward to reduce the clearance (see Figure 6–41).

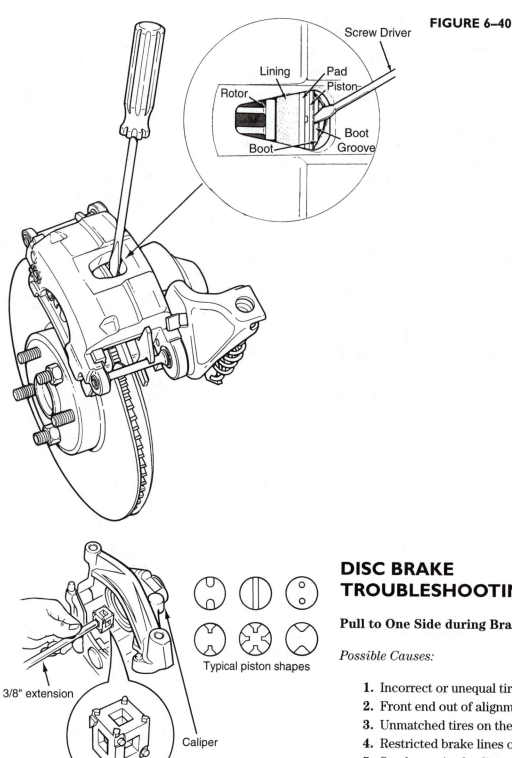

FIGURE 6–40

Typical piston shapes

3/8" extension

Caliper

FIGURE 6–41 Determine which face of the special tool best fits the holes or slots in the piston. Sometimes needle-nose pliers can be used to rotate the piston back into the caliper bore.

DISC BRAKE TROUBLESHOOTING GUIDE

Pull to One Side during Braking

Possible Causes:

1. Incorrect or unequal tire pressures
2. Front end out of alignment
3. Unmatched tires on the same axle
4. Restricted brake lines or hoses
5. Stuck or seized caliper or caliper piston
6. Defective or damaged shoe and lining (grease or brake fluid on the lining, or a bent shoe)
7. Malfunctioning rear brakes
8. Loose suspension parts
9. Loose calipers

Brake Roughness or Chatter (Pedal Pulsates)

Possible Causes:

1. Excessive lateral runout of rotor
2. Parallelism of the rotor not within specifications
3. Wheel bearings not adjusted correctly
4. Rear drums out of round
5. Brake pads worn to metal backing plate

Excessive Pedal Effort

Possible Causes:

1. Binding or seized caliper suspension
2. Binding brake pedal mechanism
3. Improper rotor surface finish
4. Malfunctioning power brake
5. Partial system failure
6. Excessively worn shoe and lining
7. Piston in the caliper stuck or sluggish
8. Fading brakes due to incorrect lining

Excessive Pedal Travel

Possible Causes:

1. Partial brake system failure
2. Insufficient fluid in the master cylinder
3. Air trapped in the system
4. Bent shoe and lining
5. Excessive pedal effort
6. Excessive parking brake travel (four-wheel disc brakes, except Corvette)

Dragging Brakes

Possible Causes:

1. Pressure trapped in the brake lines (to diagnose, open the caliper bleeder valve momentarily to relieve the pressure)
2. Restricted brake tubes or hoses
3. Improperly lubricated caliper suspension system
4. Improper clearance between the caliper and torque abutment surfaces
5. Check valve installed the outlet of the master cylinder to the disc brakes

Front Disc Brakes Very Sensitive to Light Brake Applications

Possible Causes:

1. Metering valve not holding off the front brake application
2. Incorrect lining material
3. Improper rotor surface finish
4. Check other causes listed under "Pulls to One Side during Braking"

Rear Drum Brakes Skidding under Hard Brake Applications

Possible Causes:

1. Proportioning valve
2. Contaminated rear brake lining
3. Caliper or caliper piston stuck or corroded

SUMMARY

1. Disc brakes are used on the front and on the rear wheels on many vehicles. Disc brake calipers are either fixed or floating.
2. When a disc brake is applied, the square-cut O-ring is deformed. When the brakes are released, the rubber O-ring returns to its original shape and draws the caliper piston back into the caliper and away from the rotor.
3. As the disc brake pad wears, the caliper piston moves through the square-cut O-ring to compensate for the wear. Because the piston is now moved outward, brake fluid fills the space and the brake fluid level in the master cylinder drops.
4. Disc brake wear indicators can be a metal tag that touches the rotor and makes noise or a groove molded into the pad itself. Some wear indicators are electrical and light a dash indicator lamp when the brakes are worn.
5. Caliper pistons are either chrome-plated steel or plastic (phenolic). Any damaged piston must be replaced. Both the square-cut O-ring and the dust boot must be replaced when the caliper is disassembled.
6. All metal-to-metal contact points of the disc brake assembly should be coated with an approved brake lubricant such as synthetic grease, "moly" grease, or antiseize compound.

7. After a brake overhaul, the brake pedal should be depressed several times until normal brake pedal action is achieved before performing a thorough test drive.

8. Many rear disc brake systems use an integral parking brake. Regular use of the parking brake helps maintain proper rear brake clearance.

REVIEW QUESTIONS

1. Describe how a single caliper works.
2. List the parts included in a typical overhaul kit for a single-piston floating caliper.
3. List three types of disc brake pad wear sensors.
4. Describe how to remove caliper pistons and perform a caliper overhaul.
5. Explain what causes disc brake squeal and list what a technician can do to reduce or eliminate the noise.

MULTIPLE-CHOICE QUESTIONS

1. Uneven disc brake pad wear is being discussed. Technician A says that the caliper piston may be stuck. Technician B says that the caliper may be stuck on the slides and unable to float. Which technician is correct?
   a. A only
   b. B only
   c. Both A and B
   d. Neither A nor B

2. A "chirping" noise is heard while a vehicle is moving forward but stops when the brakes are applied. Technician A says that the noise is probably caused by the disc brake pad wear sensors. Technician B says that the noise is probably a wheel bearing because the noise stops when the brakes are applied. Which technician is correct?
   a. A only
   b. B only
   c. Both A and B
   d. Neither A nor B

3. What part causes the disc brake caliper piston to retract when the brakes are released?
   a. Return (retracting) spring
   b. Rotating rotor (disc)
   c. Square-cut O-ring
   d. Caliper bushings

4. Two technicians are discussing the reason that the brake fluid level in the master cylinder drops. Technician A says that it may be normal due to the wear of the disc brake pads. Technician B says that a low brake fluid level may indicate a hydraulic leak somewhere in the system. Which technician is correct?
   a. A only
   b. B only
   c. Both A and B
   d. Neither A nor B

5. Technician A says that disc brake pads should be replaced when worn to the thickness of the steel backing. Technician B says that the pads should be removed and inspected whenever there is a brake performance complaint. Which technician is correct?
   a. A only
   b. B only
   c. Both A and B
   d. Neither A nor B

6. A typical disc brake caliper overhaul (OH) kit usually includes what parts?
   a. Square-cut O-ring seal and dust boot
   b. Replacement caliper piston and dust boot
   c. Dust boot, return spring, and caliper seal
   d. Disc brake pad clips, dust boot, and caliper piston assembly

7. Technician A says that a lack of lubrication on the back of the disc brake pads can cause brake noise. Technician B says that pads that are not correctly crimped to the caliper housing can cause brake noise. Which technician is correct?
   a. A only
   b. B only
   c. Both A and B
   d. Neither A nor B

8. Two technicians are discussing ways of removing a caliper piston. Technician A says to use compressed air. Technician B says to use large pliers. Which technician is correct?
   a. A only
   b. B only
   c. Both A and B
   d. Neither A nor B

9. Which is *not* a recommended type of grease to use on brake parts?
   a. Silicone grease
   b. Wheel bearing (chassis) grease
   c. Synthetic grease
   d. Antiseize compound

10. Technician A says that many rear disc brake caliper pistons must be turned to retract before installing replacement pads. Technician B says that some vehicles equipped with rear disc brakes use a small drum brake as the parking brake. Which technician is correct?
    a. A only
    b. B only
    c. Both A and B
    d. Neither A nor B

◀ Chapter 7 ▶

PARKING BRAKE OPERATION, DIAGNOSIS, AND SERVICE

OBJECTIVES

After studying Chapter 7, the reader will be able to:

1. Describe what is required of a parking brake.
2. Describe the parts and operation of the parking brake as used on a rear drum brake system.
3. Describe how a parking brake functions when the vehicle is equipped with rear disc brakes.
4. Explain how to adjust a parking brake properly.

Before 1967, most vehicles had only a single master cylinder operating all four brakes. If the fluid leaked at just one wheel, the operation of all brakes were lost. This required the use of a separate method to stop the vehicle in case of an emergency. This alternative method required that a separate mechanical method be used to stop the vehicle using two of the four wheel brakes. After 1967, federal regulations required the use of dual- or tandem-master cylinders, where half of the braking system has its own separate hydraulic system. In case one-half of the system fails, a dash brake warning lamp lets the driver know that a failure has occurred. The term *parking brake* has replaced the term *emergency brake* since the change to dual-master cylinder design.

PARKING BRAKE STANDARDS

According to Federal Motor Vehicle Safety Standard (FMVSS) 105, the parking brake must hold a fully loaded (laden) vehicle stationary on a slope of 30 percent for a manual transmission–equipped vehicle or a slope of 20 percent if equipped with an automatic transmission. The hand force required cannot exceed 80 lb (18 N) or a foot force greater than 100 lb (22 N). See Figure 7–1 for a typical parking brake system.

PARKING BRAKE LEVER

Parking brakes can be applied using either a hand lever or foot-operated pedal (see Figure 7-2). Some foot-operated parking brakes use a ratchet mechanism that requires that the driver push the pedal down several times to apply. This type of parking brake is commonly called *pump to set*. The lever or foot-pedal mechanism is designed to apply the required force on the parking brake using normal driver effort (see Figure 7–3).

All parking brakes lock into a slot or notch that keeps the parking brake applied until it is released. Some vehicles are equipped with a mechanism connected to the shifter mechanism that automatically releases the parking brake when the transmission is moved from park to a drive gear (either forward or reverse).

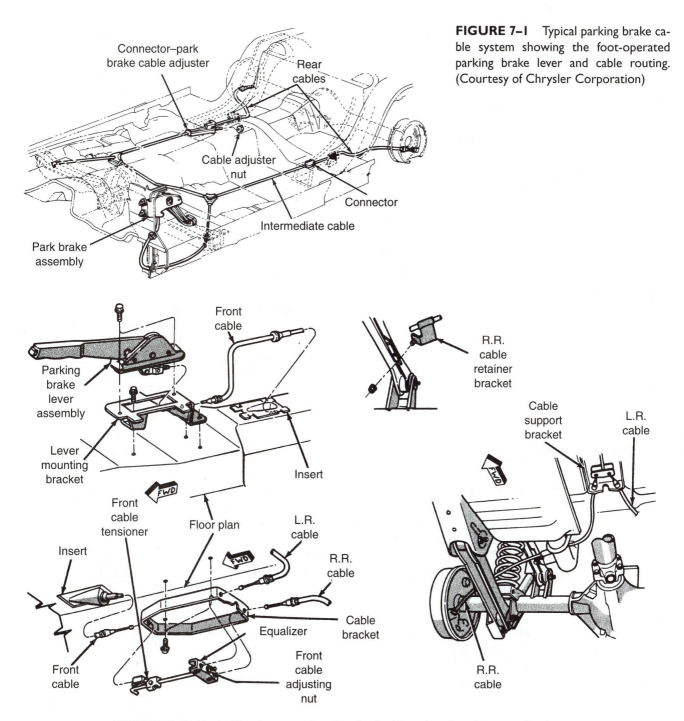

FIGURE 7–1 Typical parking brake cable system showing the foot-operated parking brake lever and cable routing. (Courtesy of Chrysler Corporation)

FIGURE 7–2 Typical hand-operated parking brake. Note that the adjustment for the cable is underneath the vehicle at the equalizer. (Courtesy of Chrysler Corporation)

PARKING BRAKE WARNING LAMP

Whenever the parking brake is engaged, a red BRAKE warning lamp lights on the dash. On most vehicles, this is the same lamp that lights when there is a hydraulic or brake fluid level problem. The warning lamp for the parking brake warns the driver that the parking brake is applied or partially applied. This warning helps prevent damage or overheating to the brake drums and linings that could occur if the vehicle was driven with the

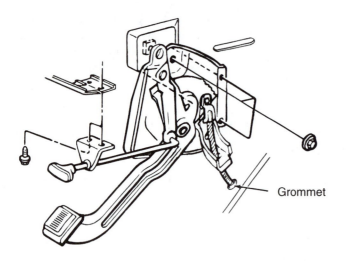

FIGURE 7–3 Foot-operated parking brake with a manual release. (Courtesy of Chrysler Corporation)

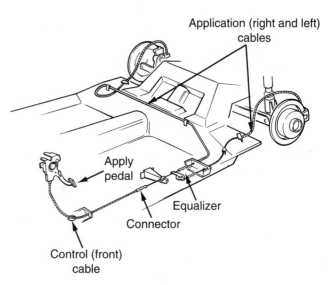

FIGURE 7–4 The cable from the activating lever to the equalizer is commonly called the *control cable*. From the equalizer, the individual brake cables are often called *application cables*. These individual cables can usually be purchased separately.

parking brake applied. If the red BRAKE warning lamp is on, check the parking brake to see if it is fully released. If the BRAKE lamp is still on, the parking brake switch may be defective, out of adjustment, or there may be a hydraulic problem.

PARKING BRAKE CABLES

Parking brake cables run through protective housing. The cable attaches to the hand lever on foot-operated pedals and runs to a junction. This front section of parking brake cable is usually called the **control cable**. The control cable then attaches to an **equalizer** to a second cable or pair of cables that run to each rear brake. These individual wheel brake cables are often called **application cables** or left or right **parking brake cables** (see Figure 7–4).

The parking brake equalizer is normally located under the vehicle. Most vehicles since the mid-1980s use wire strand cables covered with nylon for corrosion resistance. The housing or conduit contains plastic seals that help keep out dirt and water to prevent the cables from sticking and to reduce parking brake effort.

PARKING BRAKE ON DRUM BRAKES

Most parking brakes move steel woven cables attached to the rear brakes only and operate the lining through the **parking brake lever,** which is attached to the cable and the secondary lining. The parking brake force is transferred to the primary lining through a steel flat bar called a **parking brake strut**. Around the end of the slotted

strut is a spring called an **antirattle spring** (or **strut spring**), which prevents the strut from rattling whenever the parking brake is not applied (see Figure 7–5).

AUXILIARY DRUM PARKING BRAKES

Equipping a vehicle with a parking brake that has rear disc brakes is expensive. One commonly used method is to use a small brake drum inside the "hat" section of the rear disc brake rotor. This method is used on many Corvettes and Toyota Supras. While this method costs more and adds weight to the vehicle, the operation of the brake shoes by the parking brake cable is independent of the rear disc brakes (see Figure 7–6). Because the parking brake is usually only used for holding a stopped vehicle, the parking brake shoes wear little, if at all.

INTEGRAL DISC BRAKE PARKING BRAKE

Many vehicles equipped with rear disc brakes use a lead screw type of mechanism to apply the piston on the rear disc brakes. Most styles use conventional parking brake cables to activate a lever on the back of the caliper. The hydraulic part of the rear disc brake operates the same as a conventional disc brake. The parking brake lever rotates an **actuator screw**. As the screw rotates, the

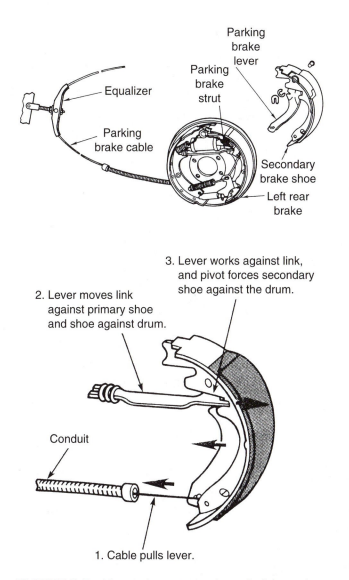

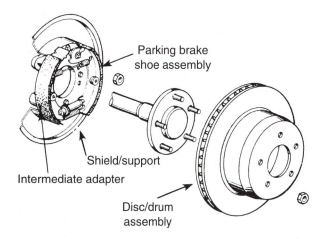

FIGURE 7–6 Many vehicles equipped with rear disc brakes use a small drum brake inside the rear rotor. When the brake shoes expand, they contact the inner surface of the rotor. (Courtesy of Allied Signal Automotive Aftermarket)

FIGURE 7–5 Notice the spring at the end of the parking brake strut. This antirattle spring keeps tension on the strut. The parking brake lever is usually attached with a pin and spring (wavy) washer and retained by a horseshoe clip.

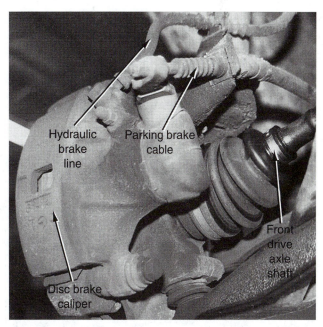

FIGURE 7–7 Subaru front disc brake caliper with parking brake.

nut of the screw does not rotate, but instead, moves against the inside of the piston. As the piston moves, it presses on the inboard brake pad. When the inboard pad contacts the disc brake rotor, the reaction force slides the caliper assembly and applies the outboard pad (see Figures 7–7 through 7–13).

PARKING BRAKE CABLE ADJUSTMENT

Most manufacturers specify a minimum of 3 or 4 and a maximum of 8 to 10 clicks when applying the parking

brake. Consult the service manual for the vehicle being serviced on the exact specification and adjustment procedures. Most vehicle manufacturers specify that the rear brakes be inspected and adjusted correctly before attempting to adjust the parking brake cable. Always follow the manufacturer's recommended procedure exactly.

Below is a general procedure for a parking brake adjusting.

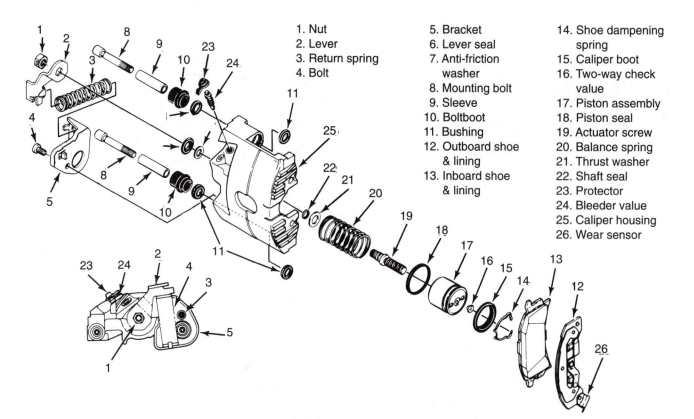

1. Nut
2. Lever
3. Return spring
4. Bolt
5. Bracket
6. Lever seal
7. Anti-friction washer
8. Mounting bolt
9. Sleeve
10. Boltboot
11. Bushing
12. Outboard shoe & lining
13. Inboard shoe & lining
14. Shoe dampening spring
15. Caliper boot
16. Two-way check value
17. Piston assembly
18. Piston seal
19. Actuator screw
20. Balance spring
21. Thrust washer
22. Shaft seal
23. Protector
24. Bleeder value
25. Caliper housing
26. Wear sensor

FIGURE 7–8 Typical General Motors rear disc brake with an integral parking brake. (Courtesy of General Motors)

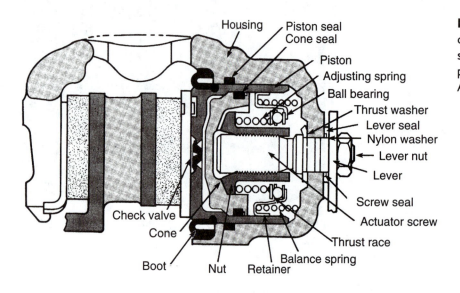

Housing
Piston seal
Cone seal
Piston
Adjusting spring
Ball bearing
Thrust washer
Lever seal
Nylon washer
Lever nut
Lever
Screw seal
Actuator screw
Thrust race
Check valve
Cone
Balance spring
Boot
Nut
Retainer

FIGURE 7–9 Cross-sectional view of a General Motors rear disc brake assembly showing the parking brake components. (Courtesy of Allied Signal Automotive Aftermarket)

1. Make certain that the rear service brakes are adjusted correctly and the lining is serviceable.
2. With the drums installed, apply the parking brake 3 or 4 clicks. There should be a slight drag on both rear wheels.
3. Adjust the cable at the equalizer (equalizes one cable's force to both rear brakes) if necessary until there is a slight drag on both rear brakes (see Figures 7–14 and 7–15).
4. Release the parking brake. Both rear brakes should be free and not dragging. Repair or replace rusted cables or readjust as necessary to ensure that the brakes are not dragging.

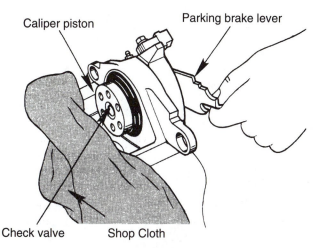

FIGURE 7–10   Removing the piston from a typical General Motors rear disc brake caliper. (Courtesy of Allied Signal Automotive Aftermarket)

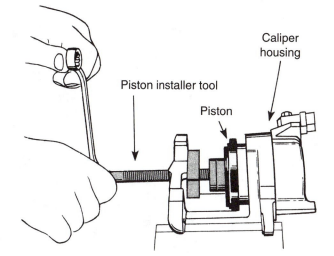

FIGURE 7–11   A piston installation tool is required to install the piston fully into a General Motors rear disc brake caliper. (Courtesy of Allied Signal Automotive Aftermarket)

◀ **TECH TIP** ▶

PARKING BRAKE CLICK TEST

When diagnosing any brake problem, apply the parking brake and count the clicks. This method works for hand- as well as foot-operated parking brakes. Most vehicle manufacturers specify a maximum of 10 clicks. If the parking brake travel exceeds this amount, the *rear* brakes may be worn or out of adjustment.

CAUTION: Do not adjust the parking brake cable until the rear brakes have been thoroughly inspected and adjusted.

If the rear brake lining is usable, check for proper operation of the self-adjustment mechanism. If the rear brakes are out of adjustment, the service brake pedal will also be low. This 10-click test is a fast and easy way to determine if the problem is due to rear brakes.

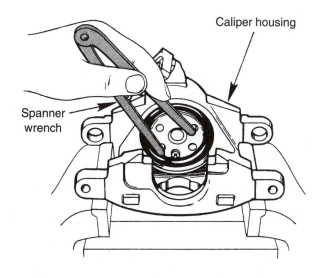

FIGURE 7–12   A spanner wrench (or needle-nose pliers) can be used to rotate the caliper piston prior to installing disc brake pads. A notch on the piston must line up with a tab on the back of the brake pad to keep the piston from rotating when the parking brake is applied. (Courtesy of Allied Signal Automotive Aftermarket)

NOTE: The rear parking brake adjustment should always be checked whenever replacing the rear brake linings. It may be necessary to loosen the parking brake cable adjustment to allow clearance to get the drum over the new linings (see Figure 7–16).

Feel the tension of the parking brake cable underneath the vehicle. It should be slightly loose (with the parking brake off). Be sure to lubricate the parking brake cable to ensure that water or ice will not cause rust or freezing of the cable.

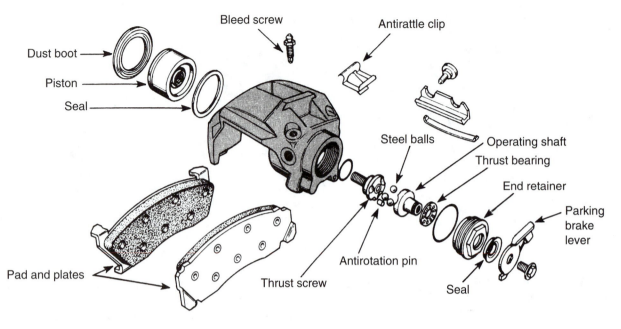

FIGURE 7–13   Typical Ford rear disc brake with integral parking brake. (Courtesy of Allied Signal Automotive Aftermarket)

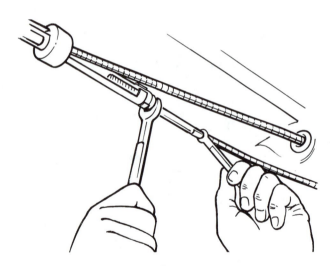

FIGURE 7–14   After checking that the rear brakes are okay and properly adjusted, the parking brake cable can be adjusted. Always follow the manufacturer's recommended procedure.

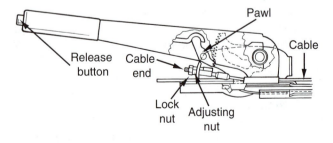

FIGURE 7–15   Many hand-operated parking brakes are adjusted inside the vehicle.

NOTE: Some vehicles are equipped with automatic adjusting parking brake lever/cable. Simply cycling the parking brake on/off/on three times is often all that is required to adjust the parking brake cable.

SUMMARY

1. Government regulations require that the parking brake be able to hold a fully loaded vehicle on a 30 percent grade.
2. The typical parking brake uses either a hand-operated lever or a foot-operated pedal to activate the parking brake.
3. On a typical drum brake system, the parking brake cable moves a parking brake lever attached to the secondary brake shoe. The primary shoe is applied through force being transferred through the strut.
4. All parking brake cables should move freely. The rear brakes should be adjusted properly before the parking brake is adjusted.

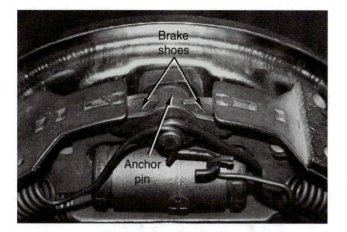

FIGURE 7–16 Always check that the brake shoes contact the anchor pin.

REVIEW QUESTIONS

1. Describe how a typical parking brake functions on a vehicle equipped with rear drum brakes.

2. Describe how a typical parking brake functions on a vehicle equipped with an integral rear disc brake system.

3. Explain how to adjust a parking brake properly.

MULTIPLE-CHOICE QUESTIONS

1. Technician A says that the parking brake cable should be adjusted at each wheel. Technician B says the parking brake cable adjustment is usually done after adjusting the rear brakes. Which technician is correct?
   a. A only
   b. B only
   c. Both A and B
   d. Neither A nor B

2. Technician A says that the parking brake hand lever can turn on the red brake warning lamp. Technician B says that a foot-operated parking brake can turn on the red brake warning lamp. Which technician is correct?
   a. A only
   b. B only
   c. Both A and B
   d. Neither A nor B

3. Technician A says that if the parking brake cable is adjusted too tight, the rear brakes may drag and overheat. Technician B says that the parking brake is properly adjusted if the cable is tight when in the released position. Which technician is correct?
   a. A only
   b. B only
   c. Both A and B
   d. Neither A nor B

4. On most drum brake systems, the parking brake lever and strut transfer the pulling force of the parking brake cable against the
   a. primary shoe
   b. secondary shoe

5. On most vehicles, the antirattle spring (strut spring) should be installed on the parking brake strut toward the _____ of the vehicle.
   a. front
   b. rear

6. In a typical integral rear disc brake caliper, the parking brake cable moves the
   a. caliper.
   b. actuator screw.
   c. auxiliary piston.
   d. rotor.

7. The rear brakes should be inspected and adjusted if necessary if the parking brake requires more than
   a. 5 clicks.
   b. 10 clicks.
   c. 15 clicks.
   d. 20 clicks.

8. A rear drum brake is being inspected. The primary shoe is not contacting the anchor pin at the top. Technician A says that this is normal. Technician B says that the parking brake cable may be adjusted too tight or is stuck. Which technician is correct?
   a. A only
   b. B only
   c. Both A and B
   d. Neither A nor B

◀ Chapter 8 ▶

MACHINING BRAKE DRUMS AND ROTORS

OBJECTIVES

After studying Chapter 8, the reader will be able to:

1. Discuss the construction of brake drums and rotors.
2. Explain the formation of hard spots in drums and rotors.
3. Describe how to measure and inspect drums and rotors before machining.
4. Discuss how surface finish is measured and its importance to satisfactory brake service.
5. Demonstrate how to machine a brake drum and rotor correctly.

Brake drums and rotors are the major energy-absorbing parts of the braking system. Friction between the friction material and the drum or rotor creates heat. As energy continues to be absorbed, the drum or rotor increases in temperature. Airflow across the drum or rotor helps to dissipate the heat and keep the temperature rise under control. See Figures 8–1 and 8–2 for examples of how drums and rotors are cooled.

BRAKE DRUMS

Brake drums are constructed of cast iron where the lining contacts the drum with mild steel centers. The drum is drilled for the lug studs. Cast iron is approximately 3 percent carbon content, which makes the drum hard,

FIGURE 8–1 An aluminum brake drum with a cast iron friction surface. The cooling fins around the outside help dissipate the heat from the friction surface to the outside air. Note the "MAX DIA 243.5 mm" cast into the drum.

yet brittle. The 3 percent carbon content of the cast iron also acts as a lubricant, which prevents noise during braking. Also, the rubbing surface can be machined without the need of a coolant (as would be required if constructed of mild steel). Because of these properties, cast iron is used on the friction surface of all drums (see Figure 8–3). Even aluminum brake drums use cast iron for the friction surface area. Besides saving weight, alu-

FIGURE 8–2 The airflow through cooling vents help brakes from overheating. (Courtesy of General Motors)

FIGURE 8–3 This air scoop is part of the water/dirt shield attached next to the rotor.

minum brake drums transfer heat to the surrounding air faster than does cast iron or steel.

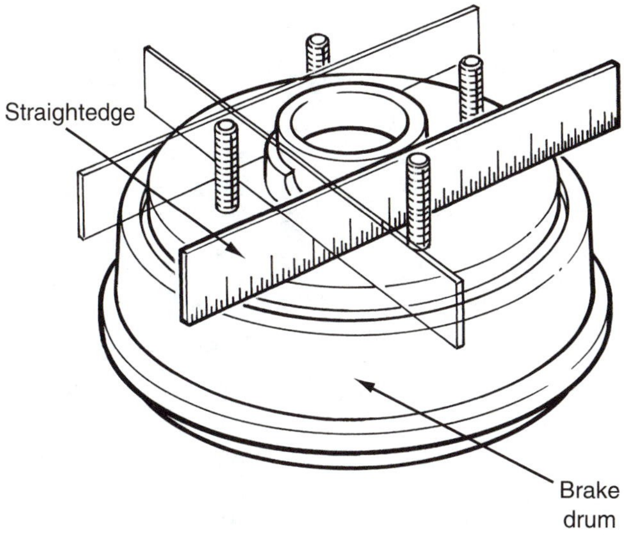

FIGURE 8–4 A straightedge can be used to check for brake drum warpage.

The first inspection step after removing a brake drum is to check it for warpage using a straightedge as shown in Figure 8–4. A warped drum is often a source of vibration. A brake drum that is out of round can cause brake pedal pulsation during braking.

HINT: To help diagnose if the front brakes or rear brakes are the cause of the vibration, try slowing the vehicle using the parking brake. If the vibration occurs, the problem is due to the rear brakes.

◀ TECH TIP ▶

MARK IT TO BE SURE

Most experts recommend that brake rotors, as well as drums and wheels, be marked *before* removing them for service. Many disc brake rotors are directional and will function correctly only if replaced in the original location. A quick and easy method is to use correction fluid. This alcohol-based liquid comes in small bottles with a small brush inside, making it easy to mark rotors with an "L" for left and an "R" for right. Correction fluid (also known by the trade names "White-Out" and "Liquid Paper") can also be used to make marks on wheel studs, wheels, and brake drums to help assure reinstallation in the same location.

FIGURE 8–5 These dark hard spots are created by heat, which actually changes the metallurgy of the cast iron drum. Most experts recommend replacement of any brake drum that has these hard spots.

HARD SPOTS

Hard spots are created by heat. Vehicles that are stopped from high speed, or long braking down a hill or mountain, can generate excessive temperatures. The metal on the surface, being at a much higher temperature, tends to expand and raise a small bump. As the spot cools, the crystallized structure of the cast iron changes to a hard steel so that the metallurgy is actually different in the hard spot than the surrounding areas (see Figure 8–5).

Some experts recommend using a grinding stone to remove hard spots. However, most experts and vehicle manufacturers agree that these hard spots have "memory" and will tend to return as soon as the brakes are subjected to severe service again and recommend that drums or rotors with hard spots be replaced.

"MACHINE TO" VERSUS "DISCARD" DIMENSION

Brake drums can usually be machined a maximum of 0.060 in. (1.5 mm) oversize (for example, a 9.500-in. drum new could wear or be machined to a maximum inside diameter of 9.560 in.) unless otherwise stamped on the drum. Most brake experts recommend that both drums on the same axle be within 0.010 in. (0.25 mm) of

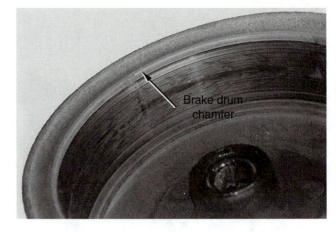

FIGURE 8–6 Most brake drums have a chamfer around the edge. If the chamfer is no longer visible, the drum is usually worn (or machined) to its maximum allowable inside diameter.

◄ TECH TIP ►

BRAKE DRUM CHAMFER

Look at the chamfer on the outer edge of most brake drums. When the chamfer is no longer visible, the brake drum is usually at or past its maximum ID (see Figure 8–6). Although this chamfer is not an accurate gauge of the inside diameter of the brake drum, it still is a helpful indicator to the technician.

each other. *The maximum specified inside diameter (ID) means the maximum wear inside diameter.* Always leave at least 0.015 in. (0.4 mm) after machining (resurfacing) for wear. Many manufacturers recommend that 0.030 in. (0.8 mm) be left for wear (see Appendix 8).

MEASURING AND INSPECTING BRAKE DRUMS

Brake drums are usually measured using a special micrometer especially designed for brake drums (see Figure 8–7). The drum should be checked for roundness, bellmouth, taper, and deep scoring (see Figure 8–8).

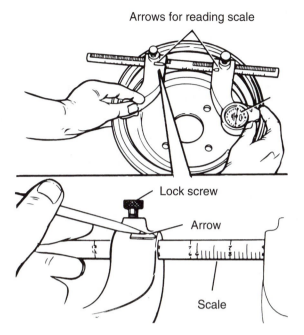

Arrows for reading scale

Lock screw

Arrow

Scale

FIGURE 8–7 Typical needle dial brake drum micrometer. The left movable arm is set to the approximate drum diameter and the right arm to the more exact drum diameter. The dial indicator (gauge) reads in thousandths of an inch. (Courtesy of AMMCO Tools, Inc.)

HINT: Hold a brake drum in the center with one hand and tap the outside of the drum with a light steel hammer. The drum should ring like a bell. If the drum sounds like a dull thud, the drum is probably cracked and needs to be replaced.

MACHINING DRUMS

If the drum has a hub with bearings, check the outer bearing races (cups) for wear and replace as necessary before placing the drum on the brake lathe. Also, carefully inspect and clean the lathe spindle shaft and cones before use. Use a **self-aligning spacer (SAS)** to be assured of even force being applied to the drum by the spindle nut. Always follow the instructions for the lathe you are using.

Hubless drums use a hole in the center of the brake drum for centering. Always check that the center hole is clean and free of burrs or nicks. Typical drum brake machining steps include:

> **Step 1.** Mount the drum on the lathe and install the silencer band as shown in Figures 8–9 and 8–10.

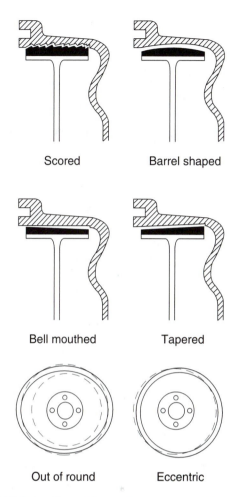

Scored | Barrel shaped

Bell mouthed | Tapered

Out of round | Eccentric

FIGURE 8–8 Typical brake drum wear problems.

> **Step 2.** Turn the drum by hand before turning on the lathe to be sure that everything is clear. Advance the tool bit manually until it just contacts the drum. This is called a **scratch cut**. See Figure 8–11.
>
> **Step 3.** Stop the lathe and back off the tool bit. Loosen the arbor nut and rotate the drum one-half turn (180°) on the arbor. Turn the lathe on and make a second scratch cut.
>
> > **a.** If the scratch cuts are side by side, the lathe is okay and machining can begin.
> >
> > **b.** If the scratch cuts are opposite, remove the drum and check for nicks, burrs, or chips on the mounting surfaces.
>
> **Step 4.** Start the lathe and set the depth of the cut (see Figures 8–12 and 8–13). The maximum rough cut depends on the lathe type. The minimum cut is usually specified as no less than 0.002 in. (0.05 mm). A shallower

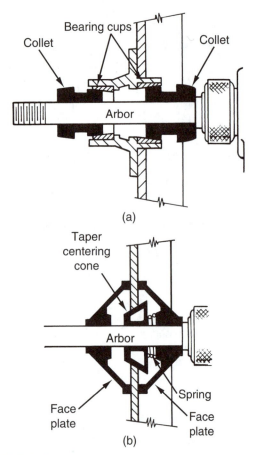

(a)

(b)

FIGURE 8–9 (a) A rotor or brake drum with a bearing hub should be installed on a brake lathe using the appropriate-size collets that fit the bearing cups (races). (b) A hubless rotor or brake drum requires a spring and a tapered centering cone. A face plate should be used on both sides of the rotor or drum to provide support.

cut usually causes the tool bit to slide over the surface of the metal rather than cut into the metal.

CAST IRON DISC BRAKE ROTORS

Disc brake rotors use cast gray iron at the area that contacts the friction pad. Rotors are made in several styles, including:

1. **Solid:** used on the rear of many vehicles equipped with rear disc brakes and on the front of some small and midsized vehicles. Solid rotors are much thinner than vented rotors.

2. **Vented:** used on the front of most vehicles. The internal vanes allow air to circulate between the two friction surfaces of the rotor. Rotors can either be

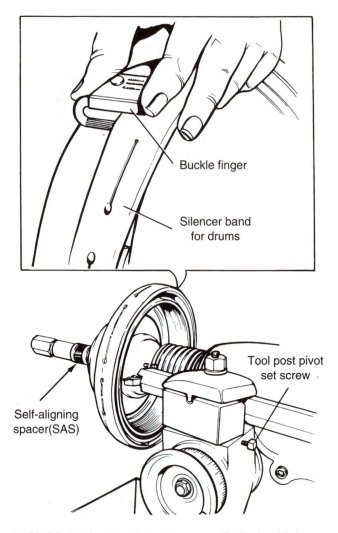

FIGURE 8–10 A self-aligning spacer (SAS) should always be used between the drum or rotor and the spindle retaining nut to help assure an even clamping force and to prevent the adapters and cone from getting into a bind. A silencer band should always be installed to prevent turning tool chatter and to assure a smooth surface finish. (Courtesy of AMMCO Tools, Inc.)

straight vane design, as shown in Figure 8–14, or directional vane design, as shown in Figure 8–15.

Composite rotors have a steel center section with a cast iron wear surface. These rotors are lighter in weight than conventional cast iron rotors (see Figure 8–16). The light weight of composite rotors makes them popular with vehicle manufacturers. However, technicians should be aware that full contact adapters that simulate the actual wheel being bolted to the rotor must be used when machining composite rotors. If composite rotors are machined incorrectly, they must usually be replaced.

FIGURE 8–11 After installing a brake drum on the lathe, turn the cutting tool outward until the tool just touches the drum. This is called a *scratch cut.* (Courtesy of AMMCO Tools, Inc.)

FIGURE 8–13 This lathe has a dial that is "diameter graduated." This means that a reading of 0.030 in. indicates a 0.015-in. cut that increases the inside diameter of the brake drum by 0.030 in.

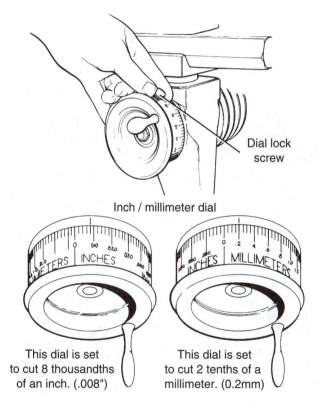

FIGURE 8–12 Set the depth of the cut indicator to zero just as the turning tool touches the drum. (Courtesy of AMMCO Tools, Inc.)

FIGURE 8–14 Severely worn vented disc brake rotor. The owner brought the vehicle to a repair shop because of a "little noise in the front." Notice the straight vane design.

ALUMINUM DISC BRAKE ROTORS

Some disc brake rotors are manufactured from an aluminum composite alloy reinforced with 20 percent silicon carbide particulate. These rotors can be distinguished from conventional cast iron rotors in several ways. These rotors will show no signs of rust and are nonmagnetic, unlike cast iron. When removed from the vehicle, aluminum composite rotors can be further distinguished by their light weight [usually, under 6 lb (2.7 kg) versus over 12 lb (5.4 kg) for cast iron rotors on the typical passenger vehicle] (see Figure 8–17).

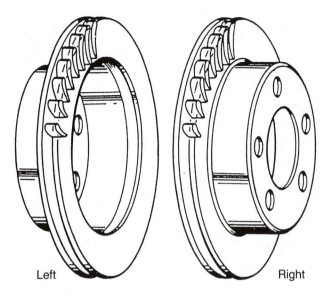

Left   Right

FIGURE 8–15   Directional vane vented disc brake rotors. Note that the fins angle toward the rear of the vehicle. It is important that this type of rotor be reinstalled on the correct side of the vehicle. (Courtesy of Allied Signal Automotive Aftermarket)

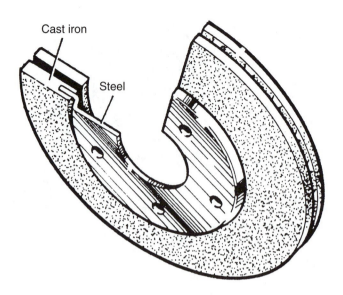

Cast iron

Steel

FIGURE 8–16   Typical composite rotor that uses cast iron friction surfaces and a steel center section.

Servicing these rotors is slightly different from servicing cast iron rotors. The dark transfer layer on the rubbing surface does not harm rotor performance and should *not* be removed unless the rotor needs to be machined for dimensional reasons (warped, etc.). *Aluminum composite disc brake rotors cannot be machined with steel cutting tools!* Carbide tools *can* be used to machine a single set of aluminum composite rotors. If a shop receives these rotors on a regular basis, a **polycrystalline diamond (PCD)**-tipped tool is a good investment. Although more expensive initially, the PCD tool can last one hundred times longer than a carbide tool.

LATERAL RUNOUT

A disc brake rotor should have a maximum of 0.002 to 0.005 in. (0.05 to 0.13 mm), depending on the manufacturer's specifications for total lateral runout.

NOTE: The diameter of a human hair is about 0.002 to 0.003 in. (0.05 to 0.08 mm). Therefore, the maximum allowable rotor runout is about equal to the thickness of a human hair. This small a measurement requires the use of an accurate measuring gauge.

See Figure 8–18.

FIGURE 8–17   Aluminum disc brake rotor. (Courtesy of Duralcan, USA)

Excessive rotor runout, is also called **wobble,** and causes disc brake rotor wear that can cause uneven thickness variations. The procedure to check lateral runout follows:

1. If the rotor is installed on a wheel where bearings are adjustable, tighten the wheel bearings temporarily to remove all end play. With a hubless rotor, install and torque the lug nuts to retain the rotor.

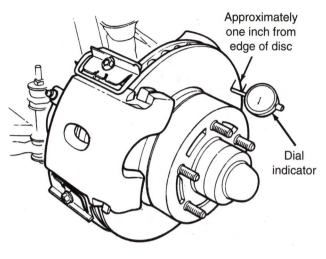

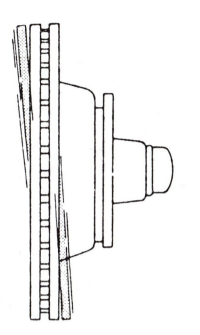

FIGURE 8–18 Excessive lateral runout can cause uneven wear. To help prevent excessive runout, always use a torque wrench or torque-absorbing adapters with an impact wrench when tightening lug nuts.

2. Attach a dial indicator to the end of the spindle and observe the reading through one complete revolution as shown in Figure 8–19.
3. Total dial indicator movement should not exceed the specifications. If greater than specifications, machine the rotor.
4. Readjust the front wheel bearing to proper end play.

ROTOR THICKNESS (PARALLELISM)

Excessive rotor runout does lead to rotor thickness variations as the rotor wears. Measure rotor thickness, using a micrometer, at four or more equally spaced locations of the rotor (see Figure 8–20). Each measurement must not vary by more than 0.0005 in. [½ of 1 thousandth of an inch (0.013 mm)] and must be greater than the minimum allowable thickness. *It is the excessive rotor thickness variation that causes brake shudder or steering wheel shimmy.*

MINIMUM THICKNESS

Most rotors have a minimum thickness cast or stamped into the rotor. This thickness is minimum wear thick-

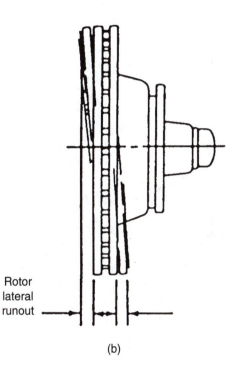

FIGURE 8–19 (a) Rotate the disc brake rotor one complete revolution while observing the dial indicator (gauge). (Courtesy of Chrysler Corporation) (b) Most vehicle manufacturers specify a maximum runout of about 0.003 in. (0.08 mm).

ness. At least 0.015 in. (0.4 mm) must remain after machining to allow for wear. [Some vehicle manufacturers, including General Motors, specify that 0.030 in. (0.8 mm) be left for wear.] Whenever machining (resurfacing) a rotor, an equal amount of material must be removed from each side.

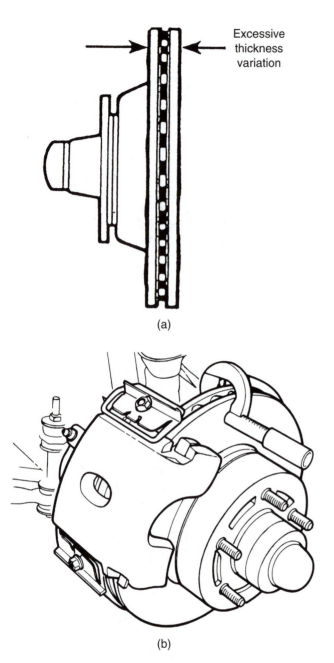

(a)

(b)

FIGURE 8–20 (a) Disc brake rotor thickness variation (parallelism). (b) The rotor should be measured with a micrometer at four or more equally spaced locations around the rotor. (Courtesy of Chrysler Corporation)

WHEN THE ROTORS SHOULD BE MACHINED

According to brake design engineers, a worn rotor has a very smooth friction surface that is ideal for replacement (new) disc brake pads. Often when the rotors are machined, the surface finish is not as smooth as speci-

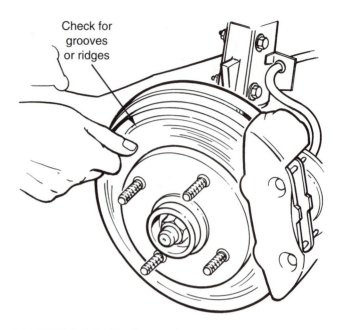

FIGURE 8–21 If a fingernail catches on a groove in the rotor, the rotor should be machined.

fied. Therefore, a rotor should **only** be machined if one of the following conditions exists:

1. Grooves are deeper than 0.060 in. (1.5 mm). This is the approximate thickness of an American nickel (see Figure 8–21).
2. Thickness variation exceeds specifications and there is a brake pedal pulsation complaint.
3. Heavy rust has corroded the friction surface of the rotor.

Therefore, if there is no complaint of a pulsating brake pedal during braking and the rotor is not deeply grooved or rusted, it should not be machined. New disc brake pads perform best against a smooth surface, and a used disc brake rotor is often smoother than a new rotor.

ROTOR FINISH

The smoothness of the rotor is called **rotor finish** or **surface finish**. Surface finish is measured in units called microinches, abbreviated μ in., where the symbol in front of the inch is the Greek lowercase letter mu. One microinch equals 0.000001 in. [0.025 micrometer (μm)]. The finish classification of microinch indicates the distance between the highest peaks and the deepest valley. The usual method of expressing surface finish is the **arithmetic average roughness height**, abbreviated **Ra** which is the average of all peaks and valleys from the mean (average) line. This surface finish is measured using a machine with a diamond stylus, as shown in Figure 8–23.

Often, a machined rotor will not be as smooth as a new rotor, resulting in a hard-stopping complaint after new brakes have been installed. Most new rotors have a surface finish of 45 to 60 μin. Ra.

◄ TECH TIP ►

BROWN = SEMI-METALLIC

Brake rotors that have used semi-metallic brake pads during operation are brown in color on the friction surface. The reason for the brown color is the rust from the steel used in the manufacture of the pads (see Figure 8–22). If the friction surface of the disc brake rotor is shiny, organic (asbestos), nonasbestos organic (NAO), or nonasbestos synthetic (NAS) pads have been used on the rotor.

This information is helpful to know, especially if the vehicle being serviced is to be equipped with semi-metallic pads as specified by the manufacturer and the rotors are shiny. In this case the incorrect lining may have been installed during previous service. The color of the friction surface is a quick and easy way to determine whether or not semi-metallic pads have been used.

MACHINING A DISC BRAKE ROTOR

Before machining a rotor, be sure that it can be machined by comparing the minimum thickness specification and the measured thickness of the rotor.

CAUTION: Some original equipment and replacement disc brake rotors are close to the minimum allowable thickness when new. Often, these rotors cannot be safely machined at all!

FIGURE 8–22 The rotor on the left is shiny, indicating that nonmetallic pads have been used, whereas the rotor on the right is brown, indicating that semi-metallic pads have been used.

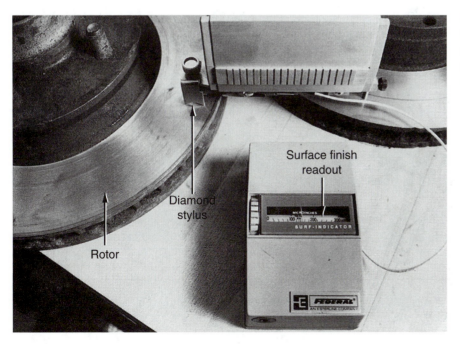

FIGURE 8–23 Electronic surface finish machine. The reading shows about 140 micro inches. This is much too rough for use but is typical for a rough-cut surface.

Typical rotor mounting configurations

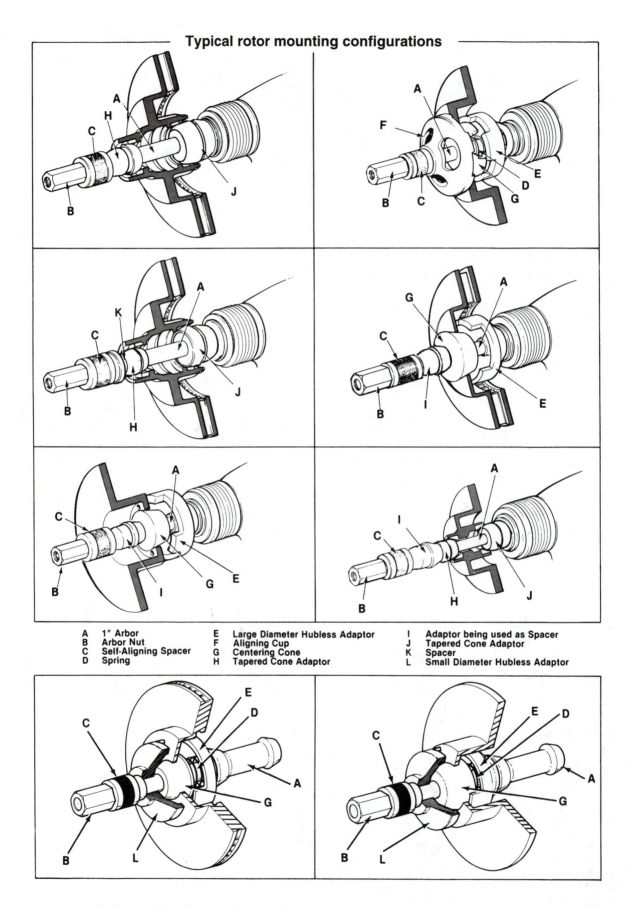

A	1″ Arbor	E	Large Diameter Hubless Adaptor	I	Adaptor being used as Spacer
B	Arbor Nut	F	Aligning Cup	J	Tapered Cone Adaptor
C	Self-Aligning Spacer	G	Centering Cone	K	Spacer
D	Spring	H	Tapered Cone Adaptor	L	Small Diameter Hubless Adaptor

FIGURE 8–24   (Courtesy of AMMCO Tools, Inc.)

The following is an example of the steps necessary to machine a disc brake rotor. Always follow the instructions for the equipment you are using.

Step 1. Mount the disc brake rotor to the spindle of the lathe using the recommended cones and adapters (see Figure 8–24).

Step 2. Install a rotor damper and position the cutting tools close to the rotor surface as shown in Figure 8–25.

NOTE: Failure to install the damper causes vibrations to occur during machining that creates a rough surface finish.

Step 3. Make a scratch cut on the rotor face as shown in Figure 8–26.

Step 4. To check that the rotor is mounted correctly, loosen the retaining nut, and turn the rotor one-half turn (180°), and retighten the nut. Make another scratch cut.

    a. The second scratch cut should be side by side with the first scratch cut if the rotor is installed properly.

    b. If the second scratch cut is on the opposite side (180°) from the first scratch cut, the rotor may not be installed correctly on the lathe.

NOTE: The runout as measured with a dial indicator on the brake lathe should be the same as the runout measured on the vehicle. If the runout is *not* the same, the rotor is not installed on the brake lathe correctly.

After proper installation of the disc brake rotor on the brake lathe, proceed with machining the rotors. For best results do not machine any more material from the rotor than is absolutely necessary. Always follow the recommendations and guidelines as specified by the vehicle manufacturer.

Rough Cut. A rough cut on a lathe involves cutting 0.005 in. per side with a feed of 0.008 in. per revolution and a 150-rpm spindle speed. This usually results in a very coarse surface finish of about 150 μin. Ra.

Finish Cut. A finish cut means removing 0.002 in. off per side, with a feed of 0.002 in. per revolution and a 150-rpm spindle speed. Although this cut usually looks

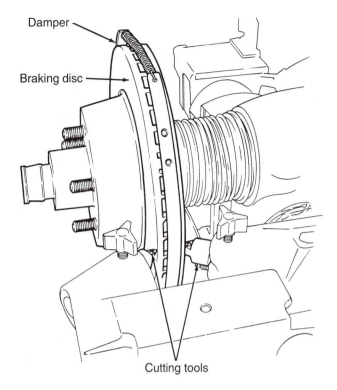

FIGURE 8–25 A damper is necessary to reduce cutting tool vibrations that can cause a rough surface finish. (Courtesy of Chrysler Corporation)

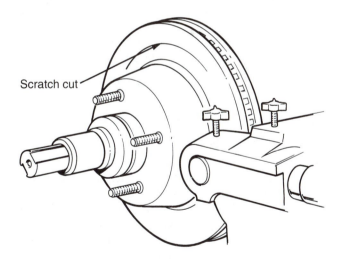
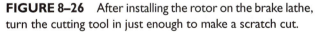

FIGURE 8–26 After installing the rotor on the brake lathe, turn the cutting tool in just enough to make a scratch cut.

smooth, the surface finish is about 90 to 100 μin. Ra. Even a typical finish cut is still not nearly as smooth as a new rotor.

Nondirectional Finish. Most vehicle and brake component manufacturers recommend a nondirectional finish to help prevent the grooves machined into the rotor from acting like record grooves that can force the pads to move outward while the rotor rotates. Some nondirectional finish tools such as those that use Scotch Brite (a registered trademark) plastic pads often do *not* make the rotor as smooth as new, even though the finish has been swirled.

Surface Finishing the Rotor. *The goal of any brake repair or service should be to restore the braking effectiveness to match new vehicle brakes.* This means that the rotor finish should be as smooth or smoother than a new rotor for maximum brake pad contact. Research conducted at Delphi (Delco) Chassis Brake Division of General Motors has shown that like-new rotor finish can be easily accomplished by using a block and sandpaper. After completing the finish cut, place 150-grit aluminum oxide sandpaper on a block and apply steady pressure against the rotor surface for 60 seconds on **each** side of the rotor (see Figure 8–27). The aluminum oxide is hard enough to remove the highest ridges left by the lathe cutting tool. This results in a surface finish ranging from 20 to 80 μin. and usually less than 40 μin., which is smoother than a new rotor.

NOTE: Many commercial rotor finish products may also give as smooth a surface finish (see Figures 8–28 and 8–29). Always compare rotor finish to the rotor finish of a *new* rotor. Microinch finish is often hard to distinguish unless you have a new rotor to which to compare.

ON-THE-VEHICLE ROTOR MACHINING

Many vehicle manufacturers recommend on-the-vehicle machining for rotors *if* the disc brake rotor *must* be machined due to deep scoring or pulsating brake pedal complaint. This is especially true of composite rotors or for vehicles such as many Honda vehicles, which require major disassembly to remove the rotors.

    Caliper mount on-the-vehicle lathes require that the disc brake caliper be removed. The cutter attaches to the steering knuckle or caliper support in the same location as the caliper (see Figure 8–30). **Hub-mount** on-the-vehicle lathes attach to the hub using the lug nuts of the vehicle. To achieve a proper cut, the hub mount *must* be calibrated for any runout caused by the hub bearings and the outside surface face of the rotor.

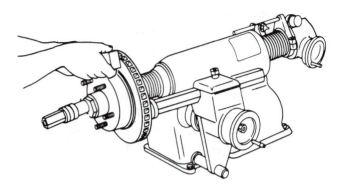

FIGURE 8–27 Sanding each side of the rotor surface for 1 min using a sanding block and 150-grit aluminum oxide sand paper after a finish cut gives the rotor the proper smoothness and finish. (Courtesy of EIS Brake Parts)

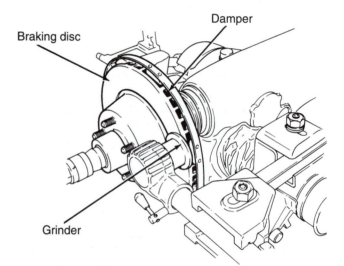

FIGURE 8–28 A grinder with sandpaper can be used to give a smooth nondirectional surface finish to the disc brake rotor. (Courtesy of Chrysler Corporation)

NOTE: All on-the-vehicle lathes require that the wheel be removed. For best results always use a torque wrench when tightening lug nuts or lathe adapters. Unequal torque on the bolts causes stress and distortion that can cause warped rotors and a pulsating brake pedal.

SUMMARY

1. Brake drums and rotors must absorb the heat generated by the friction of slowing and stopping a vehicle.

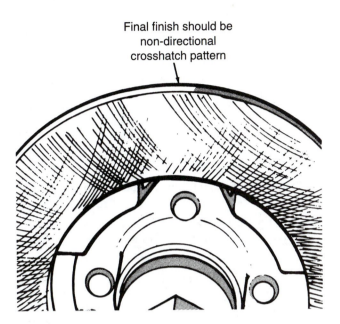

FIGURE 8–29 The correct final surface finish should be smooth and nondirectional. (Courtesy of Chrysler Corporation)

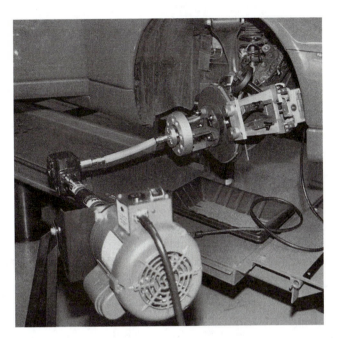

FIGURE 8–30 Typical on-the-vehicle disc brake rotor lathe.

2. All rotors should be marked before removing them from the vehicle to be assured that they will be reinstalled in the same position and on the same side of the vehicle.

3. All brake drums should be machined only enough to restore proper braking action. Brake drums should be the same size on the same axle to help prevent unequal braking.

4. Disc brake rotors should be machined and allow up to 0.030 in. (0.8 mm) for wear.

5. To assure proper braking, all rotors should be machined to a very smooth surface of less than 60 μin. finish.

REVIEW QUESTIONS

1. Explain the difference between "machine to" specifications and "discard" dimension.

2. List the steps for machining a brake drum.

3. Describe how to measure a disc brake rotor for lateral runout and thickness variation.

4. List the steps for machining a disc brake rotor.

5. Describe what is necessary to achieve like-new disc brake rotor finish.

MULTIPLE-CHOICE QUESTIONS

1. Technician A says that aluminum brake drums use cast iron friction surfaces. Technician B says that up to 0.030 in. (0.8 mm) should be left after machining a drum to allow for wear. Which technician is correct?
   - **a.** A only
   - **b.** B only
   - **c.** Both A and B
   - **d.** Neither A nor B

2. Technician A says that hard spots in a brake drum should be removed using a carbide-tip machining tool. Technician B says the drum should be replaced if hard spots are discovered. Which technician is correct?
   - **a.** A only
   - **b.** B only
   - **c.** Both A and B
   - **d.** Neither A nor B

3. Technician A says that brake drums on the same axle should be close to the same inside diameter for best brake balance. Technician B says that a brake drum may be cracked if it rings like a bell when tapped with a light steel hammer. Which technician is correct?
   a. A only
   b. B only
   c. Both A and B
   d. Neither A nor B

4. A hubless brake drum cannot be machined because it cannot be held in a lathe.
   a. True
   b. False

5. The major reason for brake pedal pulsation during braking is due to excessive rotor thickness variation.
   a. True
   b. False

6. Rotor finish is measured in
   a. millimeters.
   b. inches.
   c. microinches.
   d. centimeters.

7. The lower the Ra value of a rotor, the _____ the surface.
   a. smoother
   b. rougher
   c. higher
   d. lower

8. A disc brake rotor is being installed on a lathe for machining. During the setup a scratch test is performed. The scratch extended all the way around the rotor. Technician A says to loosen the rotor and rotate it 180° and retighten. Technician B says that the rotor is not warped. Which technician is correct?
   a. A only
   b. B only
   c. Both A and B
   d. Neither A nor B

9. Typical maximum rotor runout specifications are
   a. 0.0003 to 0.0005 in. (0.008 to 0.013 mm).
   b. 0.003 to 0.005 in. (0.08 to 0.13 mm).
   c. 0.030 to 0.050 in. (0.8 to 1.3 mm).
   d. 0.300 to 0.500 in. (8.0 to 13 mm).

10. Typical maximum rotor thickness variation (parallelism) specifications are
   a. 0.0003 to 0.0005 in. (0.008 to 0.013 mm).
   b. 0.003 to 0.005 in. (0.08 to 0.13 mm).
   c. 0.030 to 0.050 in. (0.8 to 1.3 mm).
   d. 0.300 to 0.500 in. (8.0 to 13 mm).

◀ Chapter 9 ▶

POWER BRAKE UNIT OPERATION, DIAGNOSIS, AND SERVICE

OBJECTIVES

After studying Chapter 9, the reader will be able to:

1. List the parts of a vacuum brake booster.
2. Describe how a vacuum brake booster operates.
3. Explain how to test a vacuum brake booster.
4. Describe how a hydraulic or electrohydraulic brake booster operates.

Power-assisted brakes reduce the effort of the driver to apply the necessary stopping forces of the vehicle. Power-assisted brakes were once considered a luxury and first available on large heavy cars and trucks. Without power-assisted brakes, the entire braking force was achieved by mechanical and hydraulic leverage as explained in Chapter 2 (see Figure 9–1).

VACUUM BOOSTER OPERATION

Most power brake boosters are vacuum operated, getting vacuum from the intake manifold of the engine. As any gasoline-powered engine runs, vacuum is created in the intake manifold. A vacuum is pressure below atmospheric pressure. Vacuum is measured in units of **inches of mercury**. The chemical symbol for mercury is Hg; therefore, the abbreviation is commonly written **in. Hg.** A well-running engine should produce between 17 and 21 in. Hg vacuum.

NOTE: Most manufacturers specify that the minimum engine vacuum necessary for the proper operation of a vacuum power-assist unit is 15 in. Hg. If the engine is producing less than 15 in. Hg. at idle, the cause of the low vacuum should be found and repaired before further brake system diagnosis is performed. See the Tech Tip "Check the Vacuum, Then the Brakes" for an example of how vacuum affects braking performance.

CHARCOAL FILTER

The vacuum hose leading from the engine to the power booster should run downward without any low places in the hose. If a dip or sag occurs in the vacuum hose, condensed fuel vapors and/or moisture can accumulate that can block or restrict the vacuum to the booster. Many manufacturers use a small charcoal filter in the

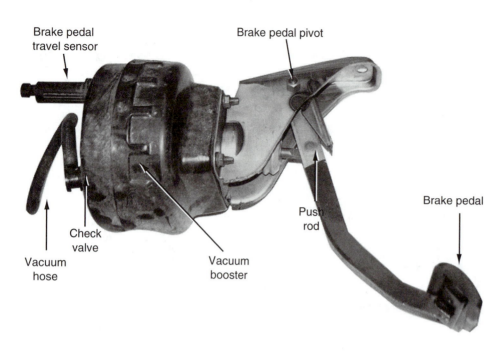

Brake pedal travel sensor

Brake pedal pivot

Check valve

Vacuum hose

Vacuum booster

Push rod

Brake pedal

FIGURE 9-1 Typical vacuum brake booster assembly. The vacuum hose attaches to the intake manifold of the engine. The brake pedal travel sensor is an input sensor for the antilock braking system.

◄ TECH TIP ►

CHECK THE VACUUM, THEN THE BRAKES

A customer complained of a very rough idle and an occasional pulsating brake pedal. The customer was certain that the engine required serious work since the vehicle had gone over 100,000 miles. During the troubleshooting procedure, a spray cleaner was used to find any vacuum (air) leaks. A large hole was found melted through a large vacuum hose next to the vacuum hose feeding the vacuum-operated power brake booster.

After repairing the vacuum leak, the vehicle was test driven again to help diagnose the cause of the pulsating brake pedal. The engine idled very smoothly after the vacuum leak was repaired *and* the brake pulsation was also cured. The vacuum leak resulted in lower-than-normal vacuum being applied to the vacuum booster. During braking, when engine vacuum is normally higher (deceleration), the vacuum booster would assist, then not assist when the vacuum was lost. This on-and-off-again supply of vacuum to the vacuum booster was noticed by the driver as a brake pulsation. Always check the vacuum at the booster whenever diagnosing any brake problems. Most vehicle manufacturers specify a maximum of 15 in. Hg of vacuum at the booster. The booster *should* be able to provide at least two or three stops even with no vacuum. The booster should also be checked to see if it can hold a vacuum after several hours. A good vacuum booster, for example, should be able to provide power assist after sitting all night without the engine being started.

vacuum line between the engine and booster as shown in Figure 9–2. The charcoal filter attracts and holds gasoline vapors and keeps the fumes from entering the vacuum booster.

VACUUM CHECK VALVE

All vacuum boosters use a one-way vacuum check valve. This valve allows air to flow in only one direction—from the booster and toward the engine. This valve prevents the loss of vacuum when the engine stops (see Figure 9–3).

CAUTION: Sometimes an engine backfire can destroy or blow the vacuum check valve out of the booster housing. If this occurs, all power assist will be lost and a much greater than normal pressure must be exerted on the brake pedal to stop the vehicle. Be sure to repair the cause of the backfire before replacing the damaged or missing check valve. Normal causes of backfire include an excessively lean air/fuel ratio or incorrect firing order or ignition timing.

FIGURE 9–2 The charcoal filter traps gasoline vapors that are present in the intake manifold and prevents them from getting into the vacuum chamber of the booster.

VACUUM BRAKE BOOSTER

A vacuum power brake booster contains a rubber diaphragm(s) which is connected to the brake pedal at one end and the master cylinder at the other end. The vacuum power unit contains the power piston assembly, which houses the control valve and reaction mechanism and the power piston return spring. The control valve is composed of the air valve (valve plunger), the floating control valve assembly, and the push rod. The reaction mechanism consists of a hydraulic piston reaction plate and a series of reaction levers. The push rod that operates the air valve projects out of the end (see Figure 9–4).

Released Position Operation. At the released position (brake pedal up), the air valve is seated on the floating control valve, which shuts off the air. The floating control valve is held away from the valve seat in the power piston insert. Vacuum from the engine is present in the space on both sides of the power piston. Any air in the system is drawn through a small passage in the power piston, over the seat in the power piston insert, and then through a passage in the power piston insert. There is a vacuum on both sides of the power piston and it is held against the rear of the housing by the power piston return spring. At rest, the hydraulic reaction plate is held against the reaction retainer. The air valve spring holds the reaction lever against the hydraulic reaction plate and also holds the air valve against its stop in the tube of the power piston. The floating control valve assembly is held against the air valve seat by the floating control valve spring (see Figure 9–5).

FIGURE 9–3 Not all check valves are located at the vacuum line to the booster housing connection. This vehicle uses an in-line check valve located between the intake manifold of the engine and the vacuum brake booster.

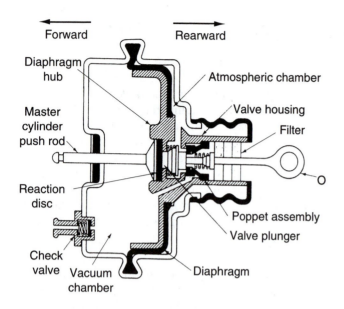

FIGURE 9–4 Cross-sectional view of a typical vacuum brake booster assembly.

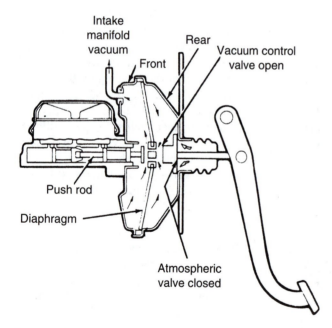

FIGURE 9–5 In the release position (brake pedal up), the vacuum is directed to both sides of the diaphragm. (Courtesy of Chrysler Corporation)

Applied Position Operation. As the brake pedal is depressed, the valve push rod moves the air valve away from the floating control valve. The floating control valve will follow until it is in contact with the raised seat in the power piston insert. When this occurs, vacuum is shut off to the rear of the power piston, and air under atmospheric pressure enters through the air fil-

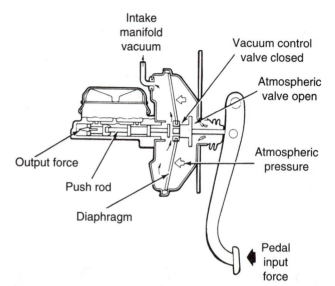

FIGURE 9–6 Simplified diagram of a vacuum brake booster in the apply position. Notice that the atmospheric valve is open and air pressure is being applied to the diaphragm. (Courtesy of Chrysler Corporation)

ter and travels past the seat of the air valve and through a passage into the housing at the rear of the power piston. Since there is still vacuum on the front side of the power piston, the atmospheric air pressure at the rear of the piston will force the power piston to travel forward.

NOTE: This movement of air into the rear chamber of the brake booster may be heard inside the vehicle as a hissing noise. The loudness of this airflow varies from vehicle to vehicle and should be considered normal.

As the power piston travels forward, the master cylinder push rod pushes the master cylinder primary and secondary pistons forward. As back-pressure builds up on the end of the master cylinder piston, the hydraulic reaction plate is moved off its seat on the power piston and presses against the reaction levers. The reaction lever pushes against the end of the air valve rod assembly. Approximately 30 percent of the load on the hydraulic master cylinder piston is transferred back through the reaction system to the brake pedal. This gives the driver a feel, which is proportional to the degree of brake application (see Figure 9–6).

Hold Position Operation. When the desired brake pedal pressure is reached, the power piston moves forward until the floating control valve, which is still on the power piston, again seats on the air valve. The power piston will now remain stationary until either pressure is applied or released at the brake pedal. As the pressure at the brake pedal is released, the air valve spring forces the air valve back to its stop on the power piston. As it returns, the air valve pushes the floating control valve off its seat on the power piston insert. The air valve seating on the floating control valve has shut off the outside air source. When it lifts the floating control valve from its seat on the power piston insert, it opens the space at the rear of the power piston insert to the vacuum source. The power piston return spring will return the piston to its released position against the rear housing, since both sides of the piston are now under a vacuum. As this occurs, the master cylinder releases its pressure and the brakes are released (see Figure 9–7).

Vacuum Failure Mode. In case of vacuum source interruption, the brake operates as a standard brake as follows: As the pedal is pushed down, the end of the air valve contacts the reaction levers and pushes, in turn, against the hydraulic reaction plate, which is fastened to the master cylinder piston rod, which applies pressure in the master cylinder. For safety in the event of a stalled engine and a loss of vacuum, a power brake should have adequate storage of vacuum for several power-assisted stops.

DUAL (TANDEM) DIAPHRAGM VACUUM BOOSTERS

To provide power assist, air pressure must work against a rubber diaphragm. The larger the area of the diaphragm, the more force can be exerted. Instead of increasing the diameter, some vacuum booster manufacturers use two smaller-diameter diaphragms and placed one in front of the other. This design increases the total area without increasing the physical diameter of the booster. This style is called a **dual diaphragm** or **tandem diaphragm** vacuum booster (see Figure 9–8).

VACUUM BOOSTER OPERATION TEST

With the engine off, apply the brakes several times to deplete the vacuum. With your foot on the brake pedal,

◀ TECH TIP ▶

A LOW SOFT BRAKE PEDAL IS NOT A POWER BOOSTER PROBLEM

Some service technicians tend to blame the power brake booster if the vehicle has a low soft brake pedal. A defective power brake booster causes a hard brake pedal, not a soft brake pedal. A soft or spongy brake pedal is usually caused by air being trapped somewhere in the hydraulic system.

Often, the technician has bled the system and therefore thinks that the system is free of any trapped air. According to remanufacturers of master cylinders and power brake boosters, most of the returned parts under warranty are not defective. Incorrect or improper bleeding procedures account for much of the problem.

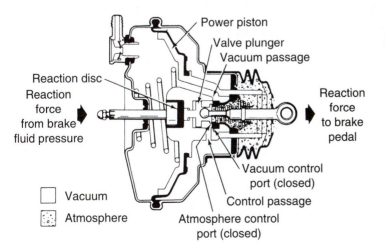

FIGURE 9–7 Cross section of a vacuum brake booster in hold position with both vacuum and atmospheric valves closed. Note that the reaction force from the brake fluid pressure is transferred back to the driver as a reaction force to the brake pedal. (Courtesy of Chrysler Corporation)

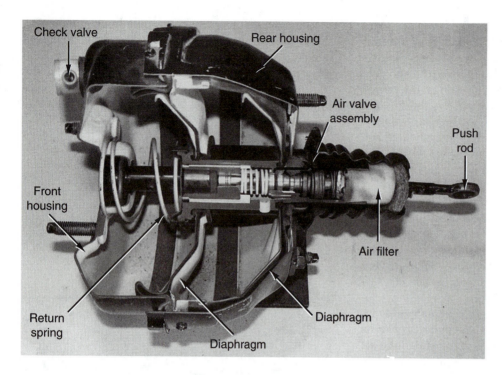

FIGURE 9–8 Cutaway showing a dual diaphragm (tandem) vacuum brake booster.

start the engine. The brake pedal *should* drop. If the brake pedal does *not* drop, check for proper vacuum source to the booster. If there is proper vacuum, repair or replacement of the power booster is required.

VACUUM BOOSTER LEAK TEST

To test if the vacuum booster can hold a vacuum, run the engine to build up a vacuum in the booster, then turn the engine off. Wait 1 minute, then depress the brake pedal several times. There should be two or more power-assisted brake applications.

If applications are not power assisted, either the vacuum check valve or the booster is leaking. To test the check valve, remove the valve from the booster and blow through the check valve. If air passes through, the valve is defective and must be replaced. If the check valve is okay, the vacuum booster is leaking and should be repaired or replaced based on the manufacturer's recommendations.

HYDRAULIC SYSTEM LEAK TEST

An internal or external hydraulic leak can also cause a brake system problem. To test if the hydraulic system (and not the booster) is leaking, depress and release the

brake pedal (service brakes) several times. This should deplete any residual power assist. On some ABS units, this may require depressing the brake pedal 20 or more times!

After depleting the power-assist unit, depress and hold the brake pedal depressed with medium pressure [20 to 35 lb (90 to 150 N)]. The brake pedal should *not* fall away. If the pedal falls, the hydraulic brake system is leaking. Check for external leakage at wheel cylinders, calipers, hydraulic lines, and hoses. If there is no external leak, there may be an internal leak inside the master cylinder. Repair or replace components as needed to correct the leakage.

PUSH ROD CLEARANCE ADJUSTMENT

Whenever the vacuum brake booster or master cylinder is replaced, the push rod length should be checked. The length of the push rod must match correctly with the master cylinder (see Figure 9–9).

If the push rod length is too long and the master cylinder is installed, the rod may be applying a force on the primary piston of the master cylinder even though the brake pedal is not applied. This can cause the brakes to overheat and then cause the brake fluid to boil. If the brake fluid boils, a total loss of braking force can occur. Obviously, this push rod clearance check and adjustment is very important. A gauge is often used to

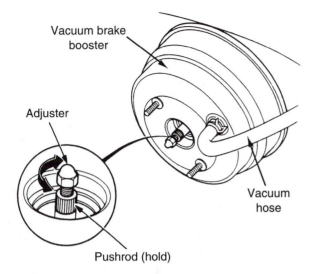

FIGURE 9–9 Typical adjustment push rod.

measure the position of the master cylinder piston and then used to determine the proper push rod clearance using the other end of the gauge (see Figure 9-10).

VACUUM BOOSTER DISASSEMBLY AND SERVICE

Some vehicle manufacturers recommend that the vacuum brake booster be disassembled and overhauled if defective.

CAUTION: Some vehicle manufacturers recommend that the vacuum brake booster be replaced as an assembly if tested to be leaking or defective. Always follow the manufacturer's recommendations.

A special holding fixture should be used before rotating (unlocking) the front and rear housing because the return spring is strong (see Figure 9–11).

Disassemble the vacuum brake booster according to the manufacturer's recommended procedures for the exact unit being serviced (see Figure 9–12). A rebuilding kit, including all necessary parts with the proper silicone grease, is available. The manufacturer usually warns that *all* parts included in the kit be replaced.

HYDRO-BOOST HYDRAULIC BRAKE BOOSTER

Hydro-boost is a hydraulically operated power-assist unit built by Bendix. The hydro-boost system uses the

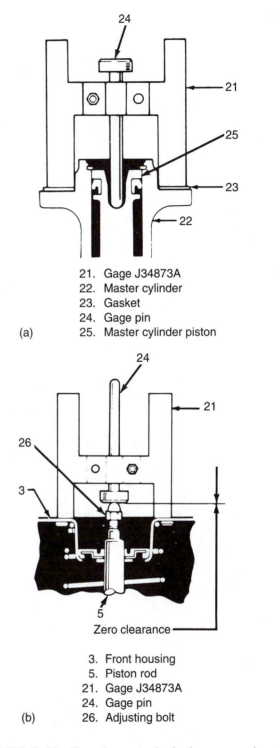

21. Gage J34873A
22. Master cylinder
23. Gasket
24. Gage pin
(a) 25. Master cylinder piston

3. Front housing
5. Piston rod
21. Gage J34873A
24. Gage pin
(b) 26. Adjusting bolt

FIGURE 9–10 Typical vacuum brake booster push rod gauging tool. (a) The tool is first placed against the mounting flange of the master cylinder and the depth of the piston determined. (b) The gauge is then turned upside down and used to gauge the push rod length. Some vacuum brake boosters do not use adjustable push rods. If found to be the incorrect length, a replacement push rod of the correct length should be installed.

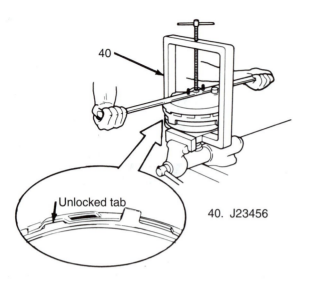

FIGURE 9–11 Holding fixture and long tool being used to rotate the two halves of a typical vacuum brake booster. (Courtesy of General Motors)

pressurized hydraulic fluid from the vehicle's power steering pump as a power source rather than using engine vacuum as is used with vacuum boosters (see Figures 9–13 and 9–14). The hydro-boost unit is used on vehicles that lack enough engine vacuum, such as turbo-charged or diesel engine vehicles.

Operation. Fluid pressure from the power steering pump enters the unit and is directed by a spool valve (see Figures 9–15 and 9–16). When the brake pedal is depressed, the lever and primary valve are moved. The valve closes off the return port, causing pressure to build in the boost pressure chamber. The hydraulic pressure pushes on the power piston, which then applies force to the output rod that connects to the master cylinder piston. In the event of a power steering pump failure, power assist is still available for several brake applications. During operation, hydraulic fluid under pressure from the power steering pump pressurizes an **accumulator**. While some units use a spring inside the

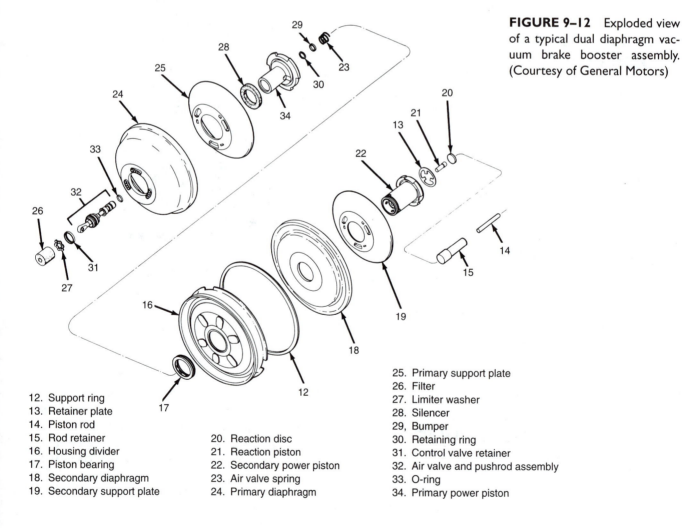

FIGURE 9–12 Exploded view of a typical dual diaphragm vacuum brake booster assembly. (Courtesy of General Motors)

12. Support ring
13. Retainer plate
14. Piston rod
15. Rod retainer
16. Housing divider
17. Piston bearing
18. Secondary diaphragm
19. Secondary support plate

20. Reaction disc
21. Reaction piston
22. Secondary power piston
23. Air valve spring
24. Primary diaphragm

25. Primary support plate
26. Filter
27. Limiter washer
28. Silencer
29. Bumper
30. Retaining ring
31. Control valve retainer
32. Air valve and pushrod assembly
33. O-ring
34. Primary power piston

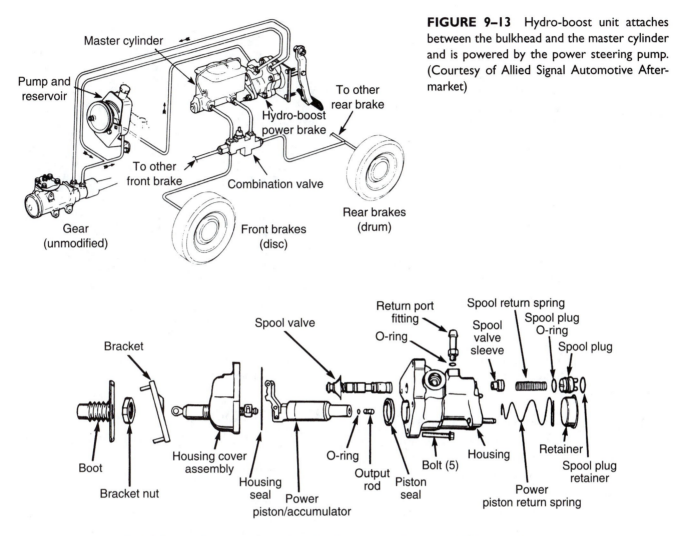

FIGURE 9–13 Hydro-boost unit attaches between the bulkhead and the master cylinder and is powered by the power steering pump. (Courtesy of Allied Signal Automotive Aftermarket)

FIGURE 9–14 Exploded view of a hydro-boost unit. (Courtesy of Allied Signal Automotive Aftermarket)

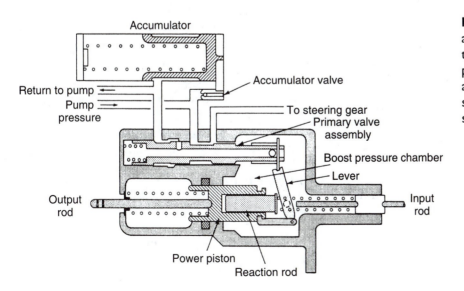

FIGURE 9–15 Simplified drawing of a hydro-boost power brake unit. Note that the hydraulic pressure from the power steering pump enters the unit and goes around the primary valve assembly, then back out and goes to the steering gear.

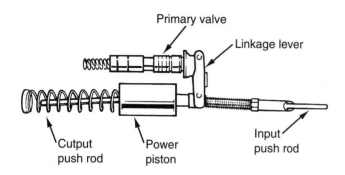

FIGURE 9–16 Operating linkages and levers used on a hydro-boost unit.

accumulator, most hydro-boost units use nitrogen gas. The fluid trapped in the accumulator under pressure is used to provide power-assisted stops in the event of a hydraulic system failure.

Diagnosis. The power source for hydro-boost units comes from the power steering pump. The first step of troubleshooting is to perform a thorough visual inspection, including:

1. Check for proper power steering fluid level.
2. Check for leaks from the unit or power steering pump.
3. Check the condition and tightness of the power steering drive belt.
4. Check for proper operation of the base brake system.

After checking all the visual components, check for proper pressure and volume from the power steering pump using a power steering pump tester, as shown in Figure 9–17. The pump should be capable of producing a minimum of 2 gallons (7.5 liters) with a *maximum* pressure of 150 psi (1000 kPa) with the steering in the straight-ahead position. With the engine off, the accumulator should be able to supply a minimum of two power-assisted brake applications.

HYDRO-BOOST FUNCTION TEST

With the engine off, apply the brake pedal several times until the accumulator is completely depleted. Depress the service brake pedal and start the engine. The pedal should fall and then push back against the driver's foot.

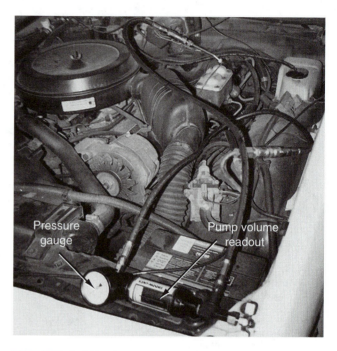

FIGURE 9–17 Power steering pump/gear tester being used on a hydro-boost power brake unit.

HYDRO-BOOST TROUBLESHOOTING GUIDE

Excessive Brake Pedal Effort

Possible Causes:

1. Loose or broken power steering pump belt
2. No fluid in the power steering reservoir
3. Leaks in the power steering, booster, or accumulator hoses
4. Leaks at tube fittings, power steering, booster, or accumulator connections
5. External leakage at the accumulator
6. Faulty booster piston seal, causing leakage at the booster flange vent
7. Faulty booster cover seal with leakage between housing and cover
8. Faulty booster spool plug seal

Slow Brake Pedal Return

Possible Causes:

1. Excessive seal friction in the booster
2. Faulty spool action

3. Broken piston return spring

4. Restriction in the return line from the booster to the pump reservoir

5. Broken spool return spring

Grabby Brakes

Possible Causes:

1. Broken spool return spring

2. Faulty spool action caused by contamination in the system

Booster Chatters—Pedal Vibrates

Possible Causes:

1. Power steering pump belt slipping

2. Low fluid level in the power steering pump reservoir

3. Faulty spool operation caused by contamination in the system

SUMMARY

1. Vacuum brake boosters use air pressure acting on a diaphragm to assist the driver's force on the brake master cylinder.

2. At rest, there is a vacuum on both sides of the vacuum booster diaphragm. When the brake pedal is depressed, atmospheric air pressure is exerted on the back side of the diaphragm.

3. The use of two diaphragms in tandem allows a smaller-diameter booster with the same area. The larger the area of the booster diaphragm, the more air pressure force can be applied to the master cylinder.

4. Hydraulic-operated brake boosters use the engine-driven power steering pump.

5. When replacing a vacuum brake booster, always check for proper push rod clearance.

6. To be assured of power-assisted brake application in the event of failure, hydraulic power-assisted brake systems use an accumulator to provide pressure to the system.

REVIEW QUESTIONS

1. Describe the purpose and function of the one-way check valve used on vacuum brake booster units.

2. Explain how vacuum is used to assist in applying the brakes.

3. Describe how to perform a vacuum booster leak test and hydraulic system leak test.

4. Explain how a hydro-boost system functions.

MULTIPLE-CHOICE QUESTIONS

1. Two technicians are discussing vacuum brake boosters. Technician A says that a low soft brake pedal is an indication of a defective booster. Technician B says that there should be at least two power-assisted brake applications after the engine stops running. Which technician is correct?

    a. A only

    b. B only

    c. Both A and B

    d. Neither A nor B

2. Technician A says that to check the operation of a vacuum brake booster, the brake pedal should be depressed until the assist is depleted and then start the engine. Technician B says that if the booster is okay, the brake pedal should *drop* when the engine starts. Which technician is correct?

    a. A only

    b. B only

    c. Both A and B

    d. Neither A nor B

3. Brake pedal feedback to the driver is provided by the

    a. vacuum check valve operation.

    b. reaction system.

    c. charcoal filter unit.

    d. vacuum diaphragm.

4. The proper operation of a vacuum brake booster requires that the engine be capable of supplying at least

    a. 15 in. Hg vacuum.

    b. 17 in. Hg vacuum.

    c. 19 in. Hg vacuum.

    d. 21 in. Hg vacuum.

5. The purpose of the charcoal filter in the vacuum hose between the engine and the vacuum brake booster is to
   a. filter the air entering the engine.
   b. trap gasoline vapors to keep them from entering the booster.
   c. act as a one-way check valve to help keep a vacuum reserve in the booster.
   d. direct the vacuum.

6. A defective vacuum brake booster will cause a
   a. hard brake pedal.
   b. soft (spongy) brake pedal.
   c. low brake pedal.
   d. slight hiss noise when the brake pedal is depressed.

7. An accumulator as is used on hydraulic brake boosters
   a. reduces brake pedal noise.
   b. provides a higher pressure being fed back to the driver's foot.
   c. provides a reserve in the event of a failure.
   d. works against engine vacuum.

8. The first step in diagnosing a hydro-boost problem is
   a. a pressure test of the pump.
   b. a volume test of the pump.
   c. to tighten the power steering drive belt.
   d. a thorough visual inspection.

◄ Chapter 10 ►

ANTILOCK BRAKING SYSTEM OPERATION, DIAGNOSIS, AND SERVICE

OBJECTIVES

After studying Chapter 10, the reader will be able to:

1. Explain the reason for ABS.
2. Describe the purpose and function of the ABS components such as wheel speed sensors, electrohydraulic unit, and electronic controller.
3. Discuss how the ABS components control wheel slippage.
4. Explain how the ABS components control acceleration traction control.
5. Describe normal ABS dash lamp operation.
6. Discuss visual inspection of the various types and brands of ABS.
7. Explain how to retrieve trouble codes.
8. Explain the various methods for bleeding ABS systems.
9. Discuss methods and tools needed to diagnose an ABS-equipped vehicle.

Antilock braking systems help prevent the wheels from locking during sudden braking, especially on slippery surfaces. This helps the driver maintain control.

THEORY OF OPERATION

When the wheels stop turning during a stop, the friction (traction) between the road surface and the tires decreases by almost 40 percent. At the surface of the road, a locked tire creates heat that generally softens the rubber of the tire. The tread rubber becomes almost liquid and loses its traction grip on the road surface.

A total loss of traction is called 100% slip where the tire is sliding across the road surface with the wheels locked (see Figure 10–1). Maximum traction occurs when the slip is controlled to between 10 and 20 percent (see Figure 10–2). This is accomplished by using electronic and hydraulic controls to monitor wheel slip and pulse the brakes on and off just the right amount to achieve the maximum possible traction without wheel lockup. A typical ABS vehicle pulses the brakes on and off between 10 and 20 times per second. This is much faster than most drivers are capable of doing.

STEERING CONTROL

When a front tire loses traction, a vehicle cannot be steered. Steering the vehicle requires that the tires have traction, and a locked wheel has little or no traction

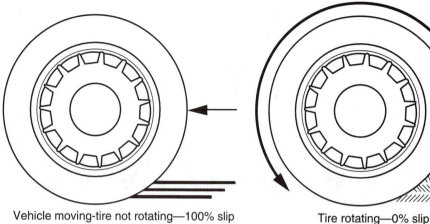

FIGURE 10–1 Maximum braking traction occurs when tire slip is between 10 and 20 percent. A rotating tire has 0% slip and a locked-up wheel has 100 percent slip.

Vehicle moving-tire not rotating—100% slip

Tire rotating—0% slip

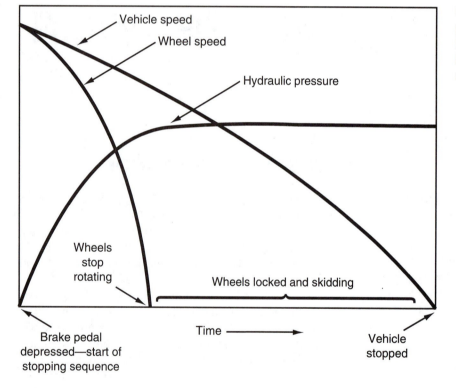

FIGURE 10–2 Typical stop on a slippery road surface without antilock brakes. Notice that the wheels stopped rotating long before the vehicle finally came to a stop.

with the road. As a result, the vehicle will continue traveling in a straight line even though the front wheels are being turned by the driver (see Figure 10–3).

SKID CONTROL

A vehicle equipped with ABS can still get into a skid; therefore, the term *skid control* is no longer used when describing antilock braking systems. ABS systems can prevent wheel lockup but cannot prevent the vehicle from skidding (see Figure 10–4).

PURPOSE AND FUNCTION OF ABS COMPONENTS

All ABS systems use three basic subsystems, including:

1. **Wheel speed sensors** (WSS). These electromagnetic sensors produce a speed signal for the electronic controller.

2. **Electronic control unit.** This is the brain, computer, or controller of any ABS unit. The controller receives wheel speed information from

the wheels and controls the hydraulic control unit if one or more wheels are slowing down at a faster rate than the other wheels. The electronic control unit is also called **controller antilock brake** (CAB) or **electronic brake control module** (EBCM).

FIGURE 10–3   Being able to steer and control the vehicle during rapid braking is one major advantage of an antilock braking system.

3. **Electromechanical hydraulic unit.** This unit does the actual work of controlling brake line pressures to keep the wheels from locking during a stop (see Figure 10–5).

ABS HYDRAULIC OPERATION

All ABS units use solenoid valves or rotary valves to control the brake fluid pressure at the wheels. There are three stages of ABS operation that are controlled by the hydraulic control unit.

Pressure Buildup (Normal Braking). The hydraulic system functions the same as any other hydraulic braking system. The driver has complete control of the pressure applied to the wheel cylinders and calipers. When greater braking force is required, the driver simply pushes the brake pedal down farther, which increases hydraulic pressure buildup in the master cylinder and in all wheel cylinders and calipers. When pressure is released from the brake pedal, the brakes release. This stage is also called **pressure buildup** (see Figure 10–6).

NOTE: Antilock braking systems cannot increase brake fluid pressure higher than the applied pressure by the driver.

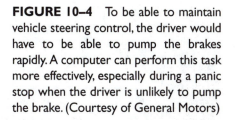

FIGURE 10–4   To be able to maintain vehicle steering control, the driver would have to be able to pump the brakes rapidly. A computer can perform this task more effectively, especially during a panic stop when the driver is unlikely to pump the brake. (Courtesy of General Motors)

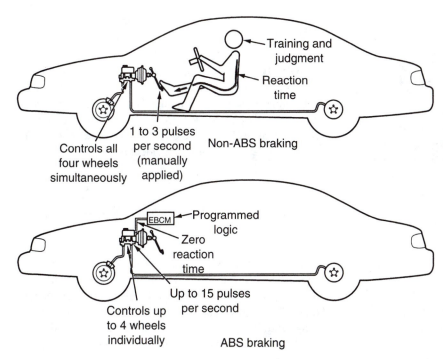

Training and judgment

Reaction time

Controls all four wheels simultaneously

1 to 3 pulses per second (manually applied)   Non-ABS braking

EBCM — Programmed logic

Zero reaction time

Controls up to 4 wheels individually

Up to 15 pulses per second

ABS braking

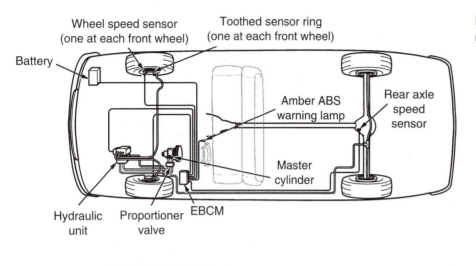

FIGURE 10–5   Typical ABS components. (Courtesy of General Motors)

Wheel speed sensor
(one at each front wheel)

Toothed sensor ring
(one at each front wheel)

Battery

Amber ABS
warning lamp

Rear axle
speed
sensor

Master
cylinder

Hydraulic
unit

Proportioner
valve

EBCM

◀ TECH TIP ▶

STOP ON A DIME?

Vehicles that are equipped with antilock braking systems help the driver avoid skidding and losing total control during braking. This is especially true for road surfaces that are slippery. However, vehicles equipped with ABS must still have traction between the tire and the road to stop.

This author had an experience with ABS on a snow-covered road. I applied the brakes while approaching a stop sign and the brake pedal started to pulsate, the electrohydraulic unit started to run, and the vehicle continued straight through the intersection! Luckily, no other vehicles were around. The vehicle did not stop for a long distance through the intersection, but it did stop straight, avoiding skidding. Because of the ice under the snow, the vehicle did not have traction between the tires and the road.

A common complaint is that the ABS did not stop the vehicle, whereas in reality, it did stop the vehicle from skidding or traveling out of control even though short stops are not always possible. The service technician should explain the purpose and function of ABS before attempting to repair a problem that may be normal on the vehicle being inspected.

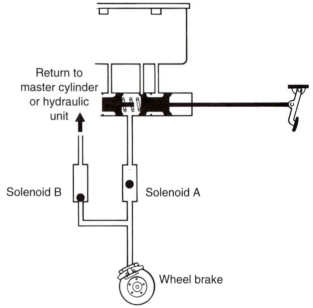

Return to
master cylinder
or hydraulic
unit

Solenoid B

Solenoid A

Wheel brake

FIGURE 10–6   In a typical antilock braking system, the flow of brake fluid is controlled by valves or solenoids. During normal braking, solenoid A is open, allowing normal pressure building to occur from the master cylinder during a normal non-ABS stop. Solenoid B is closed to maintain master cylinder pressure at the wheel brake.

the increase in pressure at the master cylinder is blocked.

Pressure Holding. Pressure holding means that the ABS controller has detected a rapid slowing of a wheel during braking. To help prevent the wheel from locking, the controller commands that a valve be shut (closed) between the master cylinder and the wheel brake drum or caliper (see Figure 10–7). By closing off the master, the pressure to the brake is held at the present level. Even if the driver pushes the break pedal down farther,

Pressure Reduction. The ABS controller can reduce the hydraulic pressure on the wheel cylinder or caliper by opening a valve and allowing the pressurized brake fluid to escape into a low-pressure area of the system. When this occurs, the pressure is reduced, called the **pressure reduction stage** (see Figure 10–8). See Figure 10–9 for a graph showing vehicle

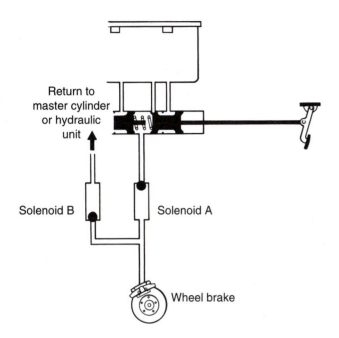

FIGURE 10–7 If a wheel is starting to slow down too fast, the ABS controller will command that solenoid A block off the master cylinder from the wheel brake. This prevents the driver from exerting additional pressure to the wheel brake. This stage is called *pressure holding.*

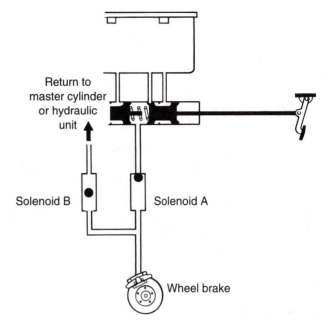

FIGURE 10–8 If a wheel is still slowing too fast and is about to lock up, solenoid B is opened, allowing the trapped pressurized brake fluid at the wheel brake to escape. This stage is called *pressure reduction (release).*

speed, wheel speed, and hydraulic pressure during a typical ABS stop.

The hydraulic pump delivers brake fluid to the master cylinder to decrease fluid pressure in the caliper. The accumulator (reservoir) temporarily stores brake fluid returning from the calipers as required to provide a smooth pressure decrease in the caliper (see Figure 10–10).

BRAKE PEDAL FEEDBACK

Many ABS units force brake fluid back into the master cylinder under pressure during an ABS stop. This pulsing brake fluid return causes the brake pedal to pulsate. Some vehicle manufacturers use the pulsation of the brake pedal to inform the driver that the wheels are tending toward lockup and that the ABS is pulsing the brakes.

NOTE: A pulsating brake pedal may be normal only during an ABS stop. It is not normal for a vehicle with ABS to have a pulsating pedal during normal braking. If the brake pedal is pulsating during a non-ABS stop, the brake drums or rotor may be warped. See Chapter 8 for details.

Some manufacturers use an **isolation valve** that prevents brake pedal pulsation even during an ABS stop.

BRAKE PEDAL TRAVEL SWITCH (SENSOR)

Some ABS systems, such as the Teves Mark IV system, use a brake pedal travel switch (sensor). The purpose of the switch is to turn on the hydraulic pump when the brake pedal has been depressed 40% of its travel. The pump runs and pumps brake fluid back into the master cylinder, which raises the brake pedal until the switch closes again turning off the pump.

NOTE: Some early ABS systems did not use a brake switch. The problem occurred when the ABS could be activated while driving over rough roads. The brake switch can be the same as the brake light switch *or* a separate switch.

The brake pedal switch is an input for the electronic controller. When the brakes are applied, the electronic controller gets ready to act if ABS needs to initialize the starting sequence of events.

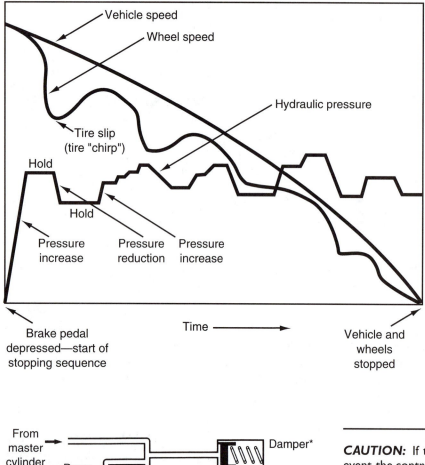

FIGURE 10–9   Typical stop with antilock brakes. Notice how the pressure increase, hold, and release are all used to bring the vehicle to a safe stop.

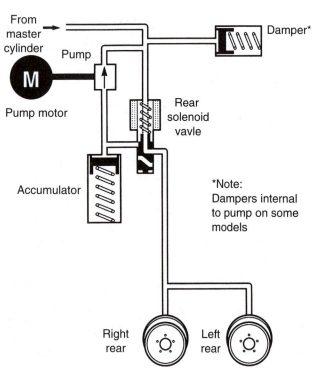

FIGURE 10–10   The brake fluid return pump sends brake fluid back to the reservoir during the pressure release stage (mode). (Courtesy of General Motors)

CAUTION: If the driver pumps the brakes during an ABS event, the controller will reset and reinitialization starts over again. This resetting process can disrupt normal ABS operation. The driver need only depress and hold the brake pedal down during a stop for best operation.

WHEEL SPEED SENSORS

Wheel speed sensors (WSS) are small electromagnetic generators. A toothed wheel is called a **tone ring, toothed ring,** or **reluctor** (see Figures 10–11 through 10–15). The speed sensor contains a coil of wire surrounding a permanent magnet. As each tooth passes by the magnet, the magnetic field around the coil is increased. Then as the tooth moves away, the magnetic field strength weakens. It is this *changing* magnetic field strength that produces a *changing* voltage (electrical pressure) in the coil of wire surrounding the magnet. This rapidly changing voltage signal is sent to the electronic controller. The electronic controller uses the **frequency** of the high and low voltages as a measure of the wheel speed. Frequency means the number of times the voltage changes per second and is measured in

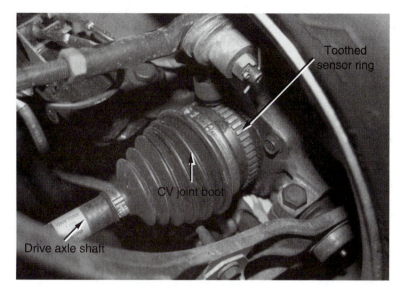

FIGURE 10–11 Typical toothed sensor ring (on the front wheel of a front-wheel-drive vehicle).

◀ **TECH TIP** ▶

KISS

KISS means "Keep It Stock, Stupid" and it is important to remember when replacing tires. Vehicles equipped with antilock brakes are "programmed" to pulse the brakes at just the right rate for maximum braking effectiveness. A larger tire rotates at a slower speed and a smaller than normal tire rotates at a faster speed. Therefore, tire size affects the speed and rate of change in speed of the wheels as measured by the wheel speed sensors.

While changing tire size will not prevent ABS operation, it *will* cause less effective braking during hard braking with the ABS activated. Using the smaller spare tire can create such a difference in wheel speed compared to the other wheels that a false wheel speed sensor code may be set and an amber ABS warning lamp on the dash may light. However, most ABS systems will still function with the spare tire installed, but the braking performance will not be as effective. For best overall performance, always replace tires with the same size and type as specified by the vehicle manufacturer.

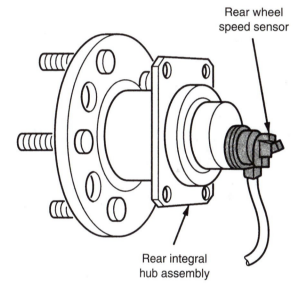

FIGURE 10–12 Many wheel speed sensors are built into the bearing assembly. This helps protect the sensor (tone) ring and sensor from possible damage from road debris. (Courtesy of General Motors)

LATERAL ACCELERATION SENSOR

Some ABS-equipped vehicles include a lateral acceleration sensor or switch that measures the vehicle's cornering force (see Figure 10–16). The signal is sent from the lateral acceleration sensor to the electronic brake control module (EBCM) to modify its control logic accounting for hard cornering conditions.

hertz. One hertz (Hz) is one cycle per second. The electronic controller looks at the frequency of all wheel speed sensors and activates the antilock control if one or more sensors indicate that a wheel is slowing down a lot faster than the other wheel sensors.

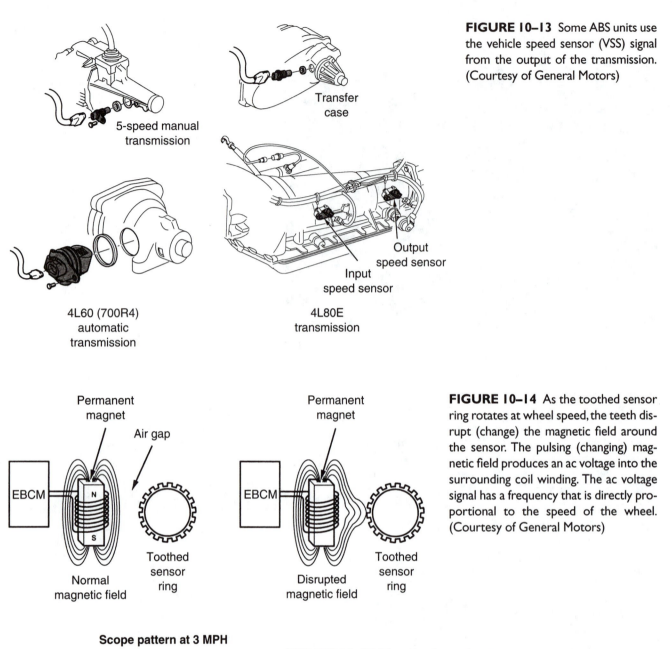

FIGURE 10–13 Some ABS units use the vehicle speed sensor (VSS) signal from the output of the transmission. (Courtesy of General Motors)

Transfer case

5-speed manual transmission

Output speed sensor

Input speed sensor

4L60 (700R4) automatic transmission

4L80E transmission

Permanent magnet

Air gap

EBCM

N

S

Toothed sensor ring

Normal magnetic field

FIGURE 10–14 As the toothed sensor ring rotates at wheel speed, the teeth disrupt (change) the magnetic field around the sensor. The pulsing (changing) magnetic field produces an ac voltage into the surrounding coil winding. The ac voltage signal has a frequency that is directly proportional to the speed of the wheel. (Courtesy of General Motors)

Permanent magnet

EBCM

Toothed sensor ring

Disrupted magnetic field

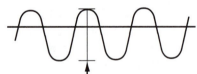

Scope pattern at 3 MPH

Approximately 5 volts

FIGURE 10–15 The wheel speed sensor generates a varying voltage as the wheel revolves. As the speed of the wheel increases, the voltage and the frequency (number of voltage cycles per second) also increase. (Courtesy of Chrysler Corporation)

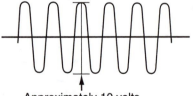

Scope pattern at 9–12 MPH

Approximately 10 volts

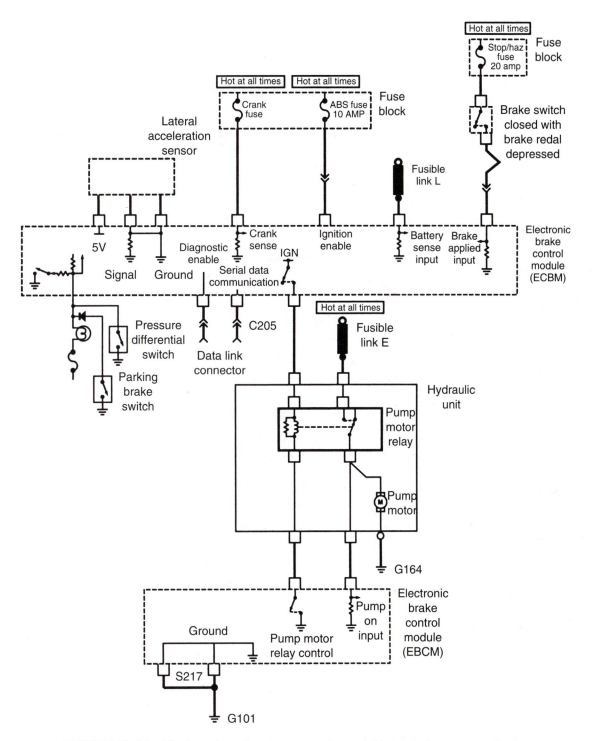

FIGURE 10–16 The lateral acceleration sensor (upper left) signals the electronic brake control module that the vehicle is cornering and changes its internal programming to allow for differences in tire loading due to cornering. (Courtesy of General Motors)

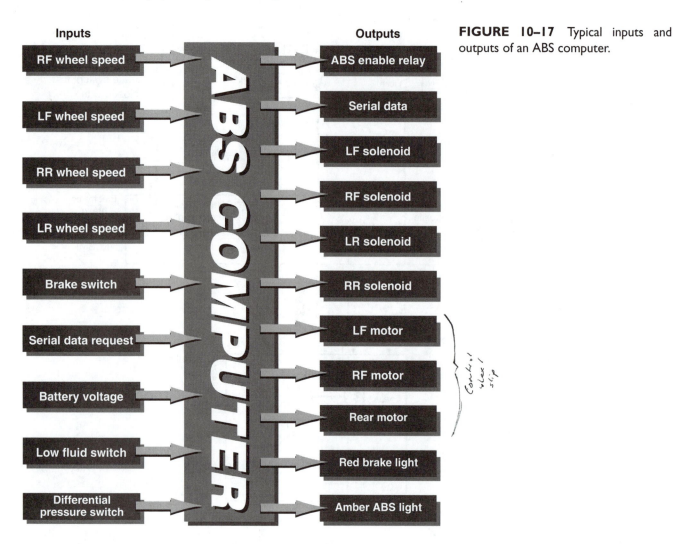

FIGURE 10-17 Typical inputs and outputs of an ABS computer.

ELECTRONIC CONTROLLER OPERATION

The electronic controller is the computer in the system that controls all parts of the ABS operation, including:

1. **Self-test.** The controller runs a self-test of all of its components every time the ignition is turned on.

NOTE: Since an antilock braking system is a safety-related system, if it malfunctions, people can be injured. This is one reason why the system does a complete "system check" every time the ignition is cycled. See the Tech Tip "Is Tire Chirp Normal with ABS?"

See Figures 10–17 through 10–19.

2. **Activation of the wheel hydraulic controls.** The controller looks at the rate of wheel deceleration and compares it to normal stopping rates using an internal computer program that is based on vehicle weight, tire size, etc. If a wheel is slowing down too fast, the controller activates the necessary hydraulic pressure controls.

REAR WHEEL ANTILOCK SYSTEMS

Antilock brakes are especially important for use on the rear wheels. Rear wheel antilock systems are commonly used on pickup trucks, sport utility vehicles (SUVs), and vans (see Figure 10–20). Rear wheel antilock systems may be abbreviated:

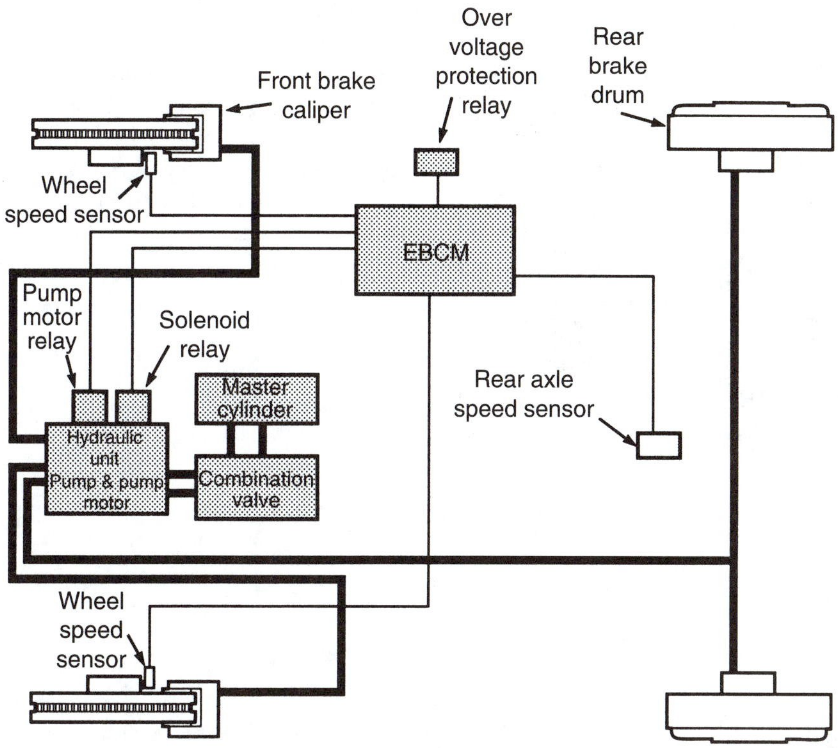

Front brake caliper

Over voltage protection relay

Rear brake drum

Wheel speed sensor

EBCM

Pump motor relay

Solenoid relay

Master cylinder

Rear axle speed sensor

Hydraulic unit Pump & pump motor

Combination valve

Wheel speed sensor

FIGURE 10–18 An overvoltage protection relay is often used to power electronic brake control modules. The purpose of the relay is to supply electrical power to the controller and protect the electronics inside from damage if a high-voltage surge were to occur in the vehicle's electrical system. (Courtesy of General Motors)

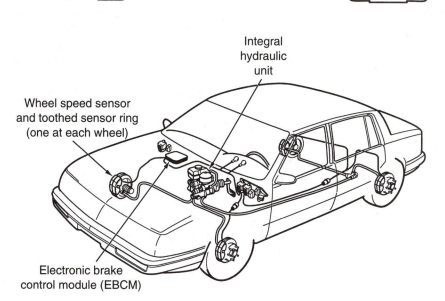

Integral hydraulic unit

Wheel speed sensor and toothed sensor ring (one at each wheel)

Electronic brake control module (EBCM)

FIGURE 10–19 A Teves Mark II ABS is equipped with four-wheel speed sensors. This integral type of ABS combines the master cylinder, hydraulic booster, and ABS hydraulic-control solenoids and valves all in one unit. (Courtesy of General Motors)

RWAL, for *rear wheel antilock*.
RABS for *rear antilock braking system*.

Rear wheel antilock systems will hold or decrease hydraulic pressure to *both* rear wheel brakes if *either* wheel starts to lock up. In other words, both rear wheels are handled as one and it does not matter to the electronic controller which wheel brake is about to lock. Rear wheel antilock braking systems are also

called **one-channel systems** because the hydraulic controls just one hydraulic circuit to both rear wheels.

THREE-CHANNEL SYSTEMS

Most rear-wheel-drive and many front-wheel-drive vehicles use a three-channel antilock braking system (see Figure 10–21). Each front wheel brake is controlled

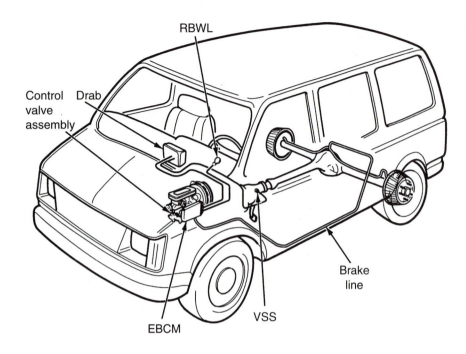

FIGURE 10–20 Typical Kelsey-Hayes rear wheel antilock (rear ABS) unit that uses the vehicle speed sensor for speed information and controls both rear wheels together if one or both tend to slow down too rapidly during a stop. (Courtesy of General Motors)

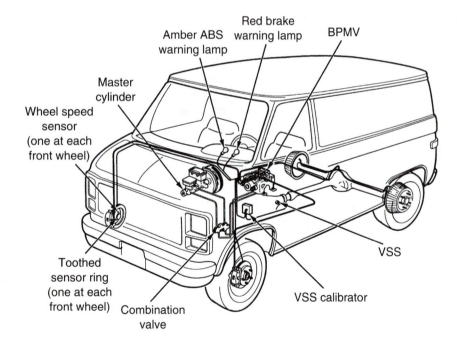

FIGURE 10–21 The typical Kelsey-Hayes four-wheel antilock (4WAL) ABS is a three-channel system that controls each front wheel individually and both rear wheels together as one. The BPMV is the *brake pressure modulator* valve, also referred to as the *electrohydraulic control unit* (EHCU). The VSS calibration can be changed if different-size tires are used on the vehicle. This helps assure maximum ABS effectiveness and correct vehicle speed being registered on the speedometer/od-ometer. (Courtesy of General Motors)

separately, and both rear brakes are controlled together.

FOUR-CHANNEL SYSTEMS

Four-channel antilock braking systems control all four wheel brakes individually. This is the most expensive type of system and requires that each wheel have a wheel speed sensor (WSS) (see Figure 10–22).

INTEGRAL ANTILOCK BRAKING SYSTEMS

Integral ABS means that the hydraulic control unit includes:

- A master cylinder with reservoir
- A hydraulic brake booster
- A brake pressure pump and motor

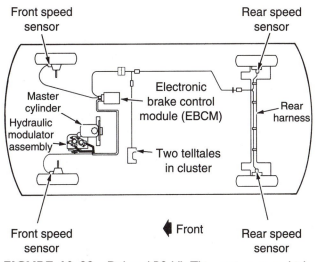

FIGURE 10–22 Delco ABS VI. This system uses high-speed bidirectional motors to reduce, hold, and apply hydraulic pressure. This four-channel system controls all four wheel brakes.

- A pressure accumulator
- Pressure monitoring switches
- Brake pressure modulator valves
- A brake fluid level sensor

Because the system functions as a brake booster as well as an antilocking braking unit, the hydraulic pump can run even in normal braking applications. If the pump fails, there will be no rear brakes and the brake pedal will be hard because of a lack of power assist (see Figures 10–23 and 10–24).

Integral ABS units are usually serviced as complete units as an assembly. The operation of each unit varies with individual vehicle and year. All integral systems are used as front-wheel-drive vehicles because if the rear brakes are lost, it is not as severe as with a rear-wheel-drive vehicle. On a front-wheel-drive vehicle, only about 20 percent of the braking force occurs on rear brakes. Teves Mark II and Delco Moraine III (Powermaster III) are examples of integral systems.

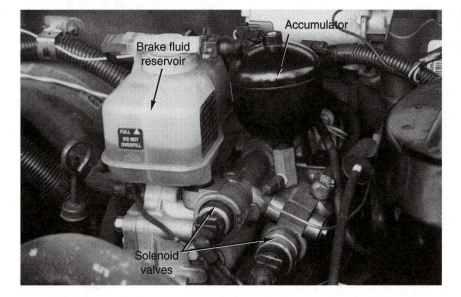

FIGURE 10–23 Typical integral ABS. (Courtesy of General Motors)

FIGURE 10–24 Integral ABS combines master cylinder, hydraulic booster, and ABS hydraulic control all in one unit.

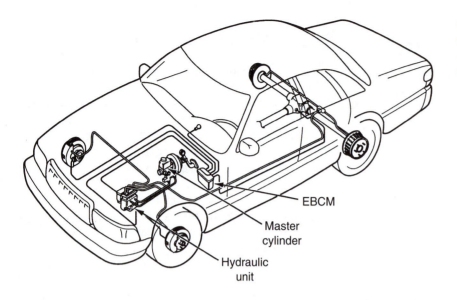

FIGURE 10–25 This nonintegral (remote) unit uses a conventional master cylinder with a separate ABS hydraulic unit. (Courtesy of General Motors)

NONINTEGRAL (REMOTE) ABS

Nonintegral systems are added to a base brake system and are often called *remote* or *add-on ABS* (see Figure 10–25). Their only function is to provide antilock braking. These systems use a hydraulic system that is remote (removed) from the master cylinder. The ABS hydraulic unit is added in series with the hydraulic brake lines. Some nonintegral systems use a hydraulic pump for fluid circulation. Bosch 2U/2S, Kelsey Hayes RWAL/RABS, Kelsey Hayes 4WAL, Teves Mark IV, and Delco VI are examples of nonintegral (remote) types of ABS.

TRACTION CONTROL

Antilock braking systems can often be used to limit wheel spin during acceleration. In a conventional antilock braking system, the wheels are kept from locking (stopping) when the electronic controller senses one or more wheels slowing too rapidly. With **traction control** (abbreviated TC), the electronic controller applies hydraulic pressure to the wheel brakes on the drive wheels if the wheels start to spin or slip during acceleration. This slipping during acceleration is referred to as **positive slip**. If the driving wheels are slipping, the vehicle will not move. If the vehicle is front-wheel drive, steering is impaired when the front wheels are spinning on slippery surfaces. Best acceleration is achieved when positive wheel slip is controlled to about 10 percent.

Low-speed traction control uses the braking system to limit positive slip up to a vehicle speed of about 30 mph (48 km/h) (see Figure 10–26). All-speed traction control systems are capable of reducing positive wheel slip at all speeds. Most speed-traction control systems use accelerator reduction and engine power reduction to limit slip. For example, if a vehicle were being driven on an icy or snow-covered road, the brakes may become overheated if heavy acceleration is attempted. To help take the load off the brakes, the acceleration and power output from the engine is reduced. Many systems use accelerator pedal reduction and fuel injector cut out or ignition timing retardation individually or in combination to help match engine power output to the available tire traction. Traction control is also called **acceleration slip regulation (ASR)** (see Figure 10–27).

***

NOTE: The ABS controller only supplies the necessary pressure to the wheel brake that is required to prevent tire slipping during acceleration. The amount of pressure varies according to the condition of the road surface and the amount of engine power being delivered to the drive wheels. A program inside the controller will disable traction control if brake system overheating is likely to occur. The driver should either wait for the brakes to cool down or use less accelerator pedal while driving.

***

BRAKE WARNING LAMP OPERATION

The first step in the correct diagnosis of an antilock braking system problem is to check the status of the brake warning lamps.

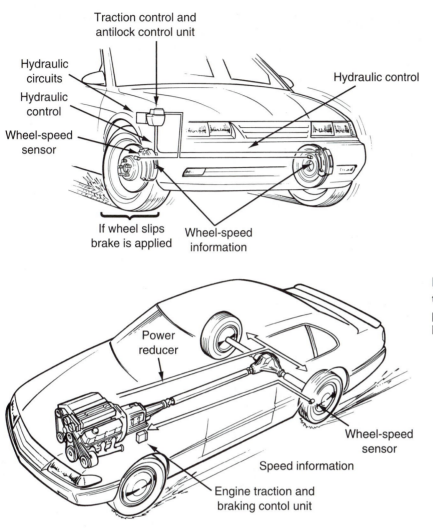

Traction control and
antilock control unit

Hydraulic
circuits

Hydraulic
control

Wheel-speed
sensor

Hydraulic control

If wheel slips
brake is applied

Wheel-speed
information

FIGURE 10–26 Brake-system-only traction control uses information from wheel speed sensors to the traction control unit. If the drive wheel is accelerating too fast, indicating wheel slip, the control unit applies hydraulic control to the wheel brake to stop the wheel slip.

Power
reducer

Wheel-speed
sensor

Speed information

Engine traction and
braking contol unit

FIGURE 10–27 All speed traction control prevents drive wheel acceleration slippage not only by applying the wheel brake(s) but also by reducing engine power.

Red Brake Warning Lamp. This lamp warns of a possible dangerous failure in the base brakes such as low brake fluid level or low pressure in half of the hydraulic system. The red brake warning lamp will also light if the parking brake is applied and may light due to an ABS failure such as low brake pressure on an integral system.

NOTE: Some antilock braking systems will light the red brake warning lamp through a resistor. This results in a dim red brake warning lamp. To check if the lamp is dim or at full brightness, simply apply the parking brake. If the warning lamp gets brighter, you know that the red brake warning lamp is indicating an ABS problem and not a hydraulic problem.

See Figure 10–28.

Amber ABS Warning Lamp. This lamp usually comes on after a start during the initialization or startup self-test sequence. The exact time the amber lamp remains on after the ignition is turned on varies with vehicle and ABS design.

THOROUGH VISUAL INSPECTION

Many ABS-related problems can be quickly diagnosed if all of the basics are carefully inspected. A thorough visual inspection should include the following items:

1. **Brake fluid level.** Check the condition and level in the reservoir.
2. **Brake fluid leaks.** Check for cracks in flexible lines or other physical damage.

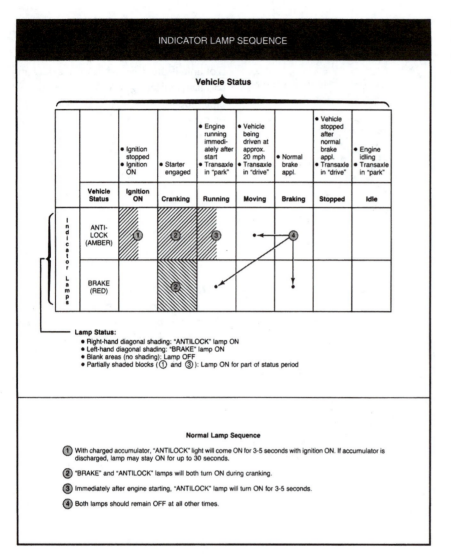

FIGURE 10–28   Another example of a normal warning lamp sequence chart during normal (no-fault) operation. (Courtesy of General Motors)

3. **Fuses and fusible links.** Check all ABS-related fuses.

4. **Wiring and connections.** Check all wiring, especially wheel speed sensor leads for damage.

5. **Wheel speed sensors.** Check that the sensor ring teeth are not damaged. Clean debris from the sensor if possible.

NOTE: Most wheel speed sensors are magnetic and therefore can attract and hold metallic particles. Be sure to remove any metallic debris from around the magnetic wheel speed sensor.

---

6. **Base brake components.** All base brake components, such as disc brake calipers, drum brake wheel cylinders, and related components, must be in proper working condition.

7. **Parking brake.** Check that the parking brake is correctly adjusted and fully released.

8. **Wheel bearings.** All wheel bearings must be free of defects and properly adjusted.

9. **Wheels and tires.** Check for correct size, proper inflation, and legal tread depth.

TEST DRIVE AND VERIFY THE FAULT

A test drive is a very important diagnostic procedure. Many ABS systems and diagnostic trouble codes (DTC) will not set unless the vehicle is moving. Often, the driver has noticed something like the self-test while driving and thought it may be a fault in the system.

◄ TECH TIP ►

IS TIRE CHIRP NORMAL WITH ABS?

Many owners of vehicles complain that their antilock braking system is not working correctly because their tires chirp and occasionally experience tire lockup during hard braking, especially at low speed. These conditions are perfectly normal, because for maximum braking between 12 and 20 percent of *slip* means that tire will slip or skid slightly during an ABS stop.

It is also normal for vehicles with ABS to have the tires lock and skid slightly when the speed of the vehicle is below 5 mph (8 km/h). This occurs because the wheel speed sensors cannot generate usable speed signals for the electronic controller. This low-speed wheel lockup seldom creates a problem. Before attempting to troubleshoot or diagnose an ABS problem, be sure that the problem is not just normal operation of the system.

NOTE: Some ABS units, such as the Delco VI, will cause the brake pedal to move up and down slightly during the cycling of the valves during the self-test. Each system has its own unique features. The service technician will have to learn to avoid attempting to repair a problem that is not a fault of the system.

Before driving, start the engine and observe the red and amber brake warning lamps. If the red brake warning lamp is on, the base brakes may not be functioning correctly. Do not drive the vehicle until the base brakes are restored to proper operation.

RETRIEVING DIAGNOSTIC TROUBLE CODES

After performing a thorough visual inspection and after verifying the customer's complaint, retrieve any stored ABS-related diagnostic trouble codes (DTCs). The *exact* procedure varies with the type of ABS and with the make, model, and year of the vehicle. *Always consult factory service information for the vehicle being diagnosed.* Some systems can only display flash codes (flashing ABS or brake lamp in sequence), whereas other systems can perform self-diagnosis and give all information to the technician through a scan tool.

◄ TECH TIP ►

WHAT'S THAT NOISE AND VIBRATION?

Many vehicle owners and service technicians have been disturbed to hear and feel an occasional groaning noise. It usually is heard and felt through the vehicle after first being started and driven. Because it occurs when first being driven in forward or reverse, many technicians have blamed the transmission or related drive line components. This is commonly heard on many ABS vehicles as part of a system check. As soon as the ABS controller senses speed from the wheel speed sensors after an ignition cycles on, the controller will run the pump either every time or whenever the accumulator pressure is below a certain level. This can occur while the vehicle is being backed out of a driveway or being driven forward because wheel sensors can only detect speed, not direction. Before serious and major repairs are attempted to "cure" a noise, make sure that it is not the normal ABS self-test activation sequence of events.

NOTE: With some antilock braking systems, the diagnostic trouble code is lost if the ignition is turned off before grounding the diagnostic connector.

GENERAL MOTORS ABS DIAGNOSTIC GUIDE

Type of ABS system	Scan Tool[a]	
	Code clearing procedure	Code retrieval
Teves II integral	Not needed	Jumper key
Teves IV nonintegral	Yes	Yes
Bosch 2U nonintegral	Yes	Yes
Delco VI	Yes	Yes
Bosch 2S	Yes	Yes
Bosch III	Yes	Yes
Delco Powermaster III	Yes	Yes

Type of ABS system	Scan Tool[a]	
	Code clearing procedure	Code retrieval
Kelsey-Hayes RWAL	Yes, or disconnect ABS fuse	Yes or flash codes
Kelsey-Hayes 4WAL	Yes	Yes

[a]A jumper key can be used for some systems. Not all scan tools can perform all functions.

FORD ABS DIAGNOSTIC GUIDE

Type of ABS system	Scan Tool[a]	
	Code clearing procedure	Code retrieval
Teves II 32 pin integral	No codes available on this system	No codes available on this system
Teves II 35 pin integral	Drive vehicle after repair	Yes
Teves IV nonintegral	Drive vehicle after repair	Yes
Kelsey-Hayes RABS	Drive vehicle after repair	Flash codes
Teves 4WABS	Drive vehicle after repair	Yes

[a]A jumper wire can be used on some systems. Not all scan tools can perform all functions.

NOTE: Some systems are diagnosed by antilock and brake warning lamps, vehicle symptoms, and the use of a breakout box.

CHRYSLER ABS DIAGNOSTIC GUIDE

Type of ABS system	Scan Tool[a]	
	Code clearing procedure	Code retrieval
Bosch III integral	Depress the brake	Turn the key off

Type of ABS system	Scan Tool[a]	
	Code clearing procedure	Code retrieval
Bendix 10 integral	Yes or disconnect the battery	Yes
Bendix 6 nonintegral	Yes or disconnect the battery	Yes
Bosch MMC	Yes	Yes
Bosch 2U nonintegral	No codes available on this system	No codes available on this system
Teves Mark II	Yes	Yes
Teves Mark IV	Yes	Yes
Kelsey-Hayes RWAL	Disconnect the battery	Flash codes
Kelsey-Hayes 4WAL	Yes	Yes

[a]A jumper wire can be used on some systems. Not all scan tools can perform all functions.

KELSEY-HAYES REAR WHEEL ANTILOCK (NONINTEGRAL)

- **GM trucks (RWAL).** Flash codes and scan data through the use of OEM scan tool (Tech I/II) or connect H to A at DLC (see Figure 10–29).

NOTE: Be sure that the brake warning lamp is on before trying to retrieve diagnostic trouble codes. If the lamp is not on, a false code 9 could be set.

- **Ford trucks (RABS).** Jumper lead flash codes only (see Figure 10–30).
- **Chrysler light trucks.** Ground diagnostic connections (see Figure 10–31).

RWAL Diagnostic Trouble Codes

NOTE: If the ignition is turned off, the failure code will be lost unless it is a hard code that will be present when the ignition is turned back on.

Code	Description
2	Open isolation valve solenoid circuit or malfunctioning EBCM/VCM
3	Open dump valve solenoid circuit or malfunctioning EBCM/VCM
4	Grounded valve reset switch circuit
5	Excessive actuations of dump valve during antilock braking
6	Erratic speed signal

Code	Description
7	Shorted isolation valve circuit or faulty EBCM/VCM
8	Shorted dump valve circuit or faulty EBCM/VCM
9	Open or grounded circuit to vehicle speed sensor
10	Brake switch circuit
12–17	Computer malfunction

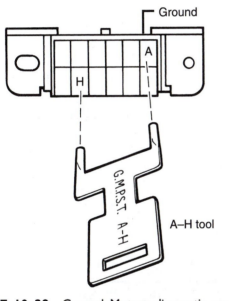

FIGURE 10–29 General Motors diagnostic connector. Flash codes are available by jumping ground (terminal A) to terminal H. This connector is located under the dash near the steering column on most General Motors vehicles.

NOTE: A scan tool may or may not be able to retrieve or display diagnostic trouble codes. Check with the technical literature for the exact vehicle being scanned.

4WAL Diagnostic Trouble Codes

Code	Description
12	System normal (2WD applications)
13	System normal—brake applied (2WD applications)
14	System normal (4WD/AWD applications)
15	System normal—brake applied (4WD/AWD applications)
21 RF, 25 LF, 31 RR, 35 LR	Speed sensor circuit open
35	VSS circuit open

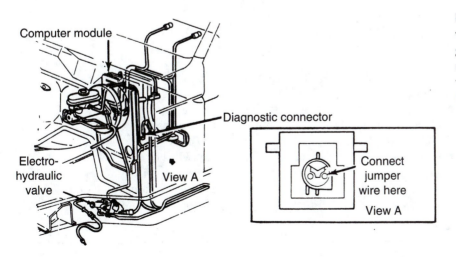

FIGURE 10–30 Connecting a jumper wire from the diagnostic connector to ground. The exact location of this diagnostic connector varies with the exact vehicle model and year. (Courtesy of Ford Motor Company)

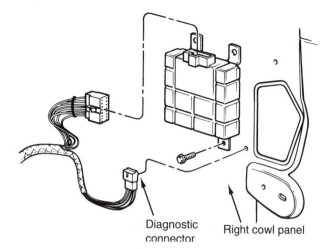

Diagnostic connector   Right cowl panel

FIGURE 10–31 The Chrysler diagnostic connector location varies with model and year. (Courtesy of Chrysler Corporation)

Code	Description
22 RF, 26 LF, 32 RR, 36 LR	Missing speed sensor signal
36	Missing VSS signal
23 RF, 27 LF, 33 RR, 37 LR	Erratic speed sensor signal

Code	Description
37	Erratic VSS signal
28	Simultaneous dropout of front wheel speed sensors
29	Simultaneous dropout of all speed sensors
35	Vehicle speed sensor circuit open
36	Missing LR or vehicle speed sensor signal
37	Erratic LR or vehicle speed sensor signal
38	Wheel speed error
41–66	Malfunctioning BPMV/EHCU
67	Open motor circuit or shorted EBCM output
68	Locked motor or shorted motor circuit
71–74	Memory failure
81	Open or shorted brake switch circuit
86	Shorted ABS warning lamp
88	Shorted red brake warning lamp (RBWL)

◄ DIAGNOSTIC STORY ►

RWAL DIAGNOSIS

The owner of an S-10 pickup truck complained that the red brake warning lamp on the dash remained on even when the parking brake was released. The problem could be:

1. A serious hydraulic problem
2. Low brake fluid
3. A stuck or defective parking brake switch

If the brake lamp is dim, RWAL trouble is indicated.

The technician found that the brake lamp was on dimly, indicating that an antilock braking problem was detected. The first step in diagnosing an antilock braking problem *with* a dash lamp on is to check for stored trouble codes. The technician used a jumper between terminals A and H on the DLC (old ALCL), and four flashes of the brake lamp indicated a code 4.

Checking a service manual, code 4 was found to be a grounded switch inside the hydraulic control unit. The hardest part about the repair was getting access to and the replacement of the defective (electrically grounded) switch. After bleeding the system and a thorough test drive, the lamp sequence and RWAL functioned correctly.

BOSCH 2 ABS (NONINTEGRAL)

The Bosch 2U/2S ABS is used on many domestic and imported brand of vehicles (see Figure 10–32).

Retrieving Diagnostic Trouble Codes. On General Motors vehicles, diagnostic trouble codes can be retrieved by connecting A to H at the diagnostic link connection (DLC) as shown in Figure 10–29. On most Bosch 2 systems, a scan tool can and should be used if available to retrieve diagnostic trouble codes.

Bosch 2U/2S ABS Diagnostic Trouble Codes

Code	Description
12	Diagnostic system operational
21 RF, 25 LF, 31 RR, 35 LR	Wheel speed sensor fault
35	Rear axle speed sensor fault
22 RF, 26 LF, 32 RR, 36 LR	Toothed wheel frequency error

Code	Description
36	Rear axle toothed wheel frequency error
41 RF, 45 LF, 55 Rear	Valve solenoid fault
61	Pump motor or motor relay fault
63	Solenoid valve relay fault
71	Electronic brake control module fault
72	Serial data link fault
74	Low voltage
75	Lateral acceleration sensor fault
76	Lateral acceleration sensor fault

BOSCH III ABS (INTEGRAL)

The Bosch III integral is used on many different makes and models of vehicles including various models of General Motors vehicles (see Figure 10–33).

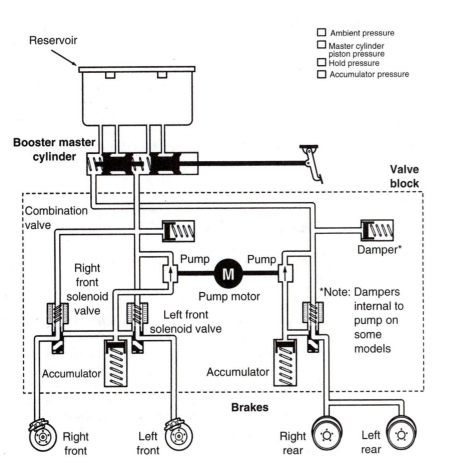

☐ Ambient pressure
☐ Master cylinder piston pressure
☐ Hold pressure
☐ Accumulator pressure

FIGURE 10–32 Bosch 2U/2S system components. (Courtesy of General Motors)

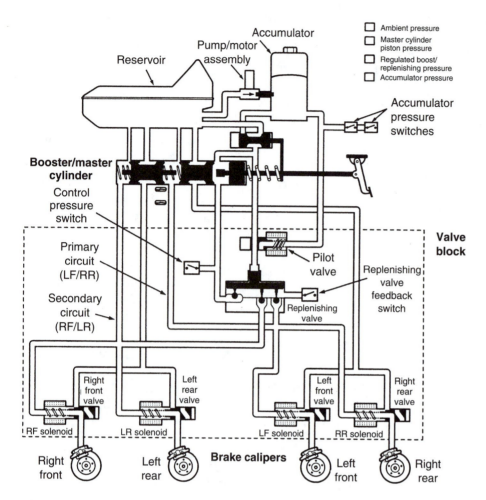

FIGURE 10–33 Bosch III system components. (Courtesy of General Motors)

Bosch III Diagnostic Trouble Codes

Code	Description
1	LF wheel circuit valve
2	RF wheel circuit valve
3	RR wheel circuit valve
4	LR wheel circuit valve
5	LF wheel speed sensor
6	RF wheel speed sensor
7	RR wheel speed sensor
8	LR wheel speed sensor
9	LF/RR wheel speed sensor
10	RF/LR wheel speed sensor
11	Replenishing valve
12	Valve relay
13	Circuit failure
14	Piston travel switches
15	Stoplight switch
16	ABCM error

Retrieving diagnostic trouble codes for a Bosch III antilock braking system varies with vehicle make, model, and year. Most codes are erased when the ignition is turned off; therefore, code retrieval must occur before key-off. If the amber ABS lamp is on, a trouble code has been set.

TEVES MARK II

Teves Mark II ABS is an integral system, as shown in Figure 10–34.

Retrieving Diagnostic Trouble Codes. On General Motors vehicles equipped with Teves Mark II ABS, diagnostic trouble codes can be retrieved by connecting terminals A and H at the data link connector (DLC). A scan tool can also be used on most General Motors vehicles to retrieve trouble codes. A scan tool is *required* to retrieve codes on other vehicle makes and models. Consult technical information for the exact vehicle being serviced.

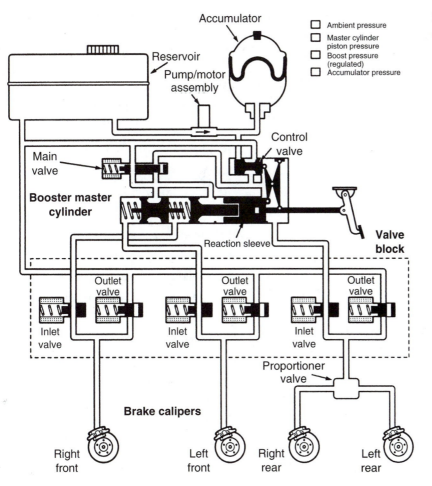

Accumulator

Reservoir

Pump/motor assembly

☐ Ambient pressure
☐ Master cylinder piston pressure
☐ Boost pressure (regulated)
☐ Accumulator pressure

Control valve

Main valve

Booster master cylinder

Reaction sleeve

Valve block

Outlet valve

Outlet valve

Outlet valve

Inlet valve

Inlet valve

Inlet valve

Proportioner valve

Brake calipers

Right front

Left front

Right rear

Left rear

FIGURE 10–34 Teves Mark II system components. (Courtesy of General Motors)

Teves Mark II Diagnostic Trouble Codes

ABS code	Description
11	EBCM
12	EBCM
21	Main valve
22	LF inlet valve
23	LF outlet valve
24	RF inlet valve
25	RF outlet valve
26	Rear inlet valve
27	Rear outlet valve
31	LF WSS
32	RF WSS
33	RR WSS
34	LR WSS
35	LF WSS

ABS code	Description
36	RF WSS
37	RR WSS
38	LR WSS
41	LF WSS
42	RF WSS
43	RR WSS
44	LR WSS
45	Two sensors (LF)
46	Two sensors (RF)
47	Two sensors (rear)
48	Three sensors
51	LF outlet valve
52	RF outlet valve
53	Rear outlet valve
54	Rear outlet valve

ABS code	Description
55	LF WSS
56	RF WSS
57	RR WSS
58	LR WSS
61	EBCM "loop" circuit
71	LF outlet valve
72	RF outlet valve
73	Rear outlet valve
74	Rear outlet valve
75	LF WSS
76	RF WSS
77	RR WSS
78	LR WSS

Clearing Diagnostic Trouble Codes. After all stored codes have been received and problems corrected, the trouble code should be erased from the computer memory by driving above 25 mph (40 km/h).

TEVES MARK IV

Teves Mark IV is a nonintegral (remote) ABS system (see Figure 10–35).

Retrieving Diagnostic Trouble Codes. Trouble codes are retrieved using only a bidirectional scan tool such as a Tech I or II to the diagnostic link connector (DLC).

Teves Mark IV ABS Diagnostic Trouble Codes

ABS code	Description
21	RF speed sensor circuit open
22	RF speed sensor signal erratic
23	RF wheel speed is 0 mph
25	LF speed sensor circuit open
26	LF speed sensor signal erratic
27	LF wheel speed is 0 mph
31	RR speed sensor circuit open
32	RR speed sensor signal erratic
33	RR wheel speed is 0 mph
35	LR speed sensor circuit open
36	LR speed sensor signal erratic
37	LR wheel speed is 0 mph
41	RF inlet valve circuit
42	RF outlet valve circuit
43	RF speed sensor noisy
45	LF inlet valve circuit

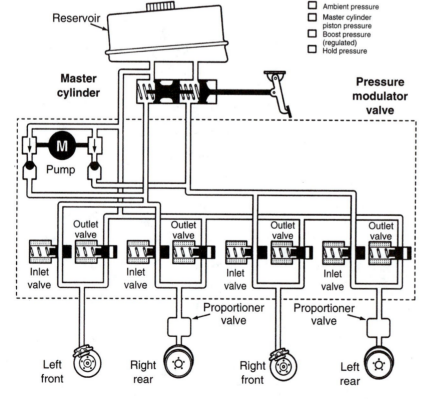

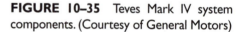

FIGURE 10–35 Teves Mark IV system components. (Courtesy of General Motors)

ABS code	Description
46	LF outlet valve circuit
47	LF speed sensor noisy
51	RR inlet valve circuit
52	RR outlet valve circuit
53	RR speed sensor noisy
55	LR inlet valve circuit
56	LR outlet valve circuit
57	LR speed sensor noisy
61	Pump motor test fault
62	Pump motor fault in ABS stop
71	EBCM check sum error
72	TCC/antilock brake switch circuit
73	Fluid level switch circuit

Clearing Diagnostic Trouble Codes. A scan tool is required to clear diagnostic trouble codes on some vehicles. Driving the vehicle over 20 mph (32 km/h) will clear the codes on some vehicles. Disconnecting the battery will also clear the codes, but will cause other "keep alive" functions of the vehicle to be lost.

DELCO ABS VI (NONINTEGRAL)

The Delco ABS VI is unique from all other antilock systems because it uses motor-driven ball screws and pistons for brake pressure reduce, hold, and apply (see Figures 10–36, 10–37 and 10–38).

Retrieving Diagnostic Codes. The Delco VI antilock braking system has extensive self-diagnostic capability. Access to this vast amount of information requires the use of a scan tool designed to interface (work) with the Delco VI system.

Delco ABS VI (Nonintegral) Diagnostic Trouble Codes

ABS code	Description
11	ABS lamp open or shorted to ground
13	ABS lamp circuit shorted to battery

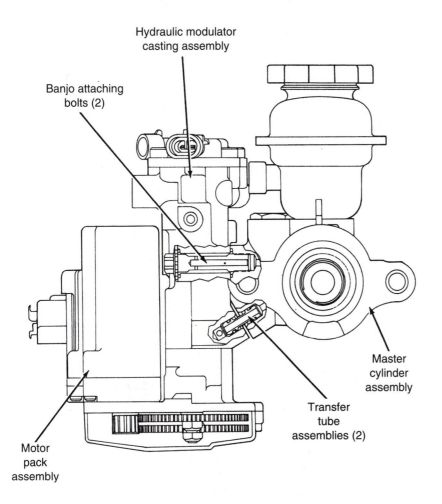

FIGURE 10–36 The Delco VI attaches to the side of the master cylinder and connects hydraulically through transfer tube assemblies. (Courtesy of General Motors)

Hydraulic modulator casting assembly

Banjo attaching bolts (2)

Motor pack assembly

Transfer tube assemblies (2)

Master cylinder assembly

ABS code	Description
14	Enable relay contacts open, fuse open
15	Enable relay contacts shorted to battery
16	Enable relay coil circuit open
17	Enable relay coil shorted to ground
18	Enable relay coil shorted to B+ or 0 ohms
21	Left front wheel speed = 0
23	Left rear wheel speed = 0
24	Right rear wheel speed = 0
25	Excessive left front wheel acceleration
26	Excessive right front wheel acceleration
27	Excessive left rear wheel acceleration
28	Excessive right rear wheel acceleration
31	Two wheel speed sensors open
36	System voltage low
37	System voltage high
38	Left front EMB will not hold motor
41	Right front EMB will not hold motor
42	Rear axle ESB will not hold motor
44	Left front EMB will not release motor, gears frozen
45	Right front EMB will not release motor, gears frozen
46	Rear axle ESB will not release motor, gears frozen
47	Left front nut failure (motor free-spins)
48	Right front nut failure (motor free-spins)
51	Rear axle nut failure (motor free-spins)
52	Left front channel in release too long
53	Right front channel in release too long
54	Rear axle in release too long
55	Motor driver interface (MDI) fault detected
56	Left front motor circuit open
57	Left front motor circuit shorted to ground
58	Left front motor circuit shorted to battery

ABS code	Description
61	Right front motor circuit open
62	Right front motor circuit shorted to ground
63	Right front motor circuit shorted to battery
64	Rear axle motor circuit open
65	Rear axle motor circuit shorted to ground
66	Rear axle motor circuit shorted to battery
67	Left front EMB release circuit open or shorted to ground
68	Left front EMB release circuit shorted to battery or driver open
71	Right front EMB release circuit open or shorted to ground
72	Right front EMB release circuit shorted to battery or driver open
76	Left front solenoid circuit open or shorted to battery
77	Left front solenoid circuit shorted to ground or driver open
78	Right front solenoid circuit open or shorted to battery
81	Right front solenoid circuit shorted to battery or driver open
82	Calibration memory failure
86	ABS controller turned "on" red brake telltale
87	Red brake telltale circuit open
88	Red brake telltale circuit shorted to battery or driver open
91	Open brake switch contacts (decel detection)
92	Open brake switch contacts
93	Test 91 or 92 failed last or current ignition cycle
94	Brake switch contacts shorted
95	Brake switch circuit open
96	Brake lamps open, brake lamp ground open, center high-mounted stop lamp open during four-way flasher operation

FIGURE 10–37

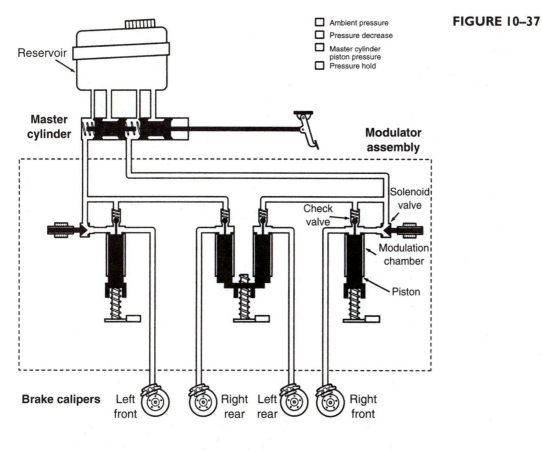

□ Ambient pressure
□ Pressure decrease
□ Master cylinder piston pressure
□ Pressure hold

Reservoir

Master cylinder

Modulator assembly

Solenoid valve

Check valve

Modulation chamber

Piston

Brake calipers   Left front   Right rear   Left rear   Right front

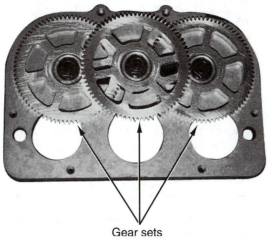

Gear sets

FIGURE 10–38 These gears are turned by high-speed electric motors and move pistons up and down to control braking during an ABS stop.

HYDRAULIC ABS SERVICE

Before doing any brake work on a vehicle equipped with antilock brakes, always consult the appropriate service information for the exact vehicle being serviced. For example, some manufacturers recommend relieving the hydraulic accumulator pressure by depressing the brake pedal many times before opening any bleeder valves. Many service checks require that a pressure gauge be installed in the system as shown in Figure 10–39.

BRAKE BLEEDING ABS

After depressurizing the unit according to the manufacturer's recommended procedures, the brakes can be bled using the same procedure as for a vehicle without ABS (see Chapter 3). Air trapped in the ABS hydraulic unit may require that a scan tool be used to cycle the valves. See the Tech Tip "Don't Forget to Bleed the E-H Unit" for details.

ABS SAFETY PRECAUTIONS

1. Avoid mounting the antenna for transmitting device near the ABS control unit. Transmitting devices include cellular (cell) telephones and citizen-band radios, among others.

2. Avoid mounting tires of diameter different from that of the original tires. Different-size tires generate different wheel speed sensor frequencies that may not be usable by the ABS controller.

◀ TECH TIP ▶

DON'T FORGET TO BLEED THE E-H UNIT

Air can easily get trapped in the ABS electronic–hydraulic (E-H) assembly whenever the hydraulic system is opened. Even though the master cylinder and all four wheel cylinders/calipers have been bled, sometimes the brake pedal will still feel spongy. Some E-H units can be bled through the use of a scan tool where the valves are pulsed in sequence by the electronic brake controller (computer). Some units are equipped with bleeder valves, whereas others must be bled by loosening the brake lines. Bleeding the E-H unit also purges out the older brake fluid that can cause rust and corrosion damage. Only DOT 3 brake fluid is specified to use in an antilock braking system. Always check the label on the brake fluid reservoir and/or service manual or owner's manual.

CAUTION: Some ABS units require that the brake pedal be depressed 20 or more times to fully discharge brake fluid from the accumulator. Failure to fully discharge the accumulator can show that the brake fluid level is too low. If additional brake fluid is added, the fluid could overflow the reservoir during an ABS stop when the accumulator discharges brake fluid back into the reservoir.

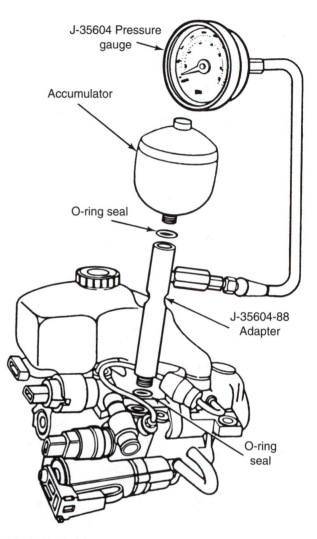

FIGURE 10–39 Installing a pressure gauge to an integral antilock system. Most service instructions of integral ABS units require that the brake pedal be depressed as many as 40 times with the ignition "off" before opening any hydraulic line or valve. (Courtesy of General Motors)

HINT: Most electrical faults are caused by heat or movement. Look closely at connectors or wiring near wheel speed sensors or wiring running near components of the exhaust system.

Look carefully at each connector for proper mating or poor terminal-to-wire connection.

3. Never open a bleeder valve or loosen a hydraulic line while the ABS is pressurized. The accumulator must be depressurized according to the manufacturer's recommended procedures.
4. If arc welding on the vehicle, *disconnect* all computers (electronic control modules) to avoid possible damage due to voltage spikes.
5. Do not pry against or hit the wheel speed sensor ring.

DIAGNOSING INTERMITTENT PROBLEMS

Problems or faults that occur only once in a while are called *intermittents*. Intermittent problems usually involve wiring or wiring connections.

SUMMARY

1. Antilock brake systems are designed to limit the amount of tire slip by pulsing the wheel brake on and off up to 20 times per second.

2. Steering control is possible during an ABS stop if the tires maintain traction with the road surface.

3. The three stages of ABS operation are pressure buildup, pressure holding, and pressure reduction.

4. The heart of an antilock braking system is the electronic controller (computer). Wheel speed sensors produce an electrical frequency that is proportional to the speed of the wheel. If a wheel is slowing down too fast, the controller controls the pressure of the wheel brake through an electrohydraulic unit.

5. Both integral and nonintegral antilock braking systems control the rear wheels only, or both fronts individually and the rear as one (three-channel), or control all four wheel brakes independently (four-channel).

6. Antilock braking systems that control the drive-wheel brakes can be used for acceleration traction control.

7. ABS diagnosis starts with checking the status of both the red brake warning lamp and the amber ABS warning lamp.

8. The second step in diagnosis of an ABS problem is to perform a thorough visual inspection.

9. The third step in diagnosis of an ABS problem is to test drive the vehicle and verify the fault.

10. Always consult the factory service information for the exact vehicle being serviced for the proper procedure to use to:

   - Retrieve diagnostic trouble codes
   - Clear diagnostic trouble codes

11. Hydraulic service on most integral-type ABS units requires that the brake pedal be depressed at least 25 times with the ignition key "off" to depressurize the hydraulic system.

REVIEW QUESTIONS

1. Describe how an antilock braking system works.

2. List the three stages of ABS operation.

3. Explain how wheel speed sensors work.

4. Describe the difference between a three- and a four-channel system.

5. Explain how ABS can be used to prevent wheel slippage during acceleration.

6. Describe the proper operation of the red and amber brake warning lamps.

7. List the items that should be checked as part of a thorough visual inspection.

8. Explain how to retrieve a diagnostic trouble code from a General Motors vehicle equipped with Kelsey-Hayes RWAL ABS.

MULTIPLE-CHOICE QUESTIONS

1. Technician A says that the ABS system is designed so that the pressure to the wheel brakes is never higher than the pressure the driver is applying through the brake pedal. Technician B says that a pulsating brake pedal during *normal* braking is a characteristic feature of most ABS-equipped vehicles. Which technician is correct?

   a. A only
   b. B only
   c. Both A and B
   d. Neither A nor B

2. A customer wanted the ABS checked because of tire chirp noise during hard braking. Technician A says that the speed sensors may be defective. Technician B says that tire chirp is normal during an ABS stop on dry pavement. Which technician is correct?

   a. A only
   b. B only

   c. Both A and B
   d. Neither A nor B

3. Two technicians are discussing ABS wheel speed sensors. Technician A says that some ABSs use a sensor located in the rear axle pinion gear area. Technician B says that all ABS systems use a wheel speed sensor at *each* wheel. Which technician is correct?

   a. A only
   b. B only
   c. Both A and B
   d. Neither A nor B

4. Technician A says that it may be normal for the hydraulic pump and solenoids to operate after the vehicle starts to move after a start. Technician B says that ABS is disabled (does not function) below about 5 mph (8 km/h). Which technician is correct?

a. A only
b. B only
c. Both A and B
d. Neither A nor B

5. Technician A says that a scan tool may be necessary to bleed some ABS hydraulic units. Technician B says that only DOT 3 brake fluid should be used with ABS. Which technician is correct?

a. A only
b. B only
c. Both A and B
d. Neither A nor B

6. The red brake warning lamp is on and the amber ABS lamp is off. Technician A says that a fault is possible in the base brake system. Technician B says that the red brake warning lamp can be turned on by a malfunction in an integral ABS with low accumulator pressure. Which technician is correct?

a. A only
b. B only
c. Both A and B
d. Neither A nor B

7. Technician A says that wheel speed sensors are magnetic. Technician B says that the toothed sensor ring is magnetic. Which technician is correct?

a. A only
b. B only
c. Both A and B
d. Neither A nor B

8. Technician A says that with some antilock braking systems, the diagnostic trouble code may be lost if the ignition is turned off before retrieving the code. Technician B says that some antilock braking systems require that a terminal be grounded to cause the amber ABS warning lamp to flash diagnostic trouble codes. Which technician is correct?

a. A only
b. B only
c. Both A and B
d. Neither A nor B

9. Technician A says that a dim red brake warning lamp could indicate a fault in the antilock braking system on some vehicles. Technician B says that the ABS fuse may have to be removed to erase some ABS diagnostic codes if a scan tool is not used. Which technician is correct?

a. A only
b. B only
c. Both A and B
d. Neither A nor B

10. Technician A says that many ABSs electrohydraulic unit can be bled using bleeder screws and the manual method. Technician B says that a scan tool is often required to bleed the ABS electrohydraulic unit. Which technician is correct?

a. A only
b. B only
c. Both A and B
d. Neither A nor B

◀ Chapter 11 ▶

TIRES AND WHEELS

OBJECTIVES

After studying Chapter 11, the reader will be able to:

1. List the various parts that make up a tire.
2. Explain how a tire is manufactured.
3. Discuss tire sizes and ratings.
4. Describe tire purchasing considerations and maintenance.
5. Explain the construction and sizing of steel and alloy wheels and attaching hardware.
6. Define sprung and unsprung weight.
7. Demonstrate the correct lug nut tightening procedure and torque.

Tires are mounted on wheels that are bolted to the vehicle to provide:

1. shock absorber action when driving over rough surfaces
2. Friction (traction) between the wheels and the road

All tires are assembled by hand from many different component parts, consisting of various rubber compounds, steel, and various types of fabric material. Tires are also available in many different designs and sizes.

PARTS OF A TIRE

Tread. The tread is the part of the tire that contacts the ground. **Tread rubber** is chemically different from other rubber parts of a tire and is compounded for a combination of traction and tire wear. **Tread depth** is usually $\frac{11}{32}$ in. on new tires (this could vary depending on the manufacturer, varying from $\frac{9}{32}$ to $\frac{15}{32}$ in.). See Figure 11–1 showing a tread depth gauge.

NOTE: A tread depth is always expressed in $\frac{1}{32}$'s of an inch even if the fraction can be reduced to $\frac{1}{16}$'s, $\frac{1}{8}$'s, or other value.

Wear indicators are also called **wear bars**. When tread depth is down to the legal limit of $\frac{2}{32}$ in., bald strips appear across the tread (see Figure 11–2).

Large, deep recesses molded in the tread and separating the tread blocks are called **circumferential grooves** or **kerfs**. Grooves running sideways across the tread of a tire are called **lateral grooves** (see Figure 11–3). Grooves in both directions are necessary for wet traction. The trapped water can actually cause the tires to ride up on a layer of water and lose contact with the ground, as illustrated in Figure 11–4. This is called **hydroplaning**. With worn tires, hydroplaning can occur when a vehicle is moving as slowly as 30 mph (50 km/h) on wet roads.

Sidewall. The sidewall is that part of the tire between the tread and the wheel. The sidewall contains all the size and construction details of the tire. White sidewall tires actually contain a strip of white rubber under the black sidewall, which is ground off at the factory to reveal the white rubber.

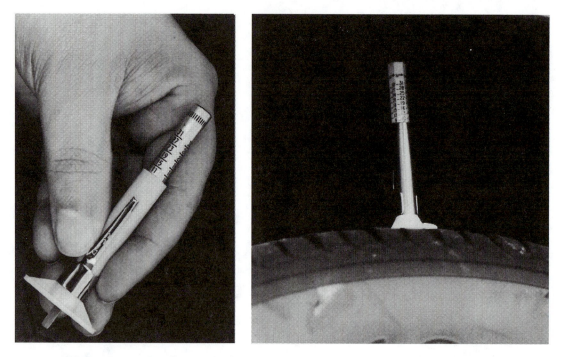

FIGURE 11–1 The center of the tread depth gauge is pushed down into the groove of the tire and the depth is read at the top edge of the sleeve. Tread depth is usually expressed in $\frac{1}{32}$'s of an inch.

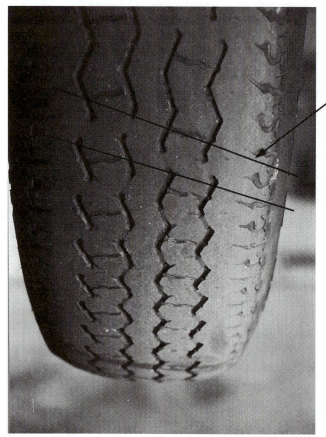

FIGURE 11–2 Wear indicators (wear bars) are strips of bald tread that show when the tread depth is down to $\frac{2}{32}$ in., the legal limit in many states.

WEAR
INDICATOR
(BALD STRIP)

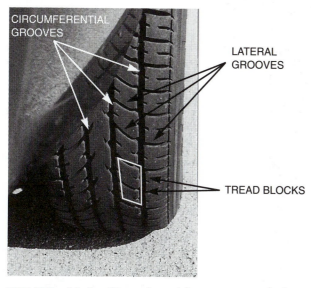

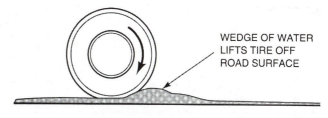

ROAD SURFACE

FIGURE 11–3 Circumferential grooves and lateral grooves are the main water channels that help evacuate water from under the tire preventing hydroplaning.

FIGURE 11–4 Hydroplaning can occur at speeds as low as 30 mph (48 km/h). If the water is deep enough and the tire tread cannot evacuate water through its grooves fast enough, the tire can be lifted off the road surface by a layer of water. Hydroplaning occurs at lower speeds as the tire becomes worn.

Bead. The bead is the foundation of the tire and is located where the tire grips the inside of the wheel rim. The bead is constructed of many turns of copper- or bronze-coated steel wire. The main body plies (layers of material) are wrapped around the bead.

CAUTION: If the bead of a tire is cut or damaged, the tire must be replaced!

Body Ply. A tire gets its strength from the layers of material wrapped around both beads under the tread and sidewall rubber. This creates the main framework or carcass of the tire. These **body plies** are often called **carcass plies**. If only one or two body plies are used and they do not cross at an angle but lay directly from bead to bead, the tire is called *radial ply*. Materials used for body plies include rayon, nylon, aramid, and polyester (see Figure 11–5).

Belt. As illustrated in Figure 11–5, a **tire belt** is two or more layers (plies) of material applied over the body plies and only under the tread area to stabilize the tread and increase tread life and handling. Belt material can be steel mesh, nylon, rayon, fiberglass, and aramid.

Inner Liner. The inner liner is the soft rubber (usually, a butyl rubber compound) lining on the inside of the tire which protects the body plies and helps provide for self-sealing of small punctures.

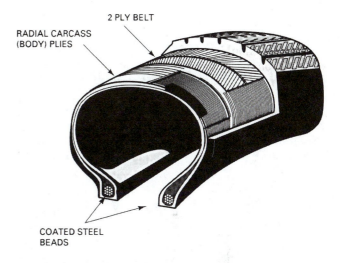

FIGURE 11–5 Typical construction of a radial tire. Some tires have only one body ply and some tires use more than two belt plies.

Major Splice. When the tire is assembled by a craftsman on a tire building machine, the body plies, belts, and tread rubber are spliced together. The fabric is overlapped. Where the majority of these overlaps occur, called the **major splice** represents the stiffest part of the tire (see Figure 11–6).

NOTE: On most new vehicles and/or new tires, the tire manufacturer paints a dot on the sidewall near the bead indicating the *largest diameter* of the tire. The largest diameter of the tire usually is near the major splice. The wheel manufacturer either marks the wheel or drills the valve core hole at the *smallest diameter* of the wheel. Therefore, the dot should be aligned with the valve core or mark for best balance and minimum radial runout.

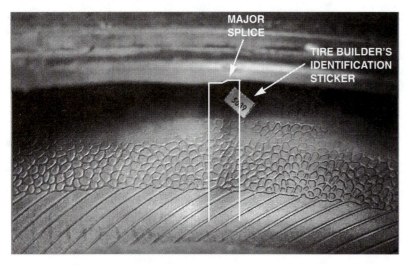

FIGURE 11–6 The major splice of a tire can often be seen and felt on the inside of the tire. The person who assembles (builds) the tire usually places a sticker near the major splice as a means of identification for quality control.

◀ TECH TIP ▶

TIRE PRESSURE AND TEMPERATURE

As the temperature of a tire increases, the pressure inside the tire also increases. The general amount of pressure gain (when temperatures increase) or loss (when temperatures decrease) is as follows:

10°F increase causes 1 psi increase.

10°F decrease causes 1 psi decrease.

For example, if a tire is correctly inflated to 35 psi when cold and then driven on a highway, the tire pressure may increase by 5 psi or more.

CAUTION: Do not let air out of a hot tire! If air is released from a hot tire to bring the pressure down to specifications, the tire will be *underinflated* when the tire has cooled. The tire pressure specification is for a cold tire.

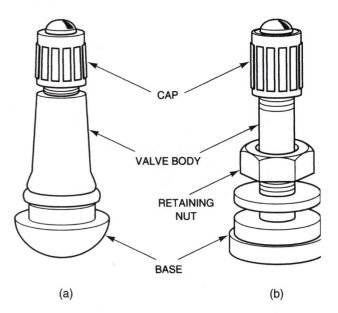

(a)   (b)

FIGURE 11–7 (a) A rubber snap-in-style tire valve assembly. (b) A metal clamp-type tire valve assembly used on most high-pressure (over 60 psi) tire applications such as is found on many trucks, RVs, and trailers. The internal Schrader valve threads into the valve itself and can be replaced individually, but most experts recommend replacing the entire valve assembly to help prevent air loss every time the tires are replaced.

TIRE VALVES

To hold air in the tire, all tires use a valve called a **Schrader valve**. The Schrader valve was invented in New York in 1844 by August Schrader. Aluminum (alloy) wheels often require special metal valve stems that use a rubber washer and are actually bolted to the wheel (see Figure 11–7).

AMERICAN METRIC TIRE SIZE DESIGNATIONS

After 1980, American tires were designated in the metric system. For example: P205/75R × 14:

P = passenger vehicle

205 = 205-mm cross-sectional width

75 = 75% aspect ratio
The height of the tire (from the wheel to the tread is 75% as great as its cross-sectional width (the width measured across its widest part). This percentage ratio of height to width is called the **aspect ratio.** (A 60 series tire is 60% as high as it is wide.)

R = radial

14 = 14-in.-diameter wheel

SERVICE DESCRIPTION

Tires built after 1990 use a "service description" method of sidewall information in accordance with ISO (International Standards Organization) standard 4000, which includes size, load, and speed ratings all together in one easy-to-read format, as illustrated in Figure 11–8.

FIGURE 11–8

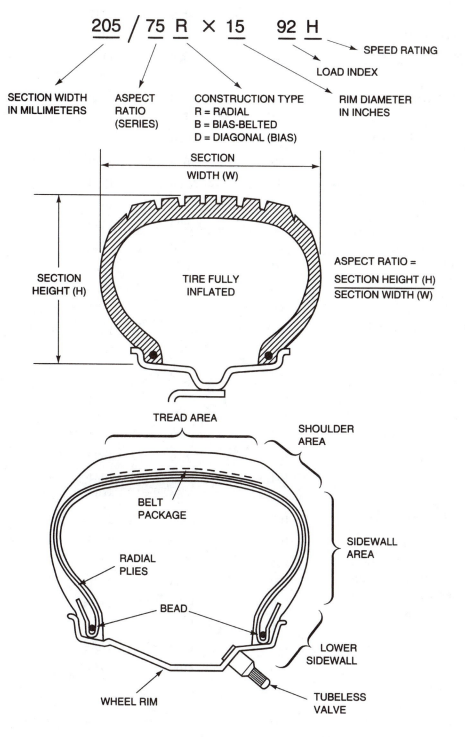

P metric designation	Service description
P 205/75HR × 15	205/75R × 15 92H

P = passenger vehicle
205 = cross-sectional
  width in millimeters
75 = aspect ratio
H = speed rating
  [130 mph (210 km/h)]
R = radial construction
15 = rim diameter in
  inches

205 = cross-sectional
  width in millimeters
75 = aspect ratio
R = radial construction
15 = rim diameter in
  inches
92 = load index
H = speed rating
  [130 mph (210 km/h)]

HIGH-FLOATATION TIRE SIZES

High-floatation light truck tires are designed to give improved off-the-road performance on sand, mud, and soft soil and still provide acceptable hard road surface performance. These tires are usually larger than conventional tires and usually require a wider than normal wheel width. High-floatation tires have a size designation such as

$$33 \times 12.50 \text{ R} \times 15 \text{ LT}$$

where 33 = approximate overall tire diameter in
  inches

12.50 = approximate cross-sectional width in
  inches

R = radial-type construction

15 = rim diameter in inches

LT = light truck designation

LOAD INDEX AND EQUIVALENT LOADS

The load index, as shown in Figure 11–9, is an abbreviated method of indicating the load-carrying capabilities of a tire. The weights listed in the chart below represent the weight that *each tire* can safely support. Multiply this amount times 4 to get the maximum that the vehicle should weigh fully loaded with cargo and passengers.

| Load index | Load | |
	kg	lb
75	387	853
76	400	882

| Load index | Load | |
	kg	lb
77	412	908
78	425	937
79	437	963
80	450	992
81	462	1019
82	475	1047
83	487	1074
84	500	1102
85	515	1135
86	530	1168
87	545	1201
88	560	1235
89	580	1279
90	600	1323
91	615	1356
92	630	1389
93	650	1433
94	670	1477
95	690	1521
96	710	1565
97	730	1609
98	750	1653
99	775	1709
100	800	1764
101	825	1819
102	850	1874
103	875	1929
104	900	1934
105	925	2039
106	950	2094
107	975	2149
108	1000	2205
109	1030	2271
110	1060	2337
111	1090	2403
112	1120	2469
113	1150	2535
114	1180	2601
115	1215	2679

FIGURE 11–9 Typical sidewall markings for load index and speed rating following the tire size.

SPEED RATINGS

Tires are rated according to the maximum *sustained* speed. A vehicle should never be driven faster than the speed rating of the tires.

CAUTION: High-speed-rated tires do not guarantee that the tires will not fail even at speeds much lower than the rating. Tire condition, inflation, and vehicle loading also affect tire performance.

Since the speed ratings were first developed in Europe, the letters correspond to metric speed in kilometers per hour with a conversion to miles per hour.

Letter	Maximum rated speed
L	120 km/h (75 mph)
M	130 km/h (81 mph)
N	140 km/h (87 mph)
P	150 km/h (93 mph)
Q	160 km/h (99 mph)
R	170 km/h (106 mph)
S	180 km/h (112 mph)
T	190 km/h (118 mph)
U	200 km/h (124 mph)
H	210 km/h (130 mph)
V	240 km/h (150 mph)

Letter	Maximum rated speed
W	270 km/h (168 mph)
Y	300 km/h (192 mph)
Z	240 km/h (150 mph and over[a])

[a]The exact speed rating for a particular Z-rated tire is determined by the tire manufacturer and may vary according to size. For example, not all brand X Z-rated tires are rated at 170 mph even though one size may be capable of these speeds.

TIRE PRESSURE AND TRACTION

All tires should be inflated to the specifications given by the vehicle manufacturer. Most vehicles have recommended tire inflation figures written in the owner's manual or on a placard or sticker on the door post or glove compartment. As a general recommendation, tires should be inflated according to the following:

1. **Maximum pressure.** The pressure number molded into the sidewall of a tire should be considered the maximum pressure when the tire is cold. (Pressures higher than that stated on the sidewall may be measured on a hot tire.) For best traction, always inflate tires to their maximum allowable pressure.
2. **Minimum pressure.** Tires should not be inflated (cold) to less than 80% of the maximum as imprinted on the sidewall of the tire.

For example:

Maximum pressure	Minimum pressure (80% of max.)
44 psi (300 kPa)	35 psi (240 kPa)
35 psi (240 kPa)	28 psi (190 kPa)

Tests done by tire engineers prove that the best traction is achieved at the maximum pressure as imprinted on the tire. Lower inflation pressure reduces ride harshness but decreases tire tread life, handling, fuel economy, and traction.

◀ TECH TIP ▶

TIRE SIZE VERSUS GEAR RATIO

Customers often ask what effect changing tire size has on fuel economy and speedometer readings. If larger (or smaller) tires are installed on a vehicle, many other factors change also. These include:

1. **Speedometer reading changes.** If larger-diameter tires are used, the speedometer will read *slower* than you are actually traveling. This can result in speeding tickets!

2. **Odometer reading changes.** Even though larger tires are said to give better fuel economy, just the opposite effect could be calculated. Since a larger-diameter tire travels farther than a smaller-diameter tire, the larger tire will cause the odometer to read a shorter distance than the vehicle actually travels. For example, if the odometer reads 100 miles traveled on tires that are 10% oversize in circumference, the actual distance traveled is 110 miles.

3. **Fuel economy.** If fuel economy is calculated on miles traveled, the result will be *lower* fuel economy than for the same vehicle with the original tires. Other tire diameter calculations include:

$$\text{mph} = \frac{\text{rpm} \times \text{diameter} \times 3.14}{\text{gear ratio}}$$

$$\text{rpm} = \frac{\text{mph} \times \text{gear ratio}}{\text{diameter} \times 3.14}$$

$$\text{Gear ratio} = \frac{\text{rpm} \times \text{diameter} \times 3.14}{\text{mph}}$$

TIRE CONICITY

Tire conicity can occur during the construction of any radial or belted tire when the parts of the tire are badly positioned, causing the tire to be smaller in diameter on one side. When this tire is installed on a vehicle, it can cause the vehicle to pull to one side of the road, due to the cone shape of the tire, as shown in Figure 11–10.

NOTE: *Radial pull* or *radial tire pull* are other terms often used to describe tire conicity.

RIM WIDTH AND TIRE SIZE

As a general rule for a given rim width, it is best not to change tire width more than 10 mm either wider or narrower. For a given tire width, it is best not to vary rim width more than ½ in. in either direction. For example, if the original tire size is 195/70 × 14, then either a 185/70 × 14 (-10 mm) or a 205/70 × 14 (+10 mm) *could* be used on the original rim (wheel) without undo harm *if* the replacement tire has proper clearance with the body and suspension components. Installing a tire on too narrow a wheel will cause the tire to wear excessively in the center of the tread. Installing a tire on too wide a wheel will cause excessive tire wear on both edges. See a tire store representative for the recommended tire size that can be safely installed on your rims.

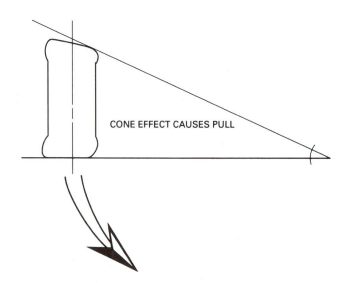

CONE EFFECT CAUSES PULL

FIGURE 11–10 Conicity is a fault in the tire that can cause the vehicle to pull to one side due to the cone effect (shape) of the tire.

UNIFORM TIRE QUALITY GRADING SYSTEM

The **U.S. Department of Transportation** (DOT) and the **National Highway Traffic Safety Administration** (NHTSA) developed a system of tire grading to help customers better judge the relative performance of tires. The three areas of tire performance are tread wear, traction, and temperature resistance, as shown in Figure 11–11.

NOTE: All tires sold in the United States must have uniform tire quality grading system ratings molded into the sidewall of the tire.

Tread Wear. The tread wear grade is a comparison rating based on the wear rate of a standardized tire tested under carefully controlled conditions which are assigned a value of 100. A tire rated "200" should have a useful life twice as long as that of the standard tire.

HINT: The standard tire has a rating for tread wear of 100. This value has generally been accepted to mean a useful life of 20,000 miles of normal driving. Therefore, a tire rated at 200 could be expected to last 40,000 miles, etc.

Tread wear rating number	Approximate tread life
100	20,000 miles (32,000 km)
150	30,000 miles (48,000 km)
200	40,000 miles (64,000 km)
250	50,000 miles (80,000 km)
300	60,000 miles (96,000 km)
400	80,000 miles (129,000 km)
500	100,000 miles (161,000 km)

The tread wear life of any tire is affected by driving habits (fast stops, starts, and cornering will decrease tread life), tire rotation (or lack of tire rotation), inflation, wheel alignment, road surfaces, and climate conditions.

Traction. Traction performance is rated by the letters A, B, or C, with A the highest rating and C the lowest rating.

IMPORTANT NOTE: The traction rating tests the tires for *wet braking* distance only! The traction rating does not include cornering traction or dry braking performance.

FIGURE 11–11 Typical uniform tire quality grading system (UTQGS) ratings as imprinted on a tire sidewall.

Temperature Resistance. Temperature resistance is rated by letters A, B, or C, with A the highest rating and C the lowest.

DOT TIRE CODE

All tires sold in the United States must be approved by the U.S. Department of Transportation (DOT). Many of the DOT tire requirements include resistance to tire damage that could be caused by curbs, chuckholes, and other common occurrences for a tire used on the public roads.

NOTE: Most race tires are *not* DOT approved and must never be used on public streets or highways.

Each tire that is DOT approved has a DOT number molded into the sidewall of the tire, as shown in Figure 11–12. This number is usually imprinted on only one side of the tire and is usually on the *opposite* side of the whitewall. The DOT code includes letters and numbers, such as

MJP2CBDX264

The first two letters identify the manufacturer name and location. For this example, the first two letters (MJ) means that the tire was made by the Goodyear Tire and Rubber Company in Topeka, Kansas.

The last three numbers are the build date code. The last of these three numbers is the year (1994), and the 26 means that it was built during the twenty-sixth week of 1994. The last letter is the same for 1984 and 2004, but

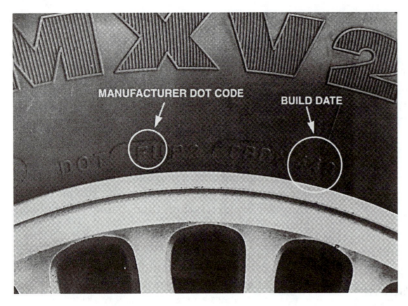

FIGURE 11–12 Typical DOT tire code. This tire has a manufacturer two-letter code of "FU" (Michelin in Bambery, Germany) and the last three numbers (440) indicate the week and year the tire was built—the 44th week of 1990).

the style and design of the tire usually change enough after 10 years to be able to identify correctly the decade in which the tire was built.

See Appendix 7 for the alphabetical listing of all manufacturers' DOT tire codes.

SPARE TIRES

Most vehicles today come equipped with space-saver spare tires that are smaller than the wheels and tires that are on the vehicle. The reason for the small size is to reduce the size and weight of the entire vehicle and to increase fuel economy by having the entire vehicle weigh less.

CAUTION: Before using a spare tire, always read the warning label (if equipped) and understand all use restrictions. For example, some spare tires are not designed to exceed 50 mph (80 km/h) or be driven more than 500 miles (800 km).

WHEELS

Today's wheels are constructed of steel or aluminum alloy. The center section of the wheel that attaches to the hub is called the **center section** or **spider** because early wheels used wooden spokes that resembled a spider's web. The rubber tire attaches to the rim of the wheel. The rim has two bead flanges where the bead of the tire is held against the wheel when the tire is inflated. The shape of this flange is very important and is designated

FIGURE 11–13 The wheel size as stamped on this old alloy wheel at the factory. This wheel is 5½ in. wide between the inside of the rim and 15 in. in diameter. The letters "JJ" refer to the contour of the inside of the rim where the wheel seals at the bead area of the tire.

by the Tire and Rim Association letters. For example, a wheel designated 14 × 6JJ means that the diameter of the wheel is 14 in. and is 6 in. wide measured from inside-to-inside flange area. The letters "JJ" indicate the *exact* shape of the flange area (see Figure 11–13).

Wheel Offset. Offset is a very important variable in wheel design. If the center section (spider) is centered on the outer rim, the offset is zero (see Figure 11–14).

Positive Offset. The wheel has a positive offset if the center section is outward from the hub mounting.

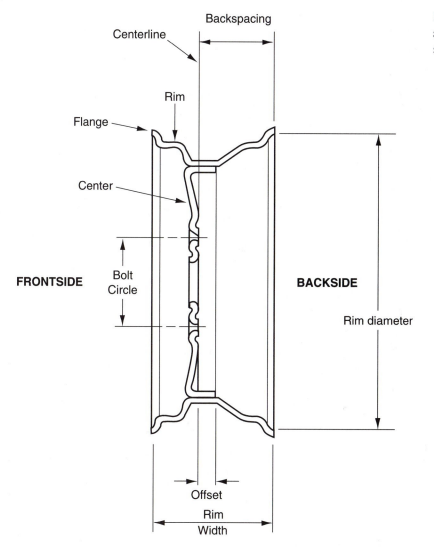

Centerline

Backspacing

Rim

Flange

Center

FRONTSIDE

Bolt
Circle

BACKSIDE

Rim diameter

Offset

Rim
Width

FIGURE 11–14 A cross-sectional view of a wheel, showing the terms and where the sizes are measured.

Front-wheel-drive vehicles commonly use positive-off-set wheels to improve the loading on the front wheels and to help provide a favorable scrub radius.

Negative Offset. The wheel has a negative offset if the center section is inboard (or "dished") from the hub mounting (see Figure 11–15). Avoid using replacement wheels that differ from the original offset. See Chapter 15 for details on the scrub radius and the effect that wheel offset has on steering and suspension geometry.

Back Spacing. Back spacing, also called **rear spacing** or **backside setting**, is the distance between the back rim edge and the wheel center section mounting pad. *This is not the same as offset.* Backspacing can be measured directly with a ruler as shown in Figure 11–16.

Determining Bolt Circle. On four-lug axles and wheels, the measurement is simply taken from center to

center on opposite studs or holes as shown in Figure 11–17. On five-lug axles and wheels, it is a little harder. One method is to measure from the far edge of one bolt hole to the center of the hole two over from the first, as shown in Figure 11–18.

Steel Wheels. Steel is the traditional wheel material. A steel wheel is very strong, due to its designed shape and the fact that it is work hardened during manufacturing. Steel wheels are formed from welded hoops flared and joined to stamped spiders.

Aluminum Wheels. Forged and cast aluminum wheels are very commonly used on cars and trucks. *Forged* means that the aluminum is hammered or forged into shape under pressure. A forged aluminum wheel is much stronger than a cast aluminum wheel. A cast aluminum wheel is constructed by pouring liquid (molten) aluminum into a mold.

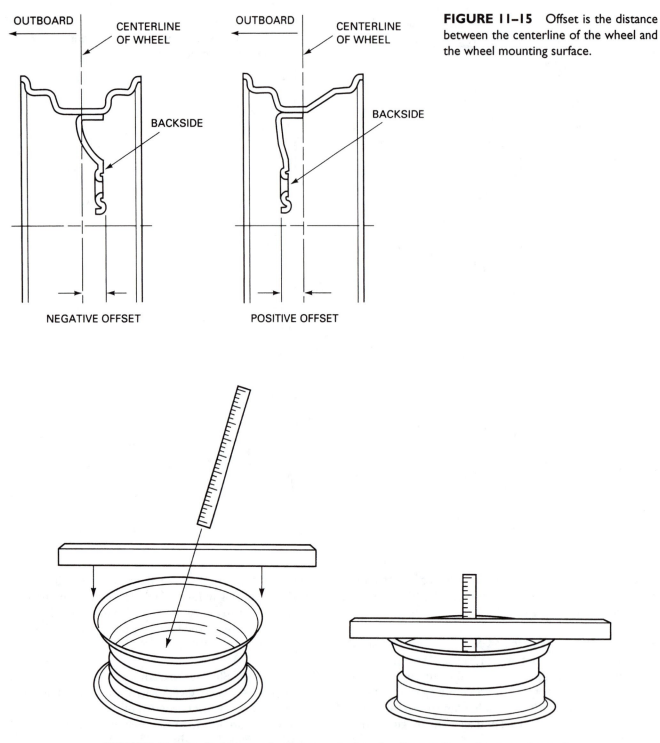

FIGURE 11–15 Offset is the distance between the centerline of the wheel and the wheel mounting surface.

FIGURE 11–16 Backing spacing (rear spacing) is the distance from the mounting pad to the edge of the rim. Most custom wheels use this measurement method to indicate the location of the mounting pad in relation to the rim.

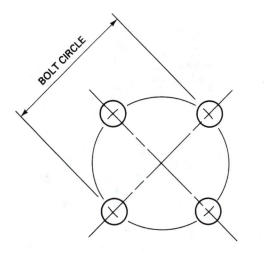

FIGURE 11–17 Bolt circle is the diameter of a circle that can be drawn through the center of each lug hole or stud.

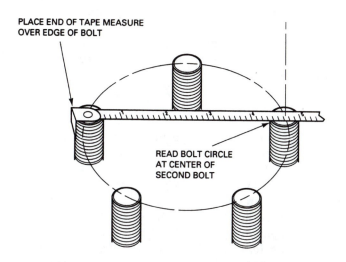

FIGURE 11–18 The easiest method to determine the approximate bolt circle of a five-lug bolt circle.

Metric Wheels. Several vehicle manufacturers, including Ford Motor Company, equip vehicles with wheels that are 365 or 390 mm in diameter rather than the more conventional 14-in., 15-in., or 16-in. wheels.

Metric Code Bead Diameter Tire and Rim

Tire size	Rim size
185/65R *365*	*365* × 135 TR
195/65R *390*	*390* × 150 TR

Note that the rim in this example is 365 or 390 mm in diameter and 135 or 150 mm wide.

WARNING: Do not attempt to mount an inch-code tire on a metric-code rim, or vice versa. A mismatch of tire size and rim size may result in tire failure and serious or fatal injury.

LUG NUTS

Lug nuts are used to hold a wheel to the brake disc, brake drum, or wheel bearing assembly. Most manufacturers use a stud in the brake or bearing assembly with a lug nut to hold the wheel. Some manufacturers, such as in some models of VW, Audi, and Mazda, use a lug *bolt* that is threaded into a hole in the brake drum or bearing assembly.

NOTE: Some aftermarket manufacturers offer a stud conversion kit to replace the lug bolt with a conventional stud and lug nut.

Typical lug nuts are tapered so that the wheel stud will center the wheel onto the vehicle. Another advantage of the taper of the lug nut and wheel is to provide a suitable surface to prevent the nuts from loosening. The taper, usually 60°, forms a wedge that helps assure that the lug nut will not loosen.

Many alloy wheels use a *shank nut* type of lug nut that has straight sides without a taper. This style of nut *must* be used with wheels designed for this nut type. If replacement wheels are used on any vehicle, check with the wheel manufacturer as to the proper type and style lug nut. See Figure 11–19, which illustrates several of the many styles of lug nuts that are available.

LUG NUT SIZE

Lug nuts are sized to the thread size of the stud onto which it screws. The diameter and the number of threads per inch are commonly stated. Since some vehicles use left-hand threads, "RH" and "LH" are commonly stated, indicating right-hand and left-hand threads. Therefore, a typical size might be $\frac{7}{16}$ – 20 RH, where $\frac{7}{16}$ indicates the diameter of the wheel stud in inches and 20 indicates that there are 20 threads per inch. Another common fractional size is $\frac{1}{2}$ × 20. Metric sizes such as M12 × 1.5 use a different sizing method.

    M = metric

    12 = 12-mm diameter of stud

    1.5 = distance from one thread peak to another in millimeters

WHEEL NUTS

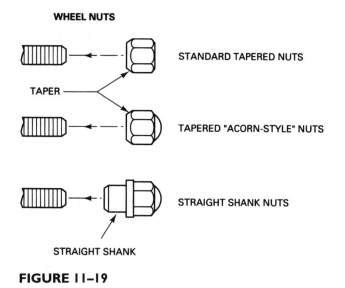

FIGURE 11–19

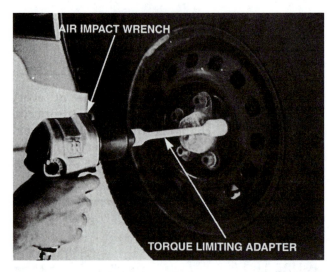

FIGURE 11–20 A torque-limiting adapter for use with an air impact wrench still requires care to prevent overtightening. The air pressure to the air impact should be limited to 125 psi (860 kPa) in most cases and the proper adapter selected for the vehicle being serviced. The torque adapter absorbs any torque beyond its designed rating. Most adapters are color coded for easy identification as to size of lug nut and torque value.

Other commonly used metric lug sizes include M12 × 1.25 and M14 × 1.5. Obviously, metric wheel studs require metric lug nuts.

TIGHTENING TORQUE

All wheels must be tightened in a star pattern to a specified torque. Always use a torque wrench or torque absorbing adapters. Proper tightening sequence and tightness is important to prevent possible damage to the wheel and/or brake rotor or drum. *Never use an air impact wrench to install wheels, except when using torque-absorbing adapters* (see Figure 11–20).

TIRE MOUNTING RECOMMENDATIONS

1. When removing a wheel from a vehicle for service, mark the location of the wheel and lug stud to ensure that the wheel can be replaced in exactly the same location. This ensures that the tire balance will be maintained if the tire–wheel assembly was balanced on the vehicle.

2. Make certain that the wheel has a good, clean metal-to-metal contact with the brake drum or rotor. Grease, oil, or dirt between these two surfaces could cause the wheel lug nuts to loosen while driving.

3. Many tires have been marked with a paint dot or sticker as shown in Figure 11–21. This mark rep-

resents the largest-diameter (high point) and/or stiffest portion of the tire. This variation is due to the overlapping of carcass and belt fabric layers as well as tread and sidewall rubber splices. The tire should be mounted to the rim with this mark lined up with the valve stem. The valve stem hole is typically drilled at the smallest diameter (low point) of the wheel. Mount the tires on the rim with the valve stem matched (lined up next to) the mark on the tire. This is called **match mounting**.

4. Never use more than 40 psi (275 kPa) to seat a tire bead.

5. Prior to mounting the tire, ensure that rim flanges are free of rust, dirt, scale, or loose or flaked rubber buildup.

6. When mounting new tires, do *not* use silicone lubricant on the tire bead. Use special rubber lubricant to help prevent tire rotation on the rim. This rubber lube is a water-based soap product that is slippery when wet (coefficient of friction less than 0.3) and almost acts as an adhesive when dry (coefficient of friction dry of over 0.5 for natural products and over 1.0 for synthetic products).

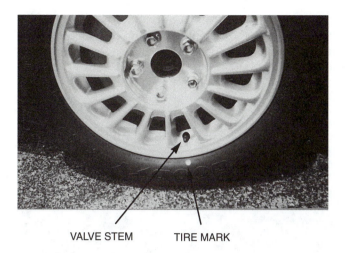

VALVE STEM   TIRE MARK

FIGURE 11–21 For the best possible ride and balance, mount the tire with the mark on the tire lined up with the valve stem. This also allows the tire assembly to be balanced with less weight than if not match mounted.

◄ **TECH TIP** ►

"I THOUGHT RADIAL TIRES COULDN'T BE ROTATED!"

When radial tires were introduced by American tire manufacturers in the 1970s, rotating tires side to side was *not* recommended because of concern of a belt or tread separation. Since the late 1980s, most tire manufacturers throughout the world, including the United States, use tire-building equipment specifically designed for radial ply tires. These newer radial tires are constructed so that the tires can now be rotated from one side of the vehicle to the other without fear of a separation being caused by the resulting reversal of the direction of rotation.

TIRE ROTATION

To assure long life and even tire wear, it is important to rotate each tire to another location. For best results, tires should be rotated every 6000 miles or 6 months (see Figure 11–22 for suggested methods of rotation).

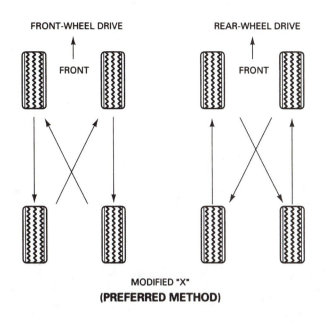

FRONT-WHEEL DRIVE   REAR-WHEEL DRIVE

FRONT   FRONT

MODIFIED "X"
(PREFERRED METHOD)

FRONT OR REAR WHEEL DRIVE   FRONT OR REAR WHEEL DRIVE

FRONT   FRONT

FULL "X"   **FRONT/REAR**
(ACCEPTABLE)   **(ACCEPTABLE)**

FIGURE 11–22 The preferred method most often recommended is the modified X method. Using this method, each tire eventually is used at each of the four wheel locations. An easy way to remember the sequence, whether front wheel drive or rear wheel drive, is to say to yourself: "Drive wheels straight, cross the nondrive wheels."

HINT: To help remind you when to rotate the tires, just remember that it should be done at every *other* oil change. Most manufacturers recommend changing the engine oil every 3000 miles (4800 km) or every 3 months, and tire rotation is recommended every 6000 miles (9600 km) or every 6 months.

TIRE INSPECTION

All tires should be inspected carefully for faults in the tire itself or for signs that something may be wrong with the steering or suspension systems of the vehicle (see Figures 11–23 through 11–25 for examples of common problems).

RADIAL RUNOUT

Even though a tire has no visible faults, it can be the cause of a vibration. If a vibration is felt above 45 mph, regardless of the engine load, the usual cause is an out-of-balance or defective out-of-round tire. Both of these problems cause a **tramp** or **up-and-down** type of vibration. This can be determined using a runout gauge to check for **radial runout** (see Figure 11–26).

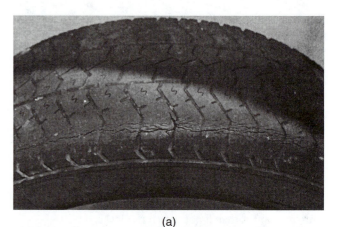

(a)

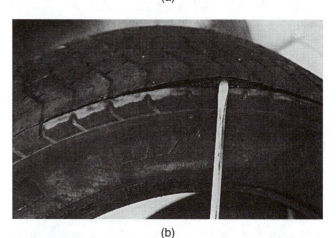

(b)

(c)

FIGURE 11–25 (a) Cracking at the edge of the belt on the tire shoulder can be an early sign of tire failure. (b) Tire failure at belt edge. The owner of the vehicle did not notice the cracking along the shoulder because it was on the inside. (c) This tire blew out as a result of a belt separation. Notice the steel wire belt sticking out.

FIGURE 11–23 Tire showing excessive shoulder wear resulting from under inflation and/or high-speed cornering.

FIGURE 11–24 Tire showing excessive wear in the center, indicating overinflation or heavy acceleration on a drive wheel.

FIGURE 11–26   For aggressive tread tires, put masking tape over the tread before measuring runout. The tape allows the dial indicator to roll smoothly over the tread of the tire.

Maximum radial runout should be less than 0.060 in. (1.5 mm). Little, if any, tramp will be noticed with less than 0.030 in. (0.8 mm) runout. If the reading is over 0.125 in. (3.2 mm), replacement of the tire is required.

CORRECTING RADIAL RUNOUT

Excessive radial runout may be corrected by one of several methods, including:

1. Try relocating the wheel on the mounting studs. Mark one stud and remount the wheel two studs away from the original position. Excessive wheel hole and/or stud tolerance may be the cause. If the radial runout is now satisfactory, re-mark the stud and wheel to prevent future occurrence of the problem.

2. Remount the tire on the wheel 180° from its original location. This can solve a runout problem, especially if the tire was not match mounted to the wheel originally.

3. If runout is still excessive, remove the tire from the wheel and check the runout of the *wheel*. If the wheel is within 0.035 in. (0.9 mm), yet the runout of the tire/wheel assembly is excessive, the problem has to be a defective tire and should be replaced.

4. Check all four tires.

LATERAL RUNOUT

Another possible problem that tires can cause is a type of vibration called **shimmy**. This rapid back-and-forth motion can be transmitted through the steering linkage to the steering wheel. Excessive runout is usually noticeable by the driver of the vehicle as a side-to-side vibration, especially at low speeds between 5 and 45 mph (8 and 72 km/h). Shimmy can be caused by an internal defect of the tire or a bent wheel. This can be checked

FIGURE 11–27 Checking the lateral runout of the wheel using a dial gauge (indicator).

using a runout gauge on the side of the tire or wheel to check for **lateral runout**.

Place the runout gauge against the side of the tire and rotate the wheel. Observe the readings. The maximum allowable reading is 0.045 in. (1.1 mm). If close to or above 0.045 in. (1.1 mm), check on the edge of the wheel to see if the cause of the lateral runout is due to a bent wheel as shown in Figure 11–27. Most manufacturers specify a maximum lateral runout of 0.035 in. (0.9 mm) for alloy wheels and 0.045 in. (1.1 mm) for steel wheels.

CORRECTING LATERAL RUNOUT

Excessive lateral runout may be corrected in one of several ways, including:

1. Retorque the wheel in the proper star pattern to the specified torque. Unequal or uneven wheel torque can cause excessive lateral runout.
2. Remove the wheel and inspect the wheel mounting flange for corrosion or for any other cause that could prevent the wheel from seating flat against the brake rotor or drum surface.
3. Check the condition of the wheel or axle bearings. Looseness in the bearings can cause the wheel to wobble.

TIRE BALANCING

Proper tire balance is important for tire life, ride comfort, and safety.

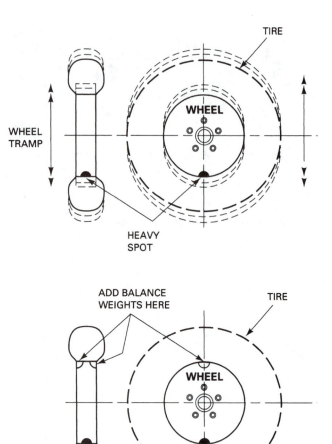

FIGURE 11–28 Weights are placed opposite the heavy spot of the tire/wheel assembly to balance the tire statically and prevent wheel tramp (up-and-down) vibration.

Static Balance. The term **static balance** means that the weight mass is evenly distributed around the axis of rotation (see Figure 11–28).

1. For example, if a wheel were spun and stopped at different places with each spin, the tire is statically balanced.
2. If the static balance is not correct, wheel tramp (vertical shake) vibration and uneven tire wear can result.

Dynamic Balance. The term **dynamic balance** means that the center line of weight mass is in the same plane as the center line of the wheel (see Figure 11–29).

1. Must be checked with the tire and the wheel being rotated to determine side-to-side out-of-balance as well as up and down.

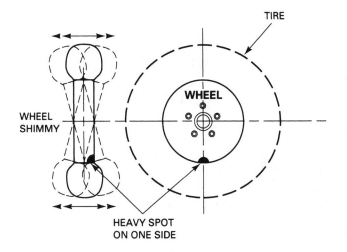

FIGURE 11–30 Coated wheel weight on an alloy wheel. Being coated prevents corrosion caused by the two different types of metals (lead weight and aluminum alloy wheel).

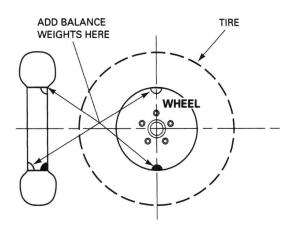

FIGURE 11–29 Weights are added to correct not only imbalance up and down but also to correct any "wobble" caused by the tire/wheel assembly being out of balance dynamically.

2. Incorrect dynamic balance causes shimmy. Shimmy-type vibration causes the steering wheel to shake from side to side.

Prebalance Checks. Before attempting to balance any tire, the following items should be checked and corrected to ensure a good tire balance:

1. Check the wheel bearing adjustment for looseness or wear.
2. Check the radial runout.
3. Check the lateral runout.
4. Remove stones from the tread.
5. Remove grease or dirt buildup on the inside of the wheel.
6. Check for proper tire pressures.
7. Remove all the old weights.

WHEEL WEIGHTS

Wheel weights are available in a variety of styles and types, including:

1. Clip-on lead weights for standard steel rims
2. Clip-on weights for alloy (aluminum) wheels
   a. **Uncoated.** These are generally *not* recommended by wheel or vehicle manufacturers because corrosion often occurs where the lead weight contacts the alloy wheel surface.
   b. **Coated.** Lead weights that are painted or coated in a plastic material are usually the *recommended* type of weight to use on alloy wheels, as shown in Figure 11–30. Weights are usually coated with a nylon or polyester-type material that often matches the color of aluminum wheels.
3. Stick-on weights that come with an adhesive backing, most often used on alloy wheels

Most wheel weights come in ¼-oz (0.25-oz) increments (ounces $\times$ 28 = grams).

0.25 oz = 7 grams	1.75 oz = 49 grams
0.50 oz = 14 grams	2.00 oz = 56 grams
0.75 oz = 21 grams	2.25 oz = 63 grams
1.00 oz = 28 grams	2.50 oz = 70 grams
1.25 oz = 35 grams	2.75 oz = 77 grams
1.50 oz = 42 grams	3.00 oz = 84 grams

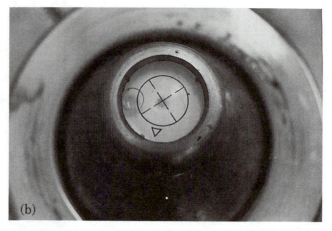

FIGURE 11–31 (a) Typical portable bubble wheel balancer. (b) When properly balanced, the bubble should be in the center of the cross hairs. The triangle mark is placed on the gauge to be used to line up the valve stem of the tire. This allows the technician to reinstall the tire/wheel assembly back onto the balancer after pounding on the weights in the same location. (c) Most bubble balancers use oil to allow the gauge to float freely. Be careful when moving this type of balance because the oil can easily spill out if tilted.

BUBBLE BALANCER

This type of static balancer is commonly used and is accurate if calibrated and used correctly. A bubble balancer is portable and can easily be stored away when not in use. It is also easy to use and relatively inexpensive (see Figure 11–31).

COMPUTER BALANCER

Most computer balancers are designed to balance wheels and tires off the vehicle. Computer dynamic balancers spin the tire at a relatively slow speed (approximately 20 mph). Sensors attached to the spindle of the balancer determine the amount and location of weights necessary to balance the tire dynamically. All computer balancers must be "programmed" or "instructed" with actual rim size and tire location for the electronic circuits to calibrate the required weight locations. Computer balancers are the most expensive type of balancer (see Figure 11–32).

Most computer balancers will be accurate to within ¼ oz (0.25 oz) while some are accurate to within ⅛ oz (0.125 oz). (Most drivers can "feel" an out-of-balance condition of 1 oz or more, but few can feel a vibration caused by just 1/4 oz.)

FIGURE 11–32 Typical computer balancer.

FIGURE 11–33 A careful visual inspection of the inside of a tire is helped by using special bead spreading clamps.

TIRE REPAIR

Tread punctures, nail holes, or cuts up to ¼ in. (2.6 mm) can be repaired. Repairs should be done from the inside of the tire using plugs or patches. The tire should be removed from the rim to make the repair. With the tire off the wheel, inspect the wheel and the tire for hidden damage. The proper steps to follow for a proper tire repair include:

1. Mark the location of the tire on the wheel.
2. Dismount the tire, inspect (see Figure 11–33), and clean the punctured area with a prebuff cleaner. *Do not use gasoline!*
3. Buff the cleaned area with sandpaper or a tire buffing tool until the rubber surface has a smooth velvet finish (see Figure 11–34).
4. Ream the puncture with a fine reamer from the inside. Cut and remove any loose wire material from the steel belts.
5. Fill the puncture with contour filling material and cut or buff the material flush with the inner liner of the tire.

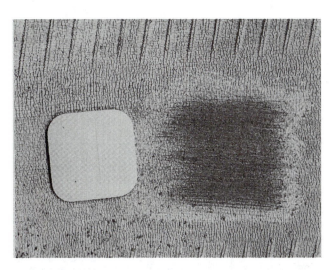

FIGURE 11–34 The area of the repair should be buffed slightly larger than the patch to be applied.

6. Apply chemical vulcanizing cement and allow to dry.

NOTE: Most vulcanizing (rubber) cement is highly flammable. Use out of the area of an open flame. Do not smoke when making a tire repair.

---

7. Apply the patch and use a stitching tool from the center toward the outside of the patch to work any air out from between the patch and the tire

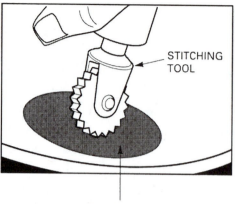

FIGURE 11–35 A stitching tool being used to force any trapped air out from under the patch.

(see Figure 11–35). Another excellent tire repair procedure uses a rubber plug. Pull the stem through the hole in the tire as shown in Figure 11–36.

8. Remount the tire on the rim aligning the marks made in step 1. Inflate to recommended pressure and check for air leaks.

There are many tire repair products on the market. Always follow the installation and repair procedures exactly according to the manufacturer's instructions.

CAUTION: Most experts agree that tire repairs should be done from the inside. Many technicians have been injured and a few killed when the tire they were repairing exploded as a steel reamer tool was inserted into the tire. The reamer can easily create a spark as it is pushed through the steel wires of a steel-belted tire. This spark can ignite a combustible mixture of gasses inside the tire that resulted from using "stop leak" or "inflator" cans. Since there is no way a technician can know if a tire has been inflated with a product that uses a combustible gas, always treat a tire as if it could explode. If a tire *has* to be repaired with a plug from the outside, remove all of the air (and trapped gas, if present), then reinflate several times. Always use a *plastic* reamer tool to reduce the chance of creating a spark.

SUMMARY

1. All tires are assembled by hand from many different materials and chemical compounds. After the "green tire" is assembled, it is placed into a mold under heat and pressure for about 30 min. The tread design and the tire shape are determined by the mold design.

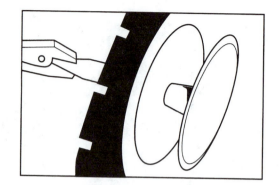

FIGURE 11–36 A rubber plug being pulled through a hole in the tire. The stem is then cut off flush with the surface of the tire tread.

2. New tires have between ⁹⁄₃₂ and ¹⁵⁄₃₂ of tread depth. Wear bars (indicators) show up as a bald strip across the tread of the tire when the tread depth gets down to ²⁄₃₂ in.

3. A 205/75R × 14 92S tire is 205 mm wide at its widest section and is 75 percent as high as it is wide. The "R" stands for radial construction and the tire is designed for a 14-in.-diameter rim. The number "92" is the load index of the tire (the higher the number, the more weight the tire can safely support). The "S" is the speed rating of this tire (S = 112 mph maximum sustained).

4. The uniform tire quality grading system is a rating for tread wear 100, 150, etc., traction (A, B, C) and temperature resistance (A, B, C).

5. Replacement wheels should have the same offset as the factory wheels to prevent abnormal tire wear and/or handling problems.

6. For safety and proper vehicle handling, all four tires of the vehicle should be of the same size, construction, and type, except where specified by the manufacturer, such as on some high-performance sports cars.

7. Tires should be rotated every 5000 to 7000 miles (8000 to 11,000 km) or at every other oil change.

8. Wheels should always be tightened with a torque wrench to the proper torque in a star pattern.

9. Properly balanced tires prolong tire life. Wheel tramp or an up-and-down type of vibration results if the tires are statically out of balance or if the tire is out of round.

10. Only coated or stick-on wheel weights should be used on alloy wheels, to prevent corrosion damage.

REVIEW QUESTIONS

1. List the various parts of a tire and explain how a tire is constructed.

2. Explain the three major areas of the uniform tire quality grading system.

3. Explain how to determine the proper tire pressure.

4. Determine the proper wheel mounting torque for your vehicle from the guidelines.

5. Determine the bolt circle, wheel diameter, and wheel width.

6. Describe the difference between static and dynamic balance.

MULTIPLE-CHOICE QUESTIONS

1. A tire labeled 215/60R × 15 92T, the "T" means
   a. its speed rating.
   b. its tread wear rating.
   c. its load rating.
   d. its temperature resistance rating.

2. The "92" in the tire designation in question 1 refers to the tire's
   a. speed rating.
   b. tread wear rating.
   c. load rating.
   d. temperature resistance rating.

3. Radial tires can cause a vehicle to pull to one side while driving. This is called *radial tire pull* and is often due to the
   a. angle of the body (carcass) plies.
   b. tire conicity.
   c. tread design.
   d. bead design.

4. Tire inflation is very important to the safe and economical operation of any vehicle. The maximum pressure should be the pressure imprinted on the sidewall of the tire and the minimum pressure should be
   a. 90 percent of maximum.
   b. 80 percent of maximum.
   c. 70 percent of maximum.
   d. 60 percent of maximum.

5. When purchasing replacement tires, do not change the tire width by more than
   a. 10 mm stock.
   b. 15 mm stock.
   c. 20 mm stock.
   d. 25 mm stock.

6. Wheel lug nuts must be tightened
   a. by hand.
   b. with a torque wrench.
   c. with an air impact wrench.
   d. hand tightened plus ¼ turn.

7. A tire is worn excessively on both edges. The most likely cause of this type of tire wear is
   a. overinflation.
   b. underinflation.
   c. excessive radial runout.
   d. excessive lateral runout.

8. When seating a bead of a tire, never exceed
   a. 30 psi.
   b. 40 psi.
   c. 50 psi.
   d. 60 psi.

9. For best tire life, most vehicle and tire manufacturers recommend tire rotation every
   a. 3000 miles.
   b. 6000 miles.
   c. 9000 miles.
   d. 12,000 miles.

10. The recommended type of wheel weight to use on aluminum (alloy) wheels is
   a. lead with plated spring steel clips.
   b. coated (painted) lead weights.
   c. lead weights with longer than normal clips.
   d. aluminum weights.

◀ Chapter 12 ▶

DRIVE AXLE SHAFTS AND CV JOINTS

OBJECTIVES

After studying Chapter 12, the reader will be able to:

1. Name drive shaft and U-joint parts, and describe their purpose, function, and operation.
2. Explain how the working angle of the U-joints are determined.
3. Describe how CV joints work.
4. Explain how to perform a U-joint inspection.
5. Describe the service procedures for replacing CV joints and boots.

A drive axle shaft transmits engine power from the transmission or transaxle (if front wheel drive) to the rear axle assembly or drive wheels (see Figures 12–1 and 12–2). **Drive shaft** is the term used by the Society of Automotive Engineers (SAE) to describe the shaft between the transmission and the rear axle assembly on a rear-wheel-drive vehicle. General Motors, Chrysler, and some other manufacturers use the term *propeller shaft* or *prop shaft* to describe this part. The SAE term is used throughout this book.

A typical drive shaft is a hollow steel tube. A splined end yoke is welded onto one end that slips over the splines of the output shaft of the transmission (see Figure 12–3). An end yoke is welded onto the other end of the drive shaft. Some drive shafts use a center support bearing. Beyond about 65 in. (165 cm) in length, a **center support bearing** must be used as shown in Figure 12–4. A center support bearing is also called a **steady bearing**.

Some vehicle manufacturers use aluminum drive shafts and they can be as long as 90 in. (230 cm) with no problem. Composite material drive shafts are also used in some vehicles. These carbon fiber–plastic drive shafts are very strong, yet lightweight and can be made in long lengths without the need for a center support bearing.

To dampen drive shaft noise, it is common to line the inside of the hollow drive shaft with cardboard. This helps eliminate the tinny sound whenever shifting between "drive" and "reverse" in a vehicle equipped with an automatic transmission.

U-JOINT DESIGN AND OPERATION

Universal joints (U-joints) are used at both ends of a drive shaft. U-joints allow the wheels and the rear axle to move up and down and remain flexible and still transfer power to the drive wheels. A simple universal joint can be made from two Y-shaped yokes connected by a cross member called a **cross** or **spider**. The four arms of the cross are called **trunnions**. See Figure 12–5 for a line drawing of a simple U-joint with all part names identified. A similar design is the common U-joint used with a socket wrench set.

Most U-joints are called **cross yoke** or **Cardan universal joints**. Cardan was a sixteenth-century Italian mathematician who worked with objects that moved freely in any direction. Torque from the engine

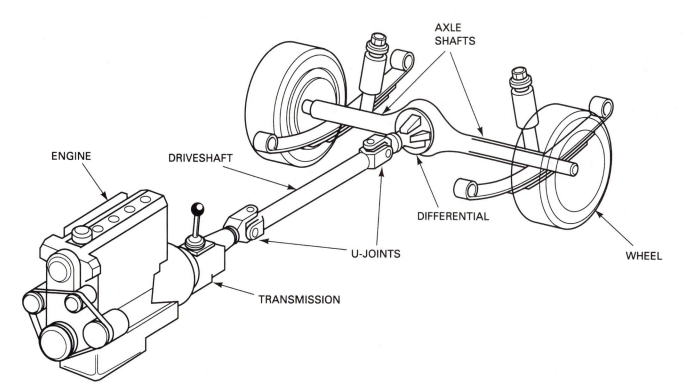

FIGURE 12–1 Typical rear-wheel-drive power train arrangement. The engine is mounted longitudinally (lengthwise) in the vehicle.

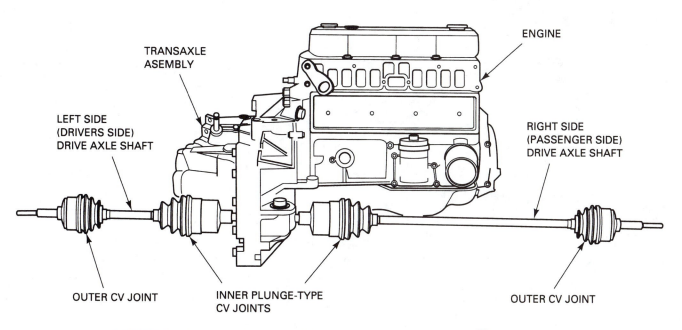

FIGURE 12–2 Typical front-wheel-drive power train arrangement. The engine is usually mounted transversely (sideways) in the vehicle.

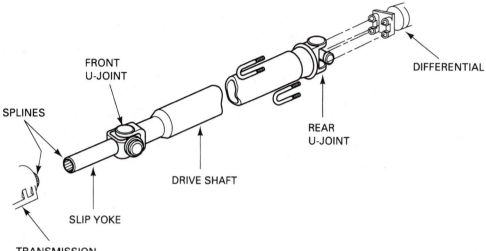

FRONT
U-JOINT

SPLINES

FIGURE 12–3 Typical drive shaft (also called a propeller shaft). The drive shaft transfers engine power from the transmission to the differential.

DIFFERENTIAL

REAR
U-JOINT

DRIVE SHAFT

SLIP YOKE

TRANSMISSION

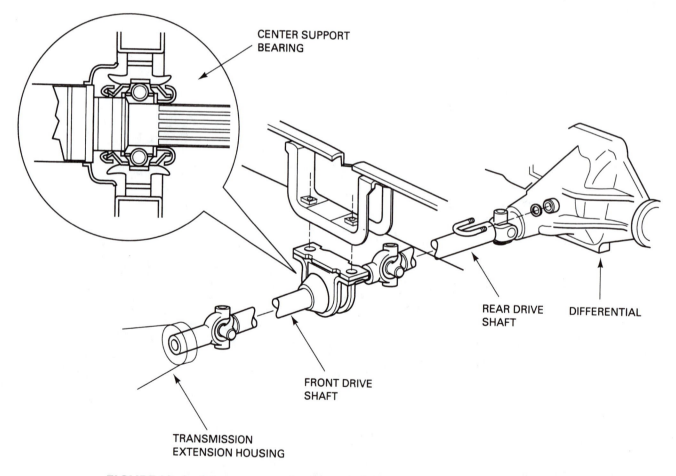

CENTER SUPPORT
BEARING

REAR DRIVE
SHAFT

DIFFERENTIAL

FRONT DRIVE
SHAFT

TRANSMISSION
EXTENSION HOUSING

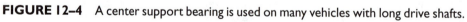

FIGURE 12–4 A center support bearing is used on many vehicles with long drive shafts.

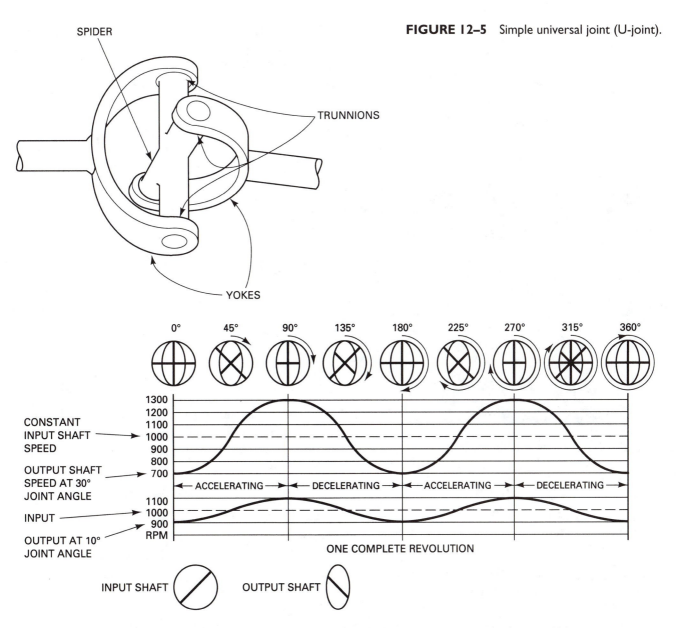

SPIDER

FIGURE 12–5 Simple universal joint (U-joint).

TRUNNIONS

YOKES

CONSTANT INPUT SHAFT SPEED

OUTPUT SHAFT SPEED AT 30° JOINT ANGLE

INPUT

OUTPUT AT 10° JOINT ANGLE

ONE COMPLETE REVOLUTION

INPUT SHAFT   OUTPUT SHAFT

FIGURE 12–6 Graphic showing how the speed difference on the output of a typical U-joint varies with the speed and the angle of the U-joint. At the bottom of the chart, the input speed is a constant 1000 rpm while the output speed varies from 900 to 1100 rpm when the angle difference in the joint is only 10°. At the top part of the chart, the input speed is a constant 1000 rpm, yet the output speed varies from 700 to 1200 rpm when the angle difference in the joint is changed to 30°. (Courtesy of Dana Corporation)

is transferred through the U-joint. The engine "drives" the U-joint at a constant speed, but the output speed of the U-joint changes because of the angle of the joint. The speed changes twice per revolution. **The greater the angle, the greater the change in speed (velocity)** (see Figure 12–6). It is very important that both U-joints are operating at about the same angle to prevent excessive drive line vibration (see Figures 12–7).

CONSTANT-VELOCITY JOINTS

Constant-velocity joints (commonly called **CV joints**) are designed to rotate without changing speed. Regular U-joints are usually designed to work up to 12° of angularity. If two Cardan style U-joints are joined together, the angle that this **double Cardan joint** can function is about 18 to 20°. See Figure 12–8 for an illustration of a double Cardan U-joint.

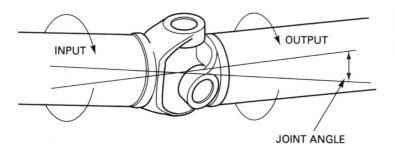

FIGURE 12–7 The joint angle is the difference between the angles of the joint. (Courtesy of Dana Corporation)

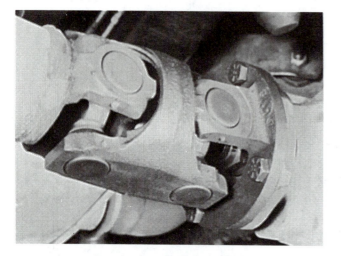

FIGURE 12–8 A double Cardan U-joint.

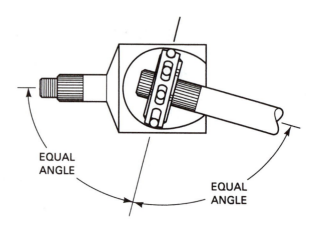

FIGURE 12–9 A constant-velocity (CV) joint can operate at high angles without a change in velocity (speed) because the joint design results in equal angles between input and output.

NOTE: Many four-wheel-drive light trucks use standard Cardan style U-joints in the front drive axles. If the front wheels are turned sharply and then accelerated, the entire truck often shakes, due to the pulsations created by the speed variations through the U-joints. This vibration is normal and cannot be corrected. It is the characteristic of this type of design and is usually not noticeable in normal driving.

The first constant-velocity joint was designed by Alfred H. Rzeppa (pronounced *shep'pa*) in the mid-1920s. The **Rzeppa joint** transfers power through six round balls that are held in position midway between the two shafts. This design resulted in the angle between the shafts being equally split regardless of the angle (see Figure 12–9). Because the angle was always split equally, power was transferred equally without the change in speed (velocity) that occurs in Cardan-style U-joints. This style of joint resulted in a constant velocity between driving and driven shafts. This style of joint could also function at greater angles than simple U-joints, up to 40°.

NOTE: CV joints are also called *Löbro joints*, after the brand name of an original equipment manufacturer.

OUTER CV JOINTS

The Rzeppa-type CV joint is most commonly used as an outer joint on most front-wheel-drive vehicles (see Figure 12–10). The outer joint must allow turning up to 40° while the front wheels move up and down and still be able to transmit engine power to drive the front wheels.

Outer CV joints are called *fixed*. The outer joints are also attached to the front wheels. They are more likely to suffer from road hazards, which often can cut through the protective outer flexible boot (see Figure 12–11). Once this boot has been split open, the special high-quality grease is thrown out and contaminants such as dirt and water can enter.

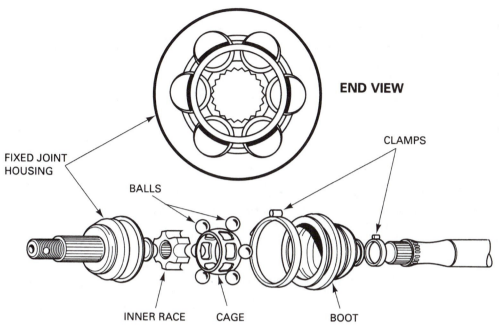

END VIEW

FIXED JOINT
HOUSING

CLAMPS

BALLS

INNER RACE   CAGE   BOOT

FIGURE 12–10 Rzeppa fixed joint. This type of CV joint is commonly used at the wheel side of the drive axle shaft. This type of joint can operate at high angles to compensate for suspension travel and steering angle changes. (Courtesy of Dana Corporation)

TORN
CV JOINT
BOOT

FIGURE 12–11 A drive axle shaft being removed from a vehicle because of the torn boot. This joint was ruined and was making a very loud clicking noise, especially when turning.

INNER CV JOINTS

Inner CV joints attach the output of the transaxle to the drive axle shaft. Inner CV joints are therefore inboard or toward the center of the vehicle (see Figure 12–12). Inner CV joints have to be able to allow the drive axle shaft to move up and down and allow the drive axle shaft to change length as required during vehicle suspension travel movements.

Unequal-length drive axle shafts (also called **half shafts**) result in unequal drive axle shaft angles to the front drive wheels (see Figure 12–13). This unequal angle often results in a pull on the steering wheel during acceleration. This pulling to one side during acceleration due to unequal engine torque being applied to the front drive wheels is called **torque steer**. To help reduce the effect of torque steer, some vehicles are manufactured with an intermediate shaft that results in equal drive axle shaft angles. Both designs use fixed outer CV joints with plunge-type inner joints (see Figures 12–14 through 12–16 for examples of each type). CV joints are also used in rear-wheel-drive vehicles and in many four-wheel-drive vehicles.

The pliable boot surrounding the CV joint must be able to remain flexible under all weather conditions and still be strong enough to avoid being punctured by road debris. There are four basic types of boot materials used over CV joints, including **natural rubber** (black), **silicone rubber** (gray), **hard thermoplastic** (black), and **urethane** (usually blue) (see Figure 12–17).

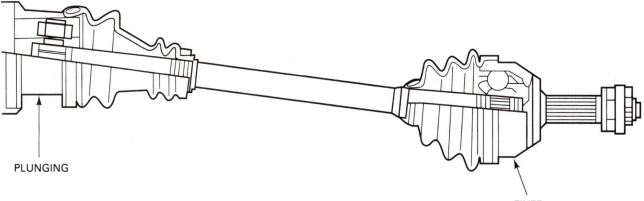

PLUNGING

FIXED

FIGURE 12–12 The fixed outer joint is required to move in all directions because the wheels must turn for steering as well as move up and down during suspension movement. The inner joint has to be able not only to move up and down, but also to plunge in and out as the suspension moves up and down. (Courtesy of Dana Corporation)

UNEQUAL LENGTH DRIVESHAFT

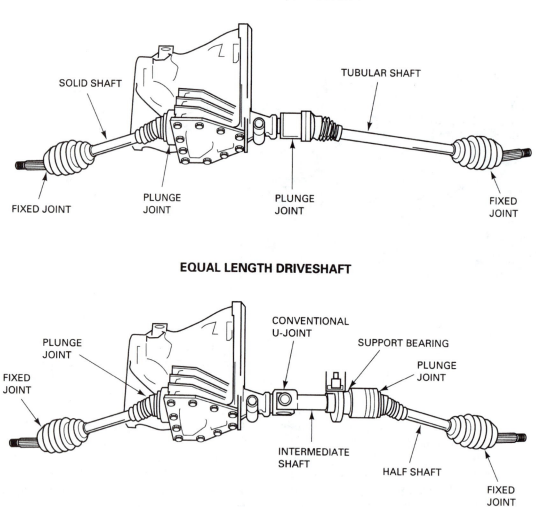

SOLID SHAFT

TUBULAR SHAFT

FIXED JOINT

PLUNGE JOINT

PLUNGE JOINT

FIXED JOINT

EQUAL LENGTH DRIVESHAFT

PLUNGE JOINT

CONVENTIONAL U-JOINT

SUPPORT BEARING

PLUNGE JOINT

FIXED JOINT

INTERMEDIATE SHAFT

HALF SHAFT

FIXED JOINT

FIGURE 12–13 Unequal-length drive shafts result in unequal drive axle shaft angles to the front drive wheels. This unequal-angle side to side often results in a steering of the vehicle during acceleration called *torque steer*. By using an intermediate shaft, both drive axles are the same angle and torque steer effect is reduced. (Courtesy of Dana Corporation)

TRIPOD TYPE PLUNGE JOINT

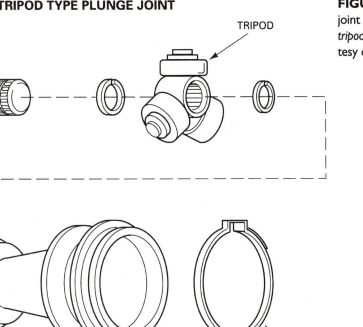

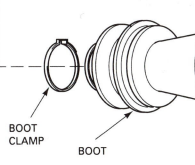

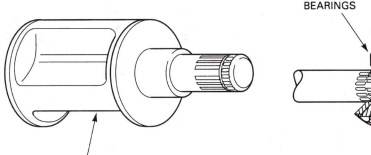

FIGURE 12–14 A tripod joint is also called a *tri-pot, tripode,* or *tulip* design. (Courtesy of Dana Corporation)

TRIPOD

BOOT CLAMP

BOOT

BOOT CLAMP

NEEDLE BEARINGS

TULIP

NOTE: *CARE MUST BE TAKEN OR TRIPOD ROLLERS MAY COME OFF TRIPOD.*

CROSS GROOVE PLUNGE JOINT

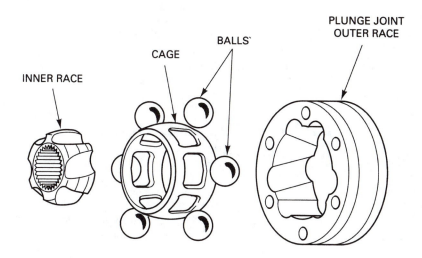

INNER RACE

CAGE

BALLS`

PLUNGE JOINT OUTER RACE

FIGURE 12–15 A cross-groove plunge joint is used on many German front-wheel-drive vehicles and as both inner and outer joints on the rear of vehicles that use an independent-type rear suspension. (Courtesy of Dana Corporation)

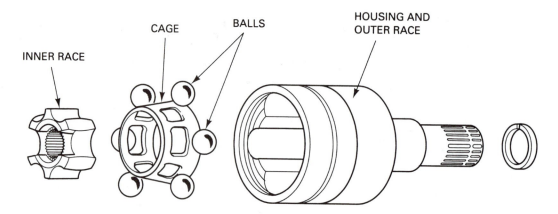

**DOUBLE OFFSET BALL
TYPE PLUNGE JOINT**

INNER RACE
CAGE
BALLS
HOUSING AND
OUTER RACE

FIGURE 12–16 Double offset ball-type plunge joint. (Courtesy of Dana Corporation)

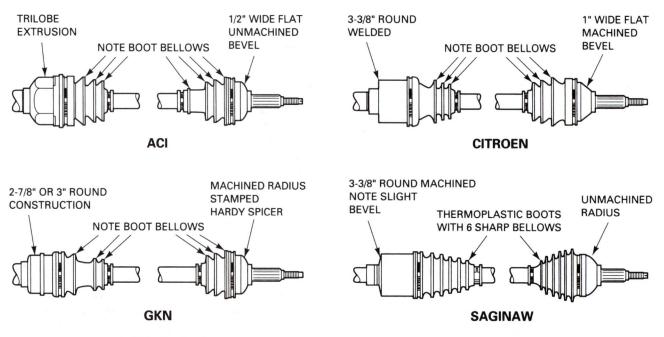

TRILOBE
EXTRUSION
NOTE BOOT BELLOWS
1/2" WIDE FLAT
UNMACHINED
BEVEL

ACI

3-3/8" ROUND
WELDED
NOTE BOOT BELLOWS
1" WIDE FLAT
MACHINED
BEVEL

CITROEN

2-7/8" OR 3" ROUND
CONSTRUCTION
NOTE BOOT BELLOWS
MACHINED RADIUS
STAMPED
HARDY SPICER

GKN

3-3/8" ROUND MACHINED
NOTE SLIGHT
BEVEL
THERMOPLASTIC BOOTS
WITH 6 SHARP BELLOWS
UNMACHINED
RADIUS

SAGINAW

FIGURE 12–17 Getting the correct boot kit or parts from the parts store is more difficult on many Chrysler front-wheel-drive vehicles because Chrysler has used four different manufacturers for their axle shaft assemblies. (Courtesy of Dana Corporation)

NOTE: Some aftermarket companies offer a split-style replacement CV joint boot. Being split means that the boot can be replaced without having to remove the drive axle shaft. Vehicle manufacturers usually do *not* recommend this type of replacement boot because the joint cannot be disassembled and cleaned properly with the drive axle still in the vehicle.

DRIVE SHAFT AND U-JOINT INSPECTION

The drive shaft should be inspected for the following:

1. Any dents or creases caused by incorrect hoisting of the vehicle or by road debris.

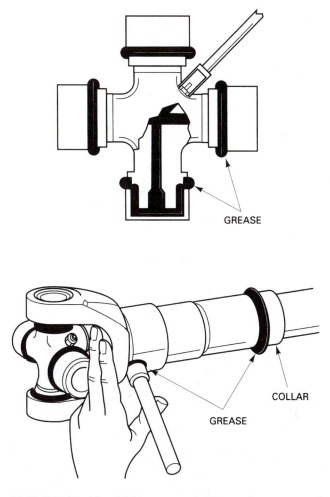

FIGURE 12-18 appears with GREASE and COLLAR labels.

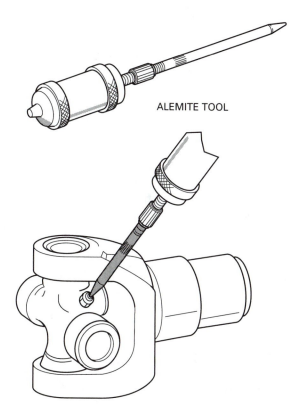

ALEMITE TOOL

FIGURE 12-18 All U-joints and spline collars equipped with a grease fitting should be greased four times a year as part of a regular lubrication service. (Courtesy of Dana Corporation)

FIGURE 12-19 Many U-joints require a special grease gun tool to reach the grease fittings. (Courtesy of Dana Corporation)

plying grease with a grease gun (see Figures 12–18 and 12–19).

Besides periodic lubrication, the drive shaft should be grabbed and moved to see if there is any movement of the U-joints. If *any* movement is noticed when the drive shaft is moved, the U-joint is worn and must be replaced.

CAUTION: A dented or creased drive shaft can collapse, especially when the vehicle is under load. This collapse of the drive shaft can cause severe damage to the vehicle and may cause an accident.

---

2. Undercoating, grease, or dirt buildup on the drive shaft can cause a vibration. Undercoating should be removed using a suitable solvent and a shop cloth. Always dispose of used shop cloths properly.

The U-joints should be inspected every time the vehicle chassis is lubricated, or four times a year. Most original equipment U-joints are permanently lubricated and have no provision for greasing. If there is a grease fitting, the U-joint should be lubricated by ap-

---

NOTE: U-joints are not serviceable items and cannot be repaired. If worn or defective, they must be replaced.

U-joints can be defective and still not show noticeable free movement. **A proper U-joint inspection can be performed only by removing the drive shaft from the vehicle.** Before removing the drive shaft, always mark the position of all mating parts to assure proper reassembly. White correction fluid, also known by the trade names White-Out and Liquid Paper, is an easy and fast-drying marking material (see Figure 12–20).

To remove the drive shaft from a rear-wheel-drive vehicle, remove the four fasteners at the rear U-joint at the differential (see Figure 12–21). Push the drive shaft forward toward the transmission and then down and

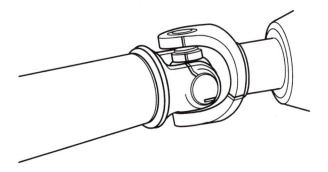

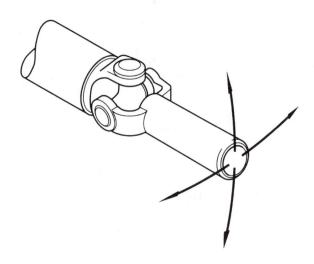

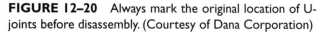

FIGURE 12–20 Always mark the original location of U-joints before disassembly. (Courtesy of Dana Corporation)

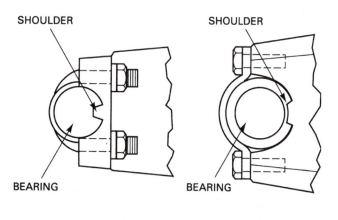

FIGURE 12–21 (Courtesy of Dana Corporation)

SHOULDER

SHOULDER

BEARING

BEARING

U-BOLT TYPE

STRAP TYPE

FIGURE 12–22 The best way to check any U-joint is to remove the drive shaft from the vehicle and move each joint in all directions. A good U-joint should be free to move without binding. (Courtesy of Dana Corporation)

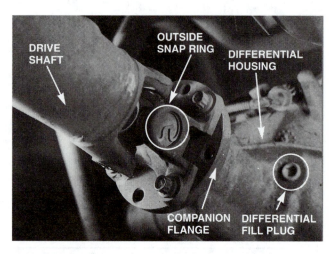

DRIVE SHAFT

OUTSIDE SNAP RING

DIFFERENTIAL HOUSING

COMPANION FLANGE

DIFFERENTIAL FILL PLUG

FIGURE 12–23 Typical U-joint that uses an outside snap ring. This style of joint bolts directly to the companion flange that is attached to the pinion gear in the differential.

toward the rear of the vehicle. The drive shaft should slip out of the transmission spline and can be removed from underneath the vehicle.

HINT: With the drive shaft removed, transmission lubricant can leak out of the rear extension housing. To prevent a mess, use an old spline the same size as the one being removed or place a plastic bag over the extension housing to hold any escaping lubricant. A rubber band can be used to hold the bag onto the extension housing.

To inspect U-joints, move each joint through its full travel, making sure that it can move (articulate) freely and equally in all directions (see Figure 12–22).

U-JOINT REPLACEMENT

All movement in a U-joint should occur between the trunnions and the needle bearings in the end caps. The end caps are press-fit to the yokes, which are welded to

the drive shaft. Three types of retainers are used to keep the bearing caps on the U-joints, including the outside snap ring (see Figure 12–23), the inside retaining ring (see Figure 12–24), or injected synthetic (usually nylon). After removing the retainer, use a press to separate the U-joint from the yoke, as shown in Figure 12–25.

U-joints that use synthetic retainers must be separated using a press and a special tool to press onto both sides of the joint in order to shear the plastic retainer as shown in Figure 12–26. Replacement U-joints use spring clips instead of injected plastic. Remove the old U-joint from the yoke as shown in Figure 12–27 and replace with a new U-joint.

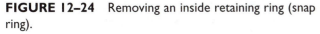

FIGURE 12–24   Removing an inside retaining ring (snap ring).

FIGURE 12–25   Pushing the lower bushing cap out of a U-joint by pushing the upper bushing cap inward using a press.

HINT: If a U-joint is slightly stiff after being installed, strike the U-joint using a brass punch and a light hammer. This often frees a stiff joint and is often called *relieving the joint.* The shock aligns the needle bearings in the end caps.

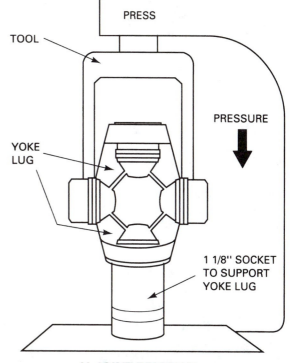

**U-JOINT REMOVAL
SYNTHETIC RETAINERS**

FIGURE 12–26   (Courtesy of Dana Corporation)

MEASURING DRIVE SHAFT ANGLES

To measure U-joint and drive shaft angles, the vehicle must be hoisted using an axle contact or drive-on-type lift so as to maintain the same drive shaft angles as the vehicle has while being driven. The working angles of the two U-joints on a drive shaft should be within ½° of each other in order to cancel out speed changes (see Figure 12–28).

To measure the working angle of a U-joint, follow these steps:

Step 1. Place an **inclinometer** (a tool used to measure angles) on the rear U-joint bearing cap. Level the bubble and read the angle (see Figure 12–29). The reading is 19.5°.

Step 2. Rotate the drive shaft 90° and read the angle of the rear yoke. For example, this reading is 17°.

Step 3. Subtract the smaller reading from the larger reading to obtain the working angle of the joint. In this example it is 2.5° ($19.5° - 17° = 2.5°$).

Repeat the same procedure for the front U-joint. The front and rear working angles should be within 0.5°. If the two working angles are not within 0.5°, shims can be added to bring the two angles closer together. The angle of the rear joint is changed by installing a tapered shim between the leaf spring and the axle as shown in Figure 12–30. The angle of the front joint is changed by adding or removing shims from the mount under the transmission.

CV JOINT DIAGNOSIS

When a CV joint wears or fails, the most common symptom is noise while driving. An outer fixed CV joint will probably be heard when turning sharply and accelerating at the same time. This noise is usually a clicking sound. While inner joint failure is less common, a defective inner

FIGURE 12–27 Removing the worn cross from the yoke.

◀ **TECH TIP** ▶

SPLINE BIND CURE

Drive line "clunk" often occurs in rear-wheel-drive vehicles when shifting between "drive" and "reverse" or when accelerating from a stop. Often the cause of this noise is excessive clearance between the teeth of the ring and pinion in the differential. Another cause is called *spline bind* and is caused by the changing rear pinion angle creating a binding in the spline when the rear springs change in height. For example, when a pickup truck stops, the weight transfers toward the front and unloads the rear springs. The front of the differential noses downward and forward as the rear springs unload. When the driver accelerates forward, the rear of the truck squats downward, causing the drive shaft to be pulled rearward when the front of the differential rotates upward. This upward movement on the spline often causes the spline to bind and make a loud clunk when the bind is finally released.

The method recommended by vehicle manufacturers to solve this noise is to follow these steps:

1. Remove the drive shaft.
2. Clean the splines on both the drive shaft yoke and the transmission output shaft.
3. Remove any burrs on the splines with a small metal file (remove all filings).
4. Apply a high-temperature grease to the spline teeth of the yoke. Apply grease to each spline but do not fill the splines. Synthetic chassis grease is preferred because of its high-temperature-resistance properties.
5. Reinstall the drive shaft.

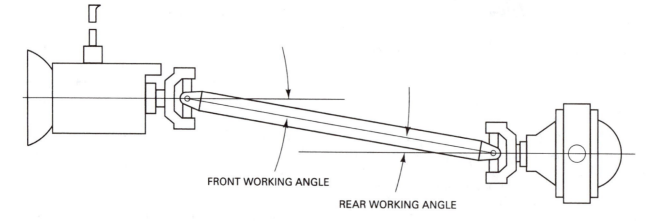

FRONT WORKING ANGLE

REAR WORKING ANGLE

FIGURE 12–28 (Courtesy of General Motors)

CV joint often creates a loud clunking noise while accelerating from rest. To help verify a defective joint, drive the vehicle in reverse while turning and accelerating. This almost always will reveal a defective outer joint.

CV JOINT SERVICE

The hub nut must be removed whenever servicing a CV joint or shaft assembly on a front-wheel-drive vehicle.

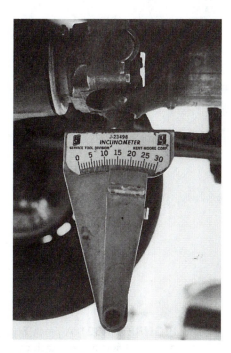

FIGURE 12–29   Inclinometer reads 19½° at this rear U-joint.

Since these nuts are usually torqued to almost 200 lb-ft (260 N-m), keep the vehicle on the ground until the hub nut is loosened and then follow these steps (see Figure 12–31):

Step 1.   Remove the front wheel and hub nut.

NOTE: Most manufacturers warn against using an air-impact wrench to remove the hub nut. The impacting force can damage the hub bearing.

Step 2.   To allow the knuckle room to move outward enough to remove the drive axle shaft, some or all of the following will have to be disconnected.
   a. Lower ball joint or **pinch bolt** (see Figure 12–32)
   b. Tie-rod end (see Figure 12–33)
   c. Stabilizer bar link
   d. Front disc brake caliper

Step 3.   Remove the splined end of the axle from the hub bearing.

Step 4.   Use a pry bar or special tool with a slide hammer as shown in Figure 12–34 and remove the inner joint from the transaxle.

Step 5.   Disassemble, clean, and inspect all components (see Figures 12–35 through 12–41).

Step 6.   Replace the entire joint if there are *any* worn parts. Pack *all* the grease that is supplied into the assembly or joint (see Figure 12–42).

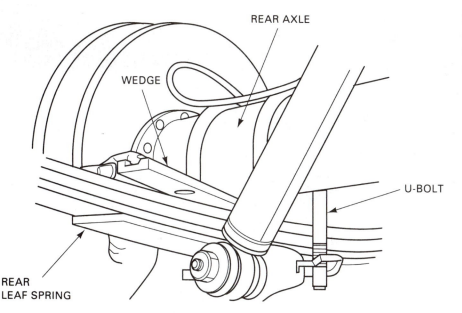

FIGURE 12–30   Placing a tapered metal wedge between the rear leaf spring and the rear axle pedestal to correct rear U-joint working angles.

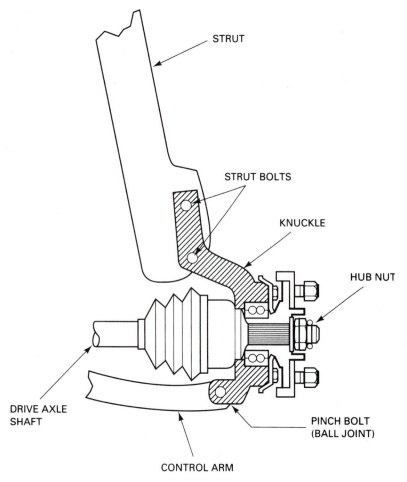

STRUT

STRUT BOLTS

KNUCKLE

HUB NUT

DRIVE AXLE
SHAFT

PINCH BOLT
(BALL JOINT)

CONTROL ARM

FIGURE 12–31 The hub nut must be removed before the hub bearing assembly or drive axle shaft can be removed from the vehicle.

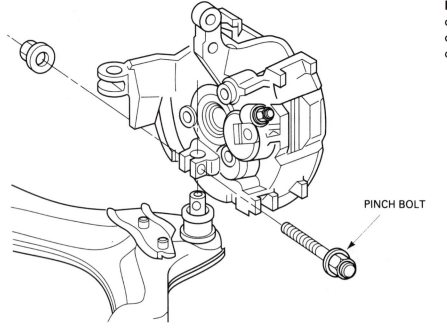

PINCH BOLT

FIGURE 12–32 A pinch bolt is used on many knuckles to attach the ball joint onto the lower control arm. (Courtesy of General Motors)

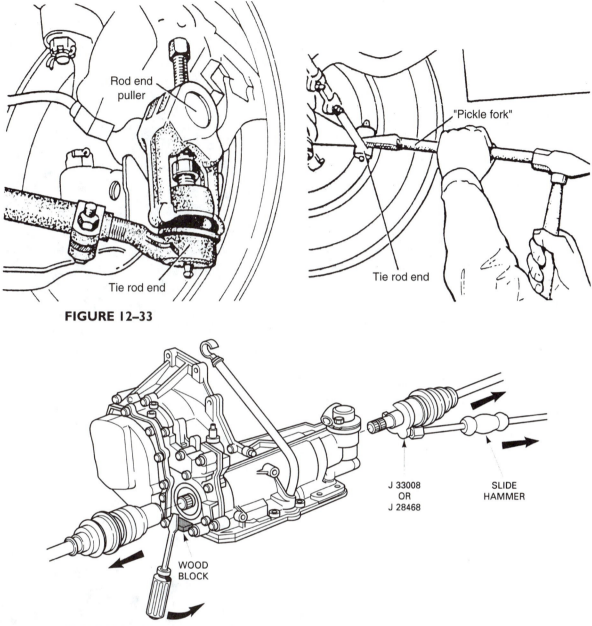

Rod end puller

"Pickle fork"

Tie rod end

Tie rod end

FIGURE 12–33

WOOD BLOCK

J 33008 OR J 28468

SLIDE HAMMER

FIGURE 12–34   Most inner CV joints can be separated from the transaxle with a pry bar. (Courtesy of General Motors)

FIGURE 12–35   If other service work requires that just one end of the drive axle shaft be disconnected from the vehicle, be sure that the free end is supported to prevent damage to the protective boots or allowing the joint to separate. The method shown using shop cloths to tie up one end may not be very pretty, but at least the technician took precautions to support the end of the shaft while removing the transaxle to replace a clutch.

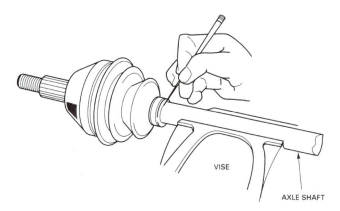

FIGURE 12–36 Mark with a scribe the location of the boots before removal. The replacement boots must be in the same location.

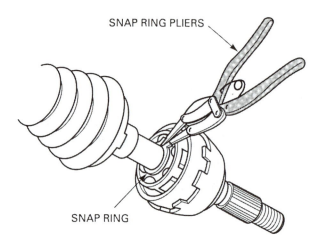

SNAP RING PLIERS

SNAP RING

FIGURE 12–37 Most CV joints use a snap ring to retain the joint to the drive axle shaft.

FIGURE 12–39 Typical outer CV joint after removing the boot and the joint from the drive axle shaft. This joint was removed from the vehicle because a torn boot was found. After disassembly and cleaning, this joint was found to be okay and was put back into service. Even though the grease looks terrible, there was enough grease in the joint to provide enough lubrication to prevent any wear from occurring.

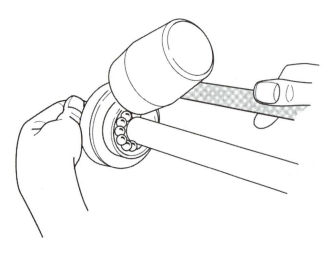

FIGURE 12–40 The cage of this Rzeppa-type CV joint is rotated so that one ball at a time can be removed. Some joints require that the technician use a brass punch and a hammer to move the cage.

FIGURE 12–38 After releasing the snap ring, most CV joints can be tapped off the shaft using a brass or shot-filled plastic (dead-blow) hammer.

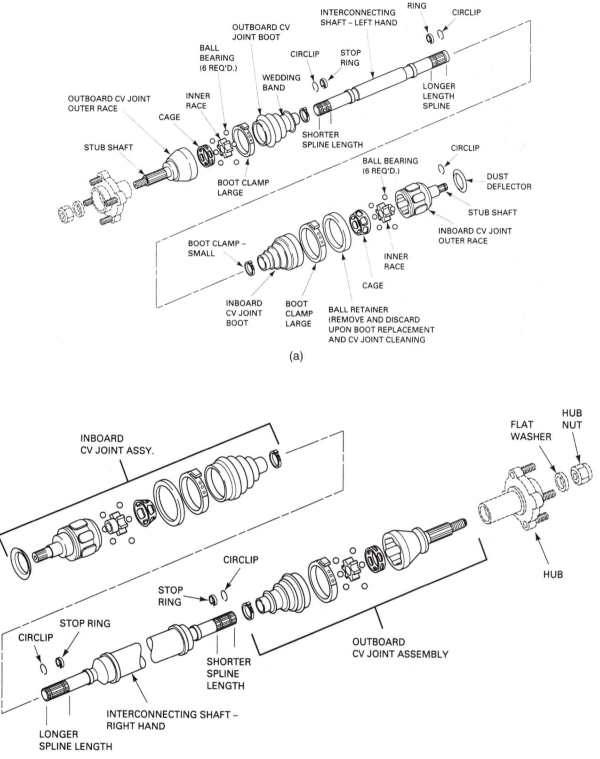

(a)

(b)

FIGURE 12–41 Typical left (a) and right (b) drive axle shaft assemblies showing all parts, including the location of stop rings and circlip. (Courtesy of Ford Motor Company)

217

FIGURE 12–42   Be sure to use *all* of the grease supplied with the replacement joint or boot kit. Use only the grease supplied and do not use substitute grease.

Most CV joint grease is molybdenum disulfide type grease, commonly referred to as *"Moly" grease.* Depending on the exact CV joint manufacturer, the exact composition of grease can vary. *The grease supplied with a replacement CV joint or boot kit should be the only grease used.* Assemble the joint and position the boot in the same location as marked before disassembly. Before clamping the last seal on the boot, be sure to release trapped air to prevent the boot from expanding when heated and collapsing when cold. This is sometimes called *burping the boot.* Clamp the boot according to the manufacturer's specifications.

Step 7.   Reinstall the drive axle shaft in the reverse order of removal.

SUMMARY

1. The drive shaft of a rear-wheel-drive vehicle transmits engine power from the transmission to the differential.

2. Universal joints (U-joints) allow the drive shaft to transmit engine power while the suspension and the rear axle assembly are moving up and down during normal driving conditions.

3. Acceptable working angles for a Cardan-type U-joint are ½ to 3°. Some angle is necessary to cause the roller bearings to rotate and a working angle of greater than 3° can lead to drive line vibrations.

4. Constant-velocity joints (CV joints) are used on all front-wheel-drive vehicles and many four-wheel-drive vehicles to provide a smooth transmission of power to the drive wheels regardless of angularity of the wheel or joint.

5. Outer or fixed CV joints commonly use a Rzeppa design, while inner CV joints are the plunging or tri-pot type.

6. A defective U-joint often makes a clicking sound when the vehicle is driven in reverse. Severely defective U-joints can cause drive line vibrations or a clunking sound when the transmission is shifted from reverse to drive or drive to reverse.

7. Incorrect drive shaft working angles can result from collapsed engine or transmission mounts.

8. Drive line clunk noise can often be corrected by applying high-temperature chassis grease to the splines of the front yoke on the drive shaft.

9. CV joints require careful cleaning, inspection, and lubrication using specific CV joint grease.

REVIEW QUESTIONS

1. Explain why Cardan-type U-joints on a drive shaft must be within ½° working angles.

2. What makes a constant-velocity joint able to transmit engine power through an angle at a constant velocity?

3. What type of grease must be used in CV joints?

4. List two items that should be checked when inspecting a drive shaft.

5. Describe how to replace a Cardan-type U-joint.

6. Explain the proper steps to perform when replacing a CV joint.

MULTIPLE-CHOICE QUESTIONS

1. A rear-wheel-drive vehicle shudders or vibrates when first accelerating from a stop. The vibration is less noticeable at higher speeds. The most likely cause is
   a. drive shaft unbalance.
   b. excessive U-joint working angles.
   c. unequal U-joint working angles.
   d. brinelling of the U-joint.

2. The owner of a full-size four-wheel-drive pickup truck complains that whenever turning sharply and accelerating rapidly, a severe vibration is created. Technician A says that the transfer case may be defective. Technician B says that this is normal if conventional Cardan U-joints are used to drive the front wheels. Which technician is correct?
   a. A only
   b. B only
   c. Both A and B
   d. Neither A nor B

3. The outer CV joints used on front-wheel-drive vehicles are
   a. fixed type.
   b. plunge type.

4. The proper grease to use with a CV joint is
   a. black chassis grease.
   b. dark blue EP grease.
   c. red moly grease.
   d. the grease that is supplied with the boot kit.

5. Two technicians are discussing U-joints. Technician A says that a defective U-joint could cause a loud clunk when the transmission is shifted between drive and reverse. Technician B says that a worn U-joint can cause a clicking sound only when driving the vehicle in reverse. Which technician is correct?
   a. A only
   b. B only
   c. Both A and B
   d. Neither A nor B

6. Incorrect or unequal U-joint working angles are most likely to be caused by
   a. a bent drive shaft.
   b. a collapsed engine or transmission mount.
   c. a dry output shaft spline.
   d. defective or damaged U-joints.

7. A defective outer CV joint will usually make a
   a. rumbling noise.
   b. growling noise.
   c. clicking noise.
   d. clunking noise.

8. The last step after installing a replacement CV joint or boot is to
   a. "burp the boot."
   b. lubricate the CV joint with chassis grease.
   c. mark the location of the boot on the drive axle shaft.
   d. separate the CV joint before installation.

9. Two technicians are discussing CV joints. Technician A says that the entire front suspension has to be disassembled to remove most CV joints. Technician B says that most CV joints are bolted onto the drive axle shaft. Which technician is correct?
   a. A only
   b. B only
   c. Both A and B
   d. Neither A nor B

10. The splines of the drive shaft yoke should be lubricated to prevent
    a. a vibration.
    b. spline bind.
    c. rust.
    d. transmission fluid leaks from the extension housing.

STEERING SYSTEM OPERATION, DIAGNOSIS, AND SERVICE

OBJECTIVES

After studying Chapter 13, the reader will be able to:

1. Identify steering system components.
2. Describe the operation of a power rack and pinion steering system.
3. Describe how to perform a dry park test to determine the condition of steering system components.
4. Perform an under-the-vehicle inspection of the steering system components.
5. Explain how to replace steering linkage parts.
6. Describe manual and power steering diagnosis and repair procedures.

The steering system consists of those parts required to turn the front wheels with the steering wheel.

STEERING COLUMNS

The typical vehicle requires about three complete revolutions (turns) of the steering wheel to rotate the front wheels from full left to full right. The front wheels rotate up to 45° while turning. The steering wheel is bolted to a splined shaft in the steering column. Since the late 1960s, the steering column and shaft are sec-

tional and are designed to collapse in the event of a collision. See Figure 13–1 for an example of the many different sections and parts contained in a typical steering column.

INTERMEDIATE SHAFT

Steering forces are transferred from the steering column to an **intermediate shaft** and a **flexible coupling**. The flexible coupling is also called a **rag joint** or **steering coupling disc** and is used to insulate noise, vibration, and harshness from being transferred from the steering components up through the steering column to the driver (see Figures 13–2 and 13–3). Many steering systems use one or more U-joints between the steering column and the steering gear as shown in Figure 13–4.

CONVENTIONAL STEERING GEAR

The rotation of the steering wheel is transferred to the front wheels through a steering gear and linkage. The intermediate shaft is splined to a **worm gear** inside a conventional steering gear. Around the worm gear is a nut with gear teeth that meshes with the teeth on a section of a gear called a **sector gear**. The sector gear is

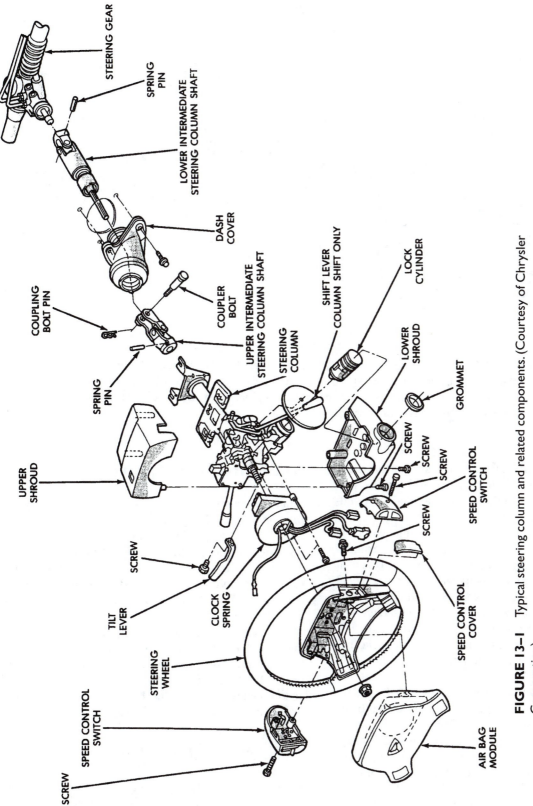

FIGURE 13–1 Typical steering column and related components. (Courtesy of Chrysler Corporation)

STEERING GEAR

SPRING PIN

LOWER INTERMEDIATE STEERING COLUMN SHAFT

DASH COVER

COUPLING BOLT PIN

COUPLER BOLT

SPRING PIN

UPPER INTERMEDIATE STEERING COLUMN SHAFT

STEERING COLUMN

SHIFT LEVER COLUMN SHIFT ONLY

LOCK CYLINDER

LOWER SHROUD

GROMMET

SCREW

SCREW

SCREW

SPEED CONTROL SWITCH

SPEED CONTROL COVER

SCREW

UPPER SHROUD

SCREW

TILT LEVER

CLOCK SPRING

STEERING WHEEL

SPEED CONTROL SWITCH

SCREW

AIR BAG MODULE

221

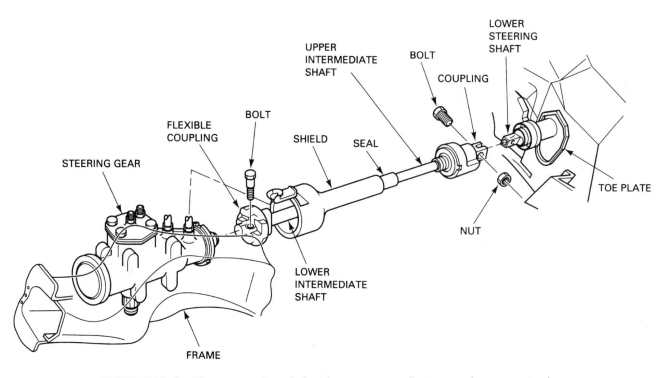

FIGURE 13–2 The intermediate shaft is the name generally given to the connection between the steering column at the bulkhead (toe plate) and the steering gear. Note that the two sections of the intermediate shaft are designed to collapse one inside the other in the event of a collision. (Courtesy of Oldsmobile Division).

part of a **pitman shaft,** also known as a **sector shaft** (see Figure 13–5). Ball, roller, or needle bearings support the sector shaft and the worm gear shaft, depending on the make and model of gear assembly.

As the steering wheel is turned, the movement is transmitted through the steering gear to an arm attached to the bottom end of the pitman shaft. This arm is called the **pitman arm**. Whenever the steering wheel is turned, the pitman arm moves. Since the late 1950s and early 1960s, most manufacturers have used a recirculating ball nut type of design. The term *recirculating ball* comes from the series of ball bearings placed between the input worm shaft and the ball nut (see Figures 13–6 and 13–7).

As the steering wheel turns, the worm shaft rolls inside a set of ball bearings. The movement of the bearings causes the ball nut to move. The ball nut has gear teeth that mesh with the gear teeth of the sector shaft. The sector gear and shaft rotate and move the pitman arm. The pitman arm is connected to steering linkage that moves and steers the front wheels. The rotating steel balls in the ball nut reduce friction by rolling in the nut along the groove in the steering shaft and through a ball guide.

STEERING GEAR RATIO

When the steering wheel is turned, the front wheels turn on their steering axis. If the steering wheel is rotated 20° and results in the front wheels rotating 1°, the steering gear ratio is 20:1 (read as "20 to 1"). The front wheels usually are able to rotate through 60 to 80° of rotation. A vehicle that turns three complete revolutions from full left to full right is said to have three turns *lock to lock.*

Most steering gears and some rack and pinion steering gears feature a **variable ratio**. This feature causes the steering ratio to decrease as the steering wheel is turned from the on-center position. The ratio change is accomplished by changing the length of the gear teeth on the sector gear.

STEERING LINKAGE

Steering linkage relays steering forces from the steering gear to the front wheels. Most conventional steering linkages use a **parallelogram**-type design. A parallelogram is a geometric box shape whose opposite sides are

(a)

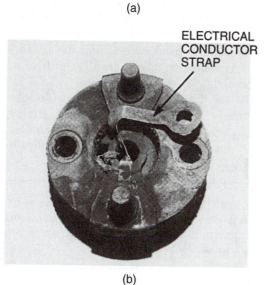

ELECTRICAL
CONDUCTOR
STRAP

(b)

FIGURE 13–4 U-joint used to change the angle of the steering shaft. Many times these U-joints become worn and can cause "jerky" or loose steering.

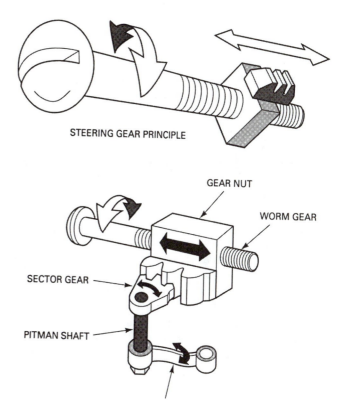

STEERING GEAR PRINCIPLE

GEAR NUT

WORM GEAR

SECTOR GEAR

PITMAN SHAFT

PITMAN ARM

FIGURE 13–5 As the steering wheel is turned, the nut moves up or down on the threads shown using a bolt to represent the worm gear and the nut representing the gear nut that meshes with the teeth of the sector gear. (Courtesy of FMC)

(c)

FIGURE 13–3 (a) Typical flexible coupling on an old pickup truck. (b) After removal of the joint, the electrical conductor strap provides the electrical ground for the horn ring. Because the flexible coupling isolates noise and vibration, the fabric also insulates the steering column shaft from the rest of the vehicle body. (c) Note the splines on the stub shaft of the steering gear. This flexible coupling was replaced because the splines in the coupling itself had rusted away, resulting in excessive play in the steering wheel.

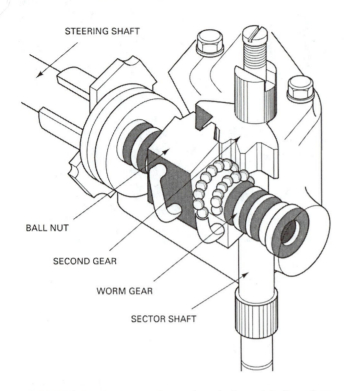

STEERING SHAFT

BALL NUT

SECOND GEAR

WORM GEAR

SECTOR SHAFT

FIGURE 13–6 Recirculating (moving) steel balls reduce the friction between the worm gear teeth and the gear nut. (Courtesy of FMC)

parallel and of equal distance. A parallelogram-type linkage uses two **tie-rods** (left and right), a **center link** (between the tie rods), an **idler arm** on the passenger side, and a **pitman arm** attached to the steering gear output shaft (pitman shaft) (see Figure 13–8).

As the steering wheel is rotated, the pitman arm is moved. The pitman arm attaches to a center link. At either end of the center link are inboard (inner) tie rods, adjusting sleeves, and outboard (outer) tie rods connecting to the steering arm which moves the front wheels. The passenger side of all these parts is supported and held horizontal by an idler arm that is bolted to the frame. The center link may be known by several names, including:

1. Center link
2. Connecting link
3. Connecting rod
4. Relay rod
5. Intermediate rod
6. Drag link (usually a truck term only)

Other types of steering linkage often used on light trucks and vans include the **cross-steer linkage** and the **Haltenberger linkage**. See Figure 13–9 for a com-

parison of parallelogram, cross-steer, and Haltenberger-type steering linkage arrangements.

NOTE: Many light trucks, vans, and some luxury cars use a steering dampener attached to the linkage. A *steering dampener* is similar to a shock absorber and it absorbs and dampens sudden motions in the steering linkage (see Figure 13–10).

Connections between all steering component parts are constructed of small ball-and-socket joints. These joints allow movement side to side to provide steering of both front wheels and the up-and-down joint movement that is required for normal suspension travel.

It is important that all these joints be lubricated with chassis grease through a **grease fitting**, also called a **zerk fitting**, at least every six months or according to the vehicle manufacturer's specifications.

NOTE: In 1922 the zerk fitting was developed by Oscar U. Zerk, an employee of the Alamite Corporation, a manufacturer of pressure lubrication equipment. A zerk or grease fitting is also known as an *Alamite fitting*.

Some vehicles come equipped with sealed joints and do not require periodic servicing.

Since the early 1980s, Ford Motor Company has used tie rod ends that use rubber bonded to the steel ball stud. Since there is no sliding friction inside the tie rod end, no lubrication is needed or required. This type of tie rod end is called a **rubber-bonded socket (RBS)** (see Figure 13–11). *RBS tie rods should never be lubricated.*

MANUAL RACK-AND-PINION STEERING

A rack-and-pinion steering unit consists of a **pinion gear** that is in mesh with a flat gear called a **rack**. The ends of the rack are connected to the front wheels through tie rods. Turning the steering wheel rotates the pinion and causes the rack to move left and right in the housing. The rack housing attaches to the body or frame of the vehicle by rubber bushings to help isolate noise and vibration. The steering forces act in a straight line and provides direct steering action with no lost motion (see Figures 13–12 and 13–13).

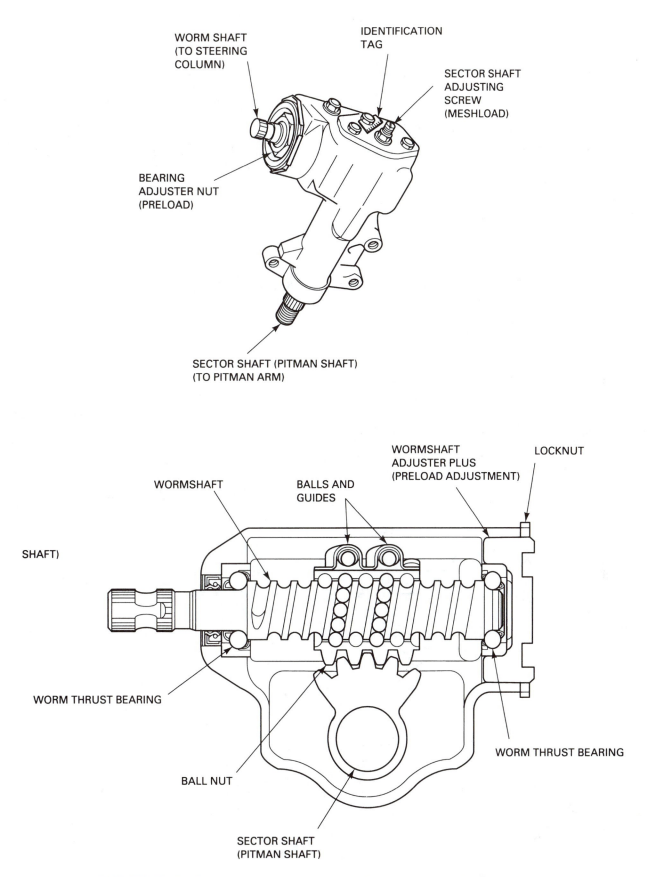

FIGURE 13–7 Recirculating ball steering gear showing worm shaft support bearings, worm shaft adjuster, and related parts. (Courtesy of Ford)

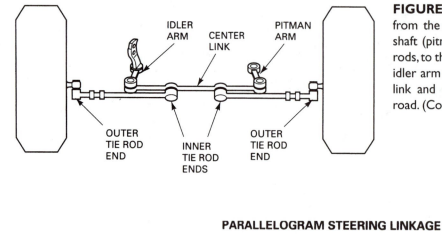

FIGURE 13–8 Steering movement is transferred from the pitman arm that is splined to the sector shaft (pitman shaft), through the center link and tie rods, to the steering knuckle at each front wheel. The idler arm supports the passenger side of the center link and keeps the steering linkage level with the road. (Courtesy of Dana Corporation)

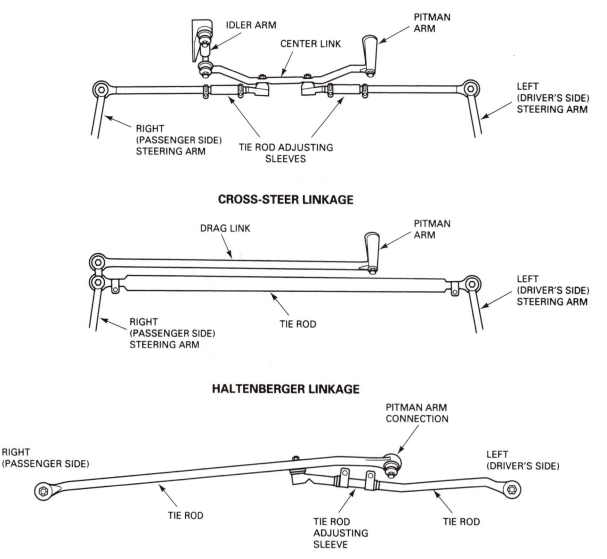

FIGURE 13–9 Types of steering linkage—parallelogram steering linkage is commonly used on most rear-wheel-drive passenger cars and light trucks. The cross-steer and Haltenberger linkage designs are used on some trucks and vans.

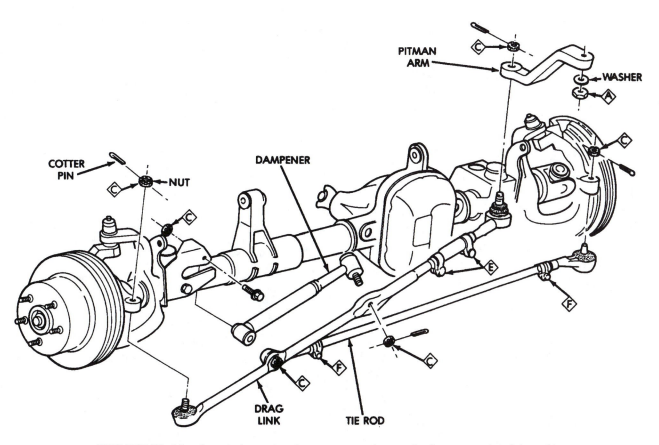

FIGURE 13–10 A typical steering dampener attaches to the frame or axle of the vehicle at one end and the steering linkage at the other end. (Courtesy of Chrysler Corporation)

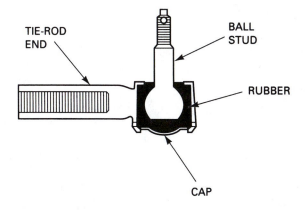

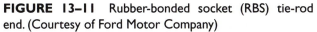

FIGURE 13–11 Rubber-bonded socket (RBS) tie-rod end. (Courtesy of Ford Motor Company)

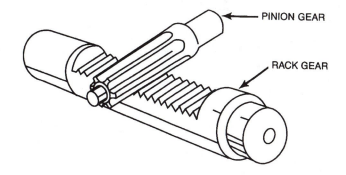

FIGURE 13–12 Basic principle and parts of a rack-and-pinion steering unit. (Courtesy of Dana Corporation)

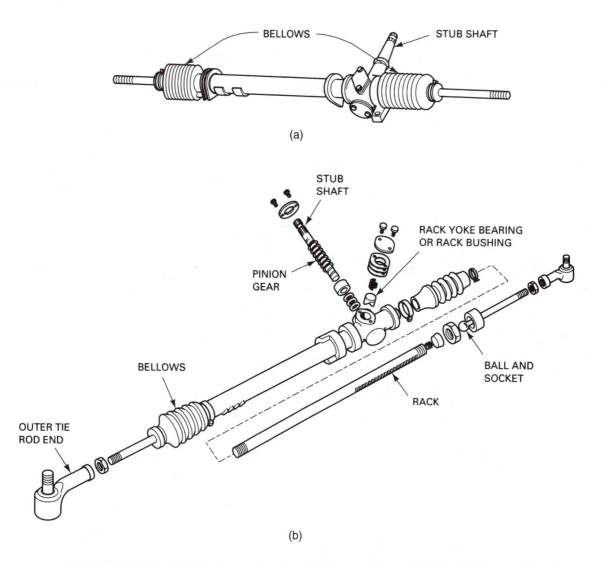

FIGURE 13–13 (a) Typical manual rack-and-pinion steering unit. (b) Exploded view of a manual rack-and-pinion assembly. The rubber bellows (boots) prevent dirt and moisture from getting around the rack-and-pinion gears. (Courtesy of Moog)

Rack-and-pinion steering is lightweight and small in size. The inner tie rod ends attach to the rack with a ball-and-socket joint. This ball and socket is retained to the end of the rack by a soft pin, a roll pin, or a stacked flange (see Figure 13–14).

POWER STEERING PUMPS

Power-assisted steering, commonly called *power steering*, allows the use of faster steering ratios and reduced steering wheel turning force. A typical power steering system requires only 2 to 3.5 lb (9 to 16 N) of effort to turn the steering wheel.

Most power steering systems use an engine-driven hydraulic pump. Power steering hydraulic pumps are usually belt driven from the front crankshaft pulley of

the engine as shown in Figure 13–15. The power steering pump drive pulley is usually fitted to a chrome-plated shaft with a press fit. The shaft is applied to a rotor with vanes that rotate between a **thrust plate** and a **pressure plate** (see Figure 13–16).

Some power steering pumps are of the slipper or roller design instead of the vane type (see Figure 13–17). When the engine starts, the drive belt rotates the power steering pump pulley and the rotor assembly inside the power steering pump (see Figure 13–18 for an example of a slipper-type pump).

With a vane-type pump, centrifugal force and hydraulic pressure push the vanes of the rotor outward into contact with the **pump ring**. The shape of the pump ring causes a change in the volume of fluid between the vanes. As the volume increases, the pressure

Inner Tie Rod End (Ball Socket)

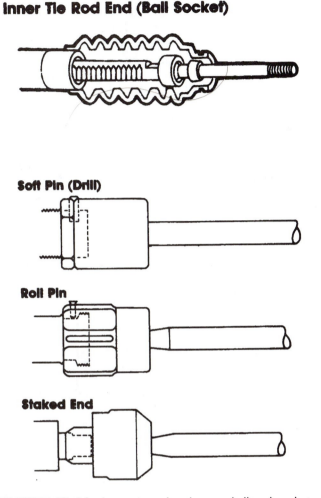

Soft Pin (Drill)

Roll Pin

Staked End

FIGURE 13-14   Inner tie-rod ends use a ball-and-socket joint to allow the tie-rod to move up and down as the wheels move up and down. Three methods used to retain the ball socket to the end of the rack include a soft metal pin (that can be easily drilled out), a roll pin (that must be pulled out), or a staked end. (Courtesy of Moog)

is decreased in the space between the vanes and draws in fluid from the pump reservoir. When the volume between the vanes decreases, the pressure is increased and flows out the pump discharge port (see Figure 13-19).

FLOW CONTROL VALVE OPERATION

Between the pump discharge ports and the outlet fitting of the pump is the **flow control valve**. Operation of the flow control valve is determined by the needs of the power steering gear or rack. When the engine speed is increased, the volume and pressure of the

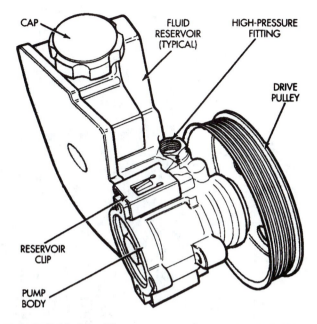

FIGURE 13-15   This power steering pump uses a reservoir that is clipped to the pump body. (Courtesy of Chrysler Corporation)

power steering pump increases (see Figure 13-20). Recirculating the power steering fluid back through the pump reduces the power required by the pump and keeps the fluid temperature from increasing while supplying adequate flow for proper power steering operation.

When the steering wheel is turned all the way to the left or right, the valve in the steering gear is closed off, shutting off flow from the pump. This causes the pressure to rise rapidly. A pressure relief valve inside the flow control valve prevents the pressure from building higher and acts as a safety valve.

INTEGRAL POWER STEERING GEAR OPERATION

With power-assisted steering, the driver effort is always in proportion to the force necessary to turn the front wheels. The hydraulic pump pressurizes the power steering fluid and sends it through a pressure hose to the steering gear. Manual steering is available if the engine stops running or if there is a loss of hydraulic assist.

Power steering gears are usually called **integral gears** because the power piston or actuator for the power assist is inside (integrated) the steering gearbox. The driver's steering effort is transferred from the steering wheel through the steering column and

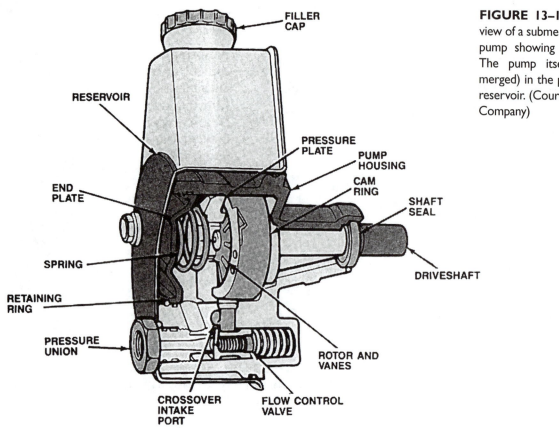

FILLER CAP

RESERVOIR

PRESSURE PLATE

PUMP HOUSING

END PLATE

CAM RING

SHAFT SEAL

SPRING

DRIVESHAFT

RETAINING RING

PRESSURE UNION

ROTOR AND VANES

CROSSOVER INTAKE PORT

FLOW CONTROL VALVE

FIGURE 13–16 Cross-sectional view of a submerged power steering pump showing the interior parts. The pump itself is inside (submerged) in the power steering fluid reservoir. (Courtesy of Ford Motor Company)

ROLLER TYPE

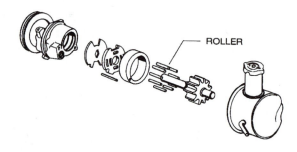

ROLLER

FIGURE 13–17 Exploded views of the three types of power steering pumps showing the design and shape of the pumps and reservoirs. (Courtesy of Moog)

VANE TYPE

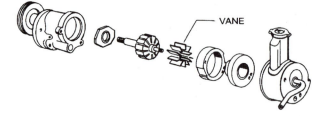

VANE

SLIPPER TYPE

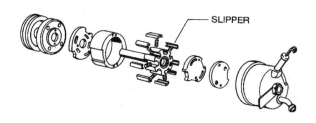

SLIPPER

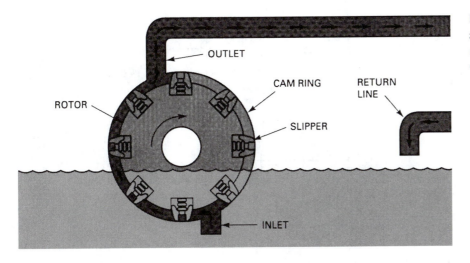

FIGURE 13–18 The operation of a slipper-type pump in a submerged reservoir. (Courtesy of Ford Motor Company)

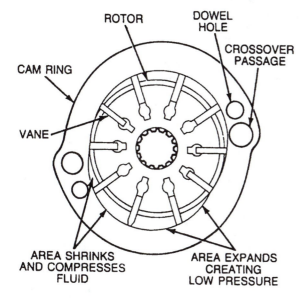

FIGURE 13–19 Fluid is being driven into the pump around the vanes as volume increases and the pressure is lowered. As the vanes rotate around toward a smaller volume, the pressure of the fluid trapped between the vanes increases as it approaches the discharge port. (Courtesy of Chrysler Corporation)

intermediate shafts to the **stub shaft** or **input shaft** of the gear assembly (see Figure 13–21). The stub shaft connects to the rotary (spool) valve inside the steering gear and directs and controls the flow of pressurized power steering fluid within the gear assembly. The stub shaft is sometimes called the **spool shaft**. The other end of the valve is connected to the worm gear through the torsion bar.

As the driver applies force to the steering wheel, the resistance of the tires on the road surface creates a resistive force and the torsion bar twists. This causes a change between the input shaft and the control sleeve which restricts the flow of power steering fluid and directs the high-pressure fluid to one end of the piston in the gear housing. This high pressure forces the piston to move the sector gear and assists the turning effort (see Figure 13–22).

POWER RACK-AND-PINION STEERING

Power rack-and-pinion steering is used in many passenger cars as well as in light trucks and vans. Its light weight and small size makes it possible to be mounted in a variety of locations. There are two basic designs of rack and pinion used today: the **end takeoff** and the **center takeoff** (see Figures 13–23 and 13–24).

The power steering pump supplies pressurized hydraulic fluid to the top or "hat" section of the unit. The steering column attaches to the stub shaft and turns a rotary spool valve just as in a conventional integral power steering gear. The spool valve assembly directs the pressurized fluid to one side of the rack piston as shown in Figure 13–25. Fluid from the other side of the rack piston returns to the spool valve area as "return" fluid (see Figure 13–26). Some power rack-and-pinion steering units use a variable-ratio rack (see Figure 13–27).

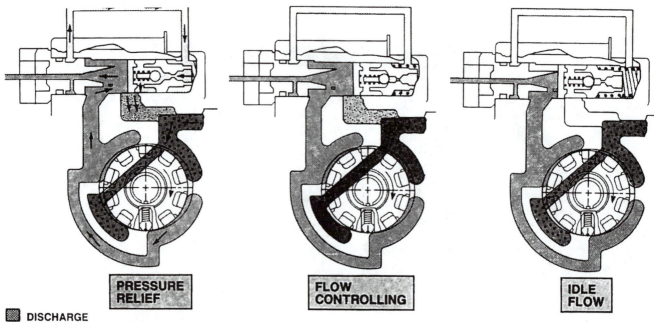

PRESSURE RELIEF

FLOW CONTROLLING

IDLE FLOW

DISCHARGE
INTAKE
STATIC
FLOW BYPASS

FIGURE 13–20 Flow control valve operation at idle while controlling the flow at higher engine speeds and during pressure relief. (Courtesy of Ford Motor Company)

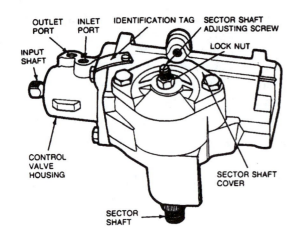

OUTLET PORT
INLET PORT
IDENTIFICATION TAG
SECTOR SHAFT ADJUSTING SCREW
LOCK NUT
INPUT SHAFT
CONTROL VALVE HOUSING
SECTOR SHAFT COVER
SECTOR SHAFT

FIGURE 13–21 Typical integral power steering gear. (Courtesy of Ford Motor Company)

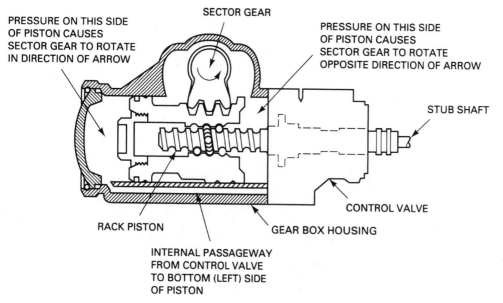

PRESSURE ON THIS SIDE OF PISTON CAUSES SECTOR GEAR TO ROTATE IN DIRECTION OF ARROW

SECTOR GEAR

PRESSURE ON THIS SIDE OF PISTON CAUSES SECTOR GEAR TO ROTATE OPPOSITE DIRECTION OF ARROW

STUB SHAFT

CONTROL VALVE

RACK PISTON

GEAR BOX HOUSING

INTERNAL PASSAGEWAY FROM CONTROL VALVE TO BOTTOM (LEFT) SIDE OF PISTON

FIGURE 13–22 Forces acting on the rack piston of an integral power steering gear.

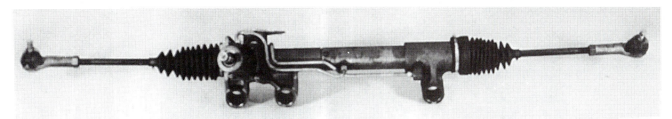

FIGURE 13–23 Typical end-takeoff power rack-and-pinion steering gear. The tie-rods connect to the ends of the rack with ball sockets.

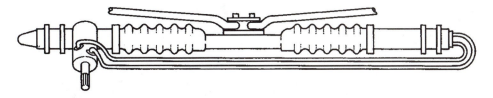

FIGURE 13–24 Typical center-takeoff power rack-and-pinion steering gear. The long tie-rods bolt to the center of the rack with step bolts that allow the outward ends of the tie-rods to move freely up and down.

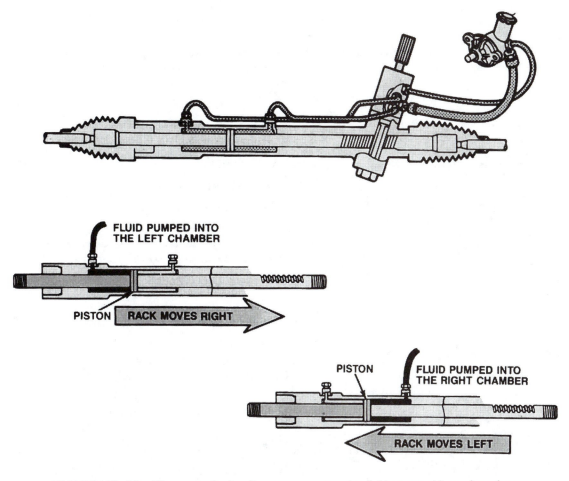

FLUID PUMPED INTO THE LEFT CHAMBER

PISTON RACK MOVES RIGHT

PISTON FLUID PUMPED INTO THE RIGHT CHAMBER

RACK MOVES LEFT

FIGURE 13–25 The control valve directs power steering fluid to one side or the other of the piston on the rack. Power steering fluid from the other side of the rack piston returns to the valve assembly, where it is directed back to the pump reservoir. (Courtesy of Ford Motor Company)

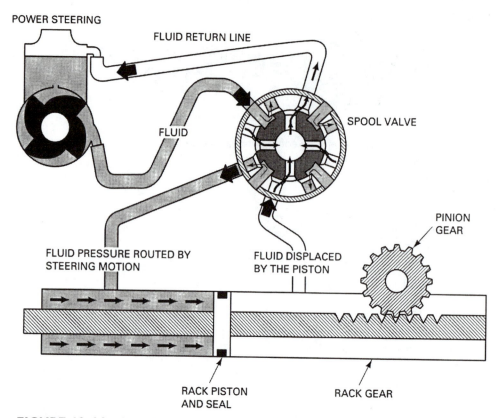

FIGURE 13–26 An end view of the control (spool) valve showing how pressurized power steering fluid is routed to one side of the rack piston. (Courtesy of FMC)

Variable Steer Rack
Gear Tooth Pitch

Variable Steer Rack &
Pinion Gear Mesh

FIGURE 13–27 Variable ratio in a rack-and-pinion steering unit is accomplished by varying the gear tooth pitch on the rack. (Courtesy of TRW Inc.)

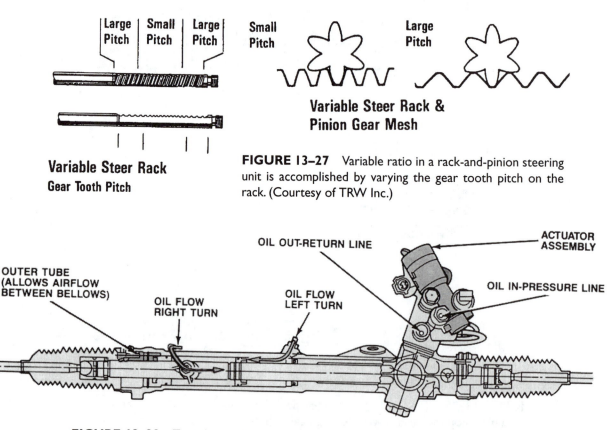

FIGURE 13–28 Typical variable-effort power rack-and-pinion steering unit with electrical actuator assembly. (Courtesy of Ford Motor Company)

VARIABLE-EFFORT POWER STEERING

Variable-effort power steering uses an electrical control solenoid or motor to control the power steering pump output volume. Two basic designs are used today: the **variable-effort** and the **two-flow** or **switched type**. See Figure 13–28, showing a variable system that uses an electric stepper motor-type actuator. Both systems control the outlet flow rate between the power steering pump and the steering gear or rack-and-pinion assembly.

Some variable-effort power steering systems *gradually* change the amount of power steering assist as the vehicle speed changes, thereby providing a less noticeable change in steering effort. Besides vehicle speed, many variable-effort systems use a **steering wheel rotation sensor** to measure rapid steering wheel movement (see Figure 13–29).

The default mode of most variable-effort power steering is to allow maximum power assist at all speeds. If, for example, the electrical connector were to become disconnected from the output actuator of the power steering pump, the orifice size is set to its largest and maximum steering assist is available.

POWER STEERING COMPUTER SENSOR

The typical power steering sensor switch is a pressure switch that completes an electrical circuit to the computer (controller) whenever the power steering pressure exceeds a certain point, usually about 300 psi (2070 kPa). As the pressure of the power steering increases, the load on the engine increases.

FOUR-WHEEL STEERING

Some vehicles are equipped with a system that steers all four wheels. Four terms are commonly used when discussing four-wheel steering:

1. **Passive rear steering.** *Passive* means that no steering wheel input is needed to cause a rear-wheel-steer effect. As the vehicle corners, forces on the suspension system allow a change in the rear toe angle. It is this slight change in toe of the rear wheels that contributes to a slight steering effect. (See Chapter 15 for details on toe and related alignment angles).

2. **Active rear steering.** *Active* means that either a mechanical, electrical, or hydraulic mechanism moves the wheels to change the rear toe.

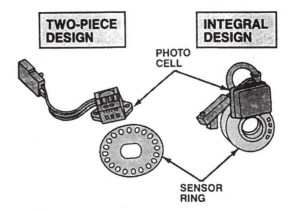

FIGURE 13–29 Two designs of steering angle sensors. These sensors can be used by the controller (computer) to vary the amount of power steering assist and for electronic ride control if the vehicle is equipped. (Courtesy of Ford Motor Company)

◀ DIAGNOSTIC STORY ▶

SIMPLE ANSWER TO A HARD PROBLEM

An owner of a Japanese sports coupe complained of stiff steering. While driving straight ahead, the steering felt stiff but would suddenly loosen when the steering wheel was turned slightly. This caused the car to dart back and forth and made it difficult to track straight ahead. After many wheel alignments and changes of tires, the real problem was discovered. The U-joint in the steering shaft between the steering column and the flexible steering shaft coupling was defective and binding. This problem was discovered after all other steering components had been replaced at great expense, including replacement of the rack-and-pinion steering assembly. It is always best to check everything, including the obvious and simplest items first before starting to replace parts, especially expensive parts.

3. **Same-phase steering.** Same-phase steering means that the front and rear wheels are steered in the same direction. Same-phase steering improves steering response, especially during rapid-lane changing (see Figure 13–30).

4. **Opposite-phase steering.** This is also called **negative-phase mode**. Opposite-phase steering is when the front wheels and rear wheels are steered in the opposite direction, as shown in Figure 13–30. Opposite-phase steering will quickly

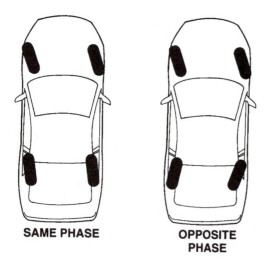

SAME PHASE OPPOSITE PHASE

FIGURE 13–30

change the vehicle's direction but may cause a feeling of oversteering.

Opposite-phase steering is best at low speeds, and same-phase steering is best for higher-speed handling and lane-change maneuvers. (See Figure 13–31 for examples of several types of four-wheel steering systems).

STEERING LINKAGE LUBRICATION

Keeping all greasable joints properly lubricated is necessary for long life and ease of steering (see Figure 13–32). During a chassis lubrication, do not forget to put grease on the **steering stop,** if there is one. Steering stops refer to the projections or built-up areas on the control arms of the front suspension to limit the steering movement at full lock (see Figure 13–33). When the steering wheel is turned as far as it can be turned, the steering should *not* stop inside the steering gear!

◀ **TECH TIP** ▶

THE KILLER B'S

The three B's that can cause steering and suspension problems are bent, broken, or binding components. Always inspect each part under the vehicle for each of the killer B's.

NOTE: Many rack-and-pinion steering units are designed with a rack travel limit internal stop and do not use an external stop on the steering knuckle or control arm.

The steering stops should be lubricated to prevent a loud grinding noise when turning while the vehicle is going over a bump. This noise is usually noticeable when turning into or out of a driveway.

DRY PARK TEST

Since many steering (and suspension) components do *not* have *exact* specifications for replacement purposes, it is extremely important that the beginning service technician work closely with an experienced veteran technician. While most technicians can determine when a steering component such as a tie rod end is definitely in need of replacement, marginally worn parts are often hard to spot and can lead to handling problems. One of the most effective, yet easy-to-perform steering component inspection methods is called the **dry park test**.

This simple test is performed with the vehicle on the ground or on a drive-on ramp-type hoist. The steering wheel is moved back and forth *slightly* while an assistant feels for movement at each section of the steering system. The technician can start checking for looseness in the steering linkage starting either at the outer tie rod ends and working toward the steering column, or from the steering column toward the outer tie rod ends. It is important to check every joint and component of the steering system (see Figure 13–34).

UNDER-VEHICLE INSPECTION

After checking the steering system components as part of a dry park test, hoist the vehicle and perform a thorough part-by-part inspection. This thorough inspection includes:

1. Inspect each part for damage due to accident or bent parts due to the vehicle hitting an object in the roadway.

CAUTION: Never straighten a bent steering linkage; always replace with new parts.

2. Idler arm inspection is performed by using *hand* force of 25 lb (110 N) up and down on the arm.

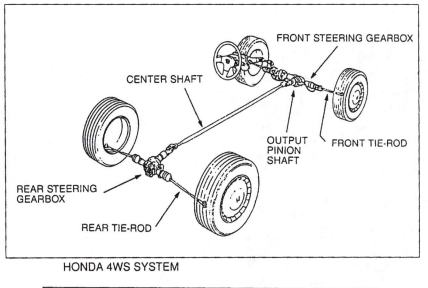

HONDA 4WS SYSTEM

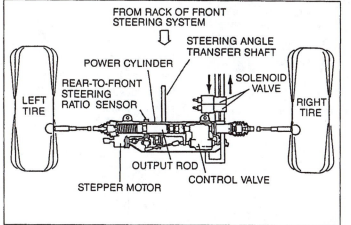

MAZDA REAR STEERING SYSTEM

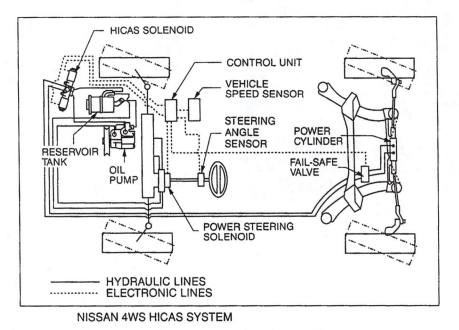

NISSAN 4WS HICAS SYSTEM

FIGURE 13–31 (Courtesy of Dana Corporation)

FIGURE 13–32 Greasing a tie-rod end. Some joints do not have a hole for excessive grease to escape, and excessive grease can destroy the seal.

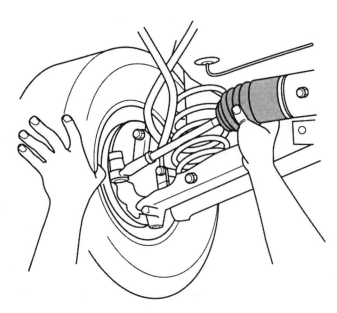

FIGURE 13–34 All joints should be felt during a dry park test. Even inner tie-rod ends (ball socket assemblies) can be felt through the rubber bellows on many rack-and-pinion steering units.

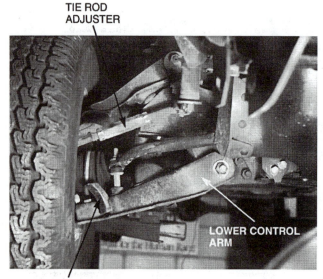

FIGURE 13–33 Steering stops should be lubricated by applying a coating of grease.

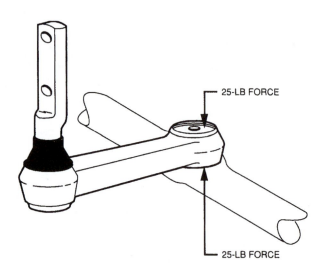

FIGURE 13–35 (Courtesy of Dana Corporation)

If the *total* movement exceeds ¼ in. (6 mm), the idler arm should be replaced (see Figure 13–35).

3. All other steering linkages should be tested *by hand* for any vertical or side-to-side looseness. It

is normal for tie rods to rotate in their sockets when the tie rod sleeve is rocked. *There should be no end play in any tie-rod.* Many tie-rods are spring loaded to help keep the ball-and-socket joint free of play as the joint wears.

4. All steering components should be tested with the wheels in the straight-ahead position. If the wheels are turned, some apparent looseness may be noticed due to the angle of the steering linkage.

◀ **DIAGNOSTIC STORY** ▶

BUMP STEER

Bump steer or **orbital steer** is a term used to describe what occurs when the steering linkage is not level, causing the front tires to turn inward or outward as the wheels and suspension move up and down. (Automotive chassis engineers call it **roll steer**.) The vehicle's direction is changed *without moving the steering wheel* whenever the tires move up and down over bumps, dips in the pavement, or even over gentle rises!

This author experienced bump steer once and will never forget the horrible feeling of not having control of the vehicle. After replacing an idler arm and aligning the front wheels, everything was okay until about 40 mph (65 km/h); then the vehicle started darting from one lane of the freeway to another. Because there were no "bumps" as such, bump steer was not considered as a cause. Even when holding the steering wheel perfectly still and straight ahead, the vehicle would go left, then right. Did a tie rod break? It certainly felt exactly like that's what happened. I slowed down to below 30 mph and returned to the shop.

After several hours of checking everything, including the alignment, I discovered that the idler arm was not level with the pitman arm. This caused a pull on the steering linkage whenever the suspension moved up and down. As the suspension compressed, the steering linkage pulled inward on the tie rod on that side of the vehicle. As the wheel moved inward (toed-in), it created a pull just as if the wheel had been turned by the driver.

This is why all steering linkage *must* be parallel with the lower control. The reason for the bump steer was that the idler arm bolted to the frame which was slotted vertically. I didn't pay any attention to the location of the original idler arm and simply bolted the replacement to the frame. After raising the idler arm back up where it belonged [about ½ in. (13 mm)] the steering problem was corrected.

Other common causes of bump steer are worn or deteriorated rack mounting bushings, noncentered steering linkage, or a bent steering linkage. If the steering components are not level, any bump or dip in the road will cause the vehicle to steer one direction or the other. Always check the steering system carefully whenever a customer complains about any "weird" handling problem.

CAUTION: Do not turn the front wheels of the vehicle while suspended on a lift to check for looseness in the steering linkage. The extra leverage of the wheel and tire assembly can cause a much greater force being applied to the steering components than can be exerted by hand alone. This extra force may cause some apparent movement in good components that may not need to be replaced.

STEERING LINKAGE REPLACEMENT

When replacing any steering system component, it is best to replace all defective and marginally good components at the same time. Replacing steering system components involves these steps:

1. Hoist the vehicle safely with the wheels in the straight-ahead position. Remove the front wheels, if necessary, to gain access to the components.

2. Loosen the retainer nut on tapered components such as tie-rod ends. Use a tie-rod removal puller or *taper breaker*, as shown in Figure 13–36, or use hammers to deform the taper slightly as shown in Figure 13–37.

CAUTION: Vehicle manufacturers often warn not to use a tapered "pickle fork" tool to separate tapered parts. The wedge tool can tear the grease seal and damage both the part being removed and the adjoining part.

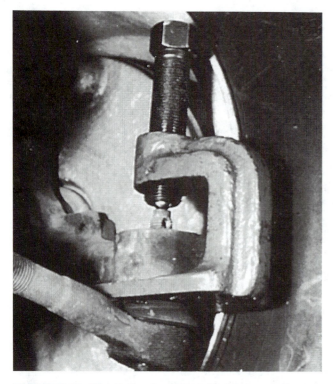

FIGURE 13–36 Puller (taper breaker) being used to remove a tie-rod end.

FIGURE 13–37 Two hammers being used to disconnect a tie-rod end from the steering knuckle. One hammer is used as a backing for the second hammer . Notice that the attaching nut has been loosened but not removed. This prevents the tie-rod end from falling when the tapered connection is knocked loose.

Pitman arms require a large puller to remove the pitman arm from the splines of the pitman shaft (see Figure 13–38).

3. Replace the part using the hardware and fasteners supplied with the replacement part. *Do not reuse precrimped torque prevailing nuts used*

FIGURE 13–38 Using a puller to remove the pitman arm from the pitman shaft.

at the factory as original equipment on many tie-rod ends.

CAUTION: Whenever tightening the nuts of tapered parts such as tie-rods, *do not* loosen after reaching the proper assembly torque to align the cotter key hole. If the cotter key does not fit, *tighten* the nut further until the hole lines up for the cotter key.

When replacing tie-rod ends, use the adjusting sleeve to adjust the total length of the tie-rod to the same position and length as the original. Measure the original length of the tie rods and assemble the replacement tie rod(s) to the same overall length (see Figures 13–39 and 13–40).

Inner tie-rod end assemblies used on rack-and-pinion steering units require special consideration and often special tools. The inner tie-rod end, also called a **ball socket assembly**, should be replaced whenever there is any noticeable free play in the ball-and-socket joint. The inner tie-rod assemblies are attached to the end of the steering rack by one of several methods including:

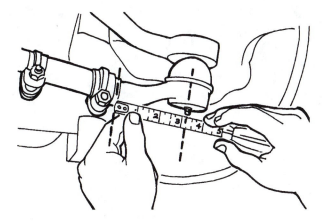

FIGURE 13–39 Replacement tie-rods should be installed the same length as the original. Measure from the edge of the tie-rod sleeve to the center of the grease fitting. When the new tie-rod is threaded to this dimension, the toe setting will be close to the original. (Courtesy of Dana Corporation)

1. **Staked.** This method is common on Saginaw-style rack-and-pinion steering units found on General Motors and many Chrysler vehicles. Use two wrenches to remove, as shown in Figure 13–41. The flange around the outer tie-rod must be restaked to the flat on the end of the rack, as shown in Figures 13–42 and 13–43.

2. **Riveted or pinned.** Commonly found on Ford vehicles (see Figure 13–44). Some roll pins require a special puller or the pin can be drilled out. Many styles use an aluminum rivet. A special very deep socket or a large open end wrench can usually be used to shear the aluminum rivet by unscrewing the socket assembly from the end of the rack while the rack-and-pinion unit is still in the vehicle.

HINT: When replacing a rack-and-pinion assembly, specify a long rack rather than a short rack. A short rack does not include the bellows (boots) or inner tie-rod ends (ball socket assemblies). The labor and cost required to exchange or replace these parts usually makes it less expensive and easier to replace the entire steering unit.

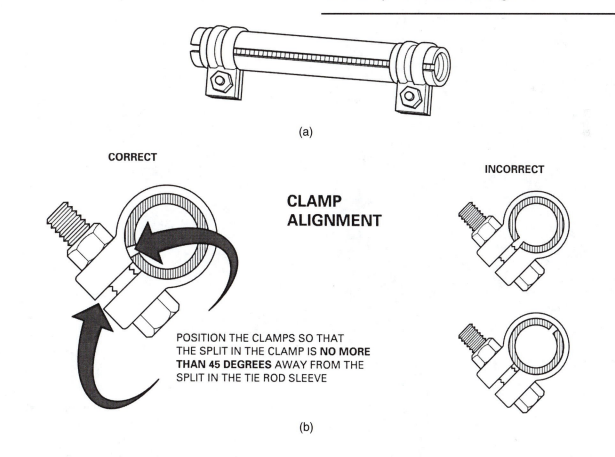

FIGURE 13–40 (a) Tie-rod adjusting sleeve. (Courtesy of Dana Corporation) (b) Be sure to position the clamp correctly on the sleeve. (Courtesy of FMC)

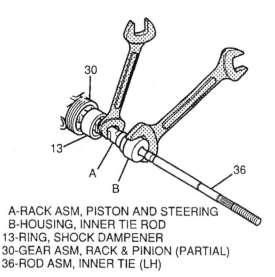

A-RACK ASM, PISTON AND STEERING
B-HOUSING, INNER TIE ROD
13-RING, SHOCK DAMPENER
30-GEAR ASM, RACK & PINION (PARTIAL)
36-ROD ASM, INNER TIE (LH)

FIGURE 13–41 Removing a staked inner tie-rod assembly requires two wrenches: one to hold the rack and the other to unscrew the joint from the end of the steering rack. (Courtesy of Oldsmobile)

◄ TECH TIP ►

JOUNCE/REBOUND TEST

All steering linkages should be level and "work" at the same angle as the suspension arms, as shown in Figure 13–45. A simple test to check these items is performed as follows:

1. Park on a level, hard surface with the wheels straight ahead and the steering wheel in the "unlocked" position.
2. Bounce or "jounce" the vehicle up and down at the front bumper while watching the steering wheel. The steering wheel should *not* move.

If the steering wheel moves while the vehicle is being bounced, look for a possible bent steering linkage, suspension arm, or steering rack.

POWER STEERING DIAGNOSIS AND TROUBLESHOOTING

Power steering systems are generally very reliable, yet many problems are caused by not correcting simple service items such as:

FIGURE 13–42 As the inner tie-rod end is unthreaded from the rack, the staked socket housing bends.

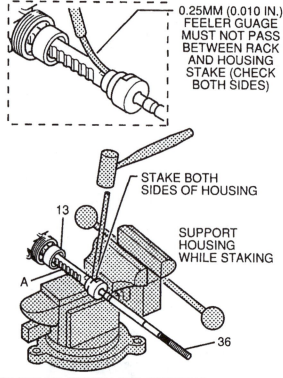

A-RACK ASM, PISTON AND STEERING
13-RING, SHOCK DAMPENER
36-ROD ASM, INNER TIE (LH)

FIGURE 13–43 When the inner tie-rod end is reassembled, both sides of the housing must be staked down onto the flat shoulder of the rack. (Courtesy of Oldsmobile)

1. **A loose, worn, or defective power steering drive belt (including serpentine belts).** This can cause "jerky" steering and belt noise especially when turning.

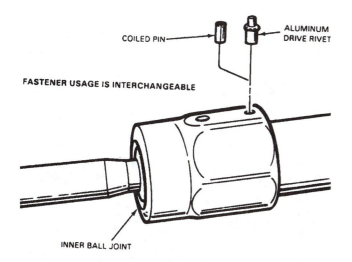

FIGURE 13–44 Many inner tie-rod ends (ball socket assemblies) are pinned or riveted to the end of the rack. (Courtesy of Dana Corporation)

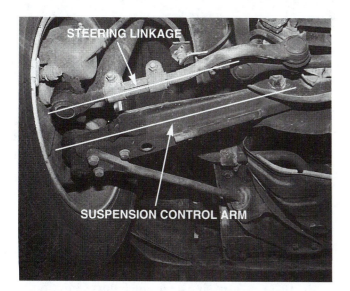

FIGURE 13–45 The steering and suspension arms must remain parallel to prevent the up-and-down motion of the suspension from causing the front wheels to turn inward or outward.

NOTE: Do not guess at the proper belt tension. Always use a belt tension gauge or observe the marks on the tensioner as shown in Figures 13–46 and 13–47. Always apply force on the pump at the proper location to prevent damage to the pump (see Figure 13–48).

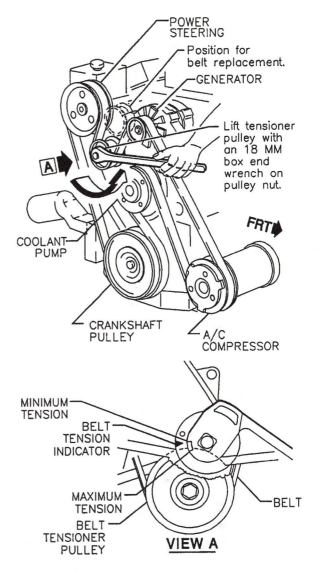

FIGURE 13–46 (Courtesy of Oldsmobile)

2. **A bent or misaligned drive pulley.** Usually this is caused by a collision or improper reassembly of the power steering pump after an engine repair procedure. This can cause a severe grinding noise whenever the engine is running and may sound like an engine problem.

3. **Low or contaminated power steering fluid.** Usually this is caused by a slight leak at the high-pressure hose or defective inner rack seals on a power rack-and-pinion power steering system. This can cause a loud whine noise and lack of normal power steering assist.

4. **Broken or loose power steering pump mounting brackets.** In extreme cases the pump

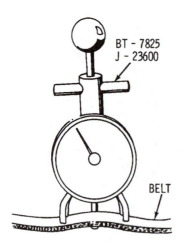

FIGURE 13–47   Belt tension gauge. Most vehicle manufacturers recommend a maximum tension of 170 lb (77 kg) for a new belt and a minimum tension of 90 lb (41 kg) for a used belt. (Courtesy of Oldsmobile)

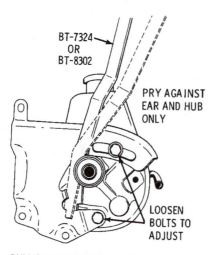

FIGURE 13–48   Most power steering pumps have a designated area to be used to pry against to tension the drive belt. Never pry against the sheet metal or plastic reservoir. (Courtesy of Oldsmobile)

mounting bolts can be broken. These problems can cause "jerky" steering. It is important to inspect the pump mounting brackets and hardware carefully when diagnosing a steering-related problem. The brackets tend to crack at the adjustment points and pivot areas. Tighten all the hardware to assure that the belt will remain tight and not slip, causing noise or a power-assist problem.

  5. **Underinflated tires.**

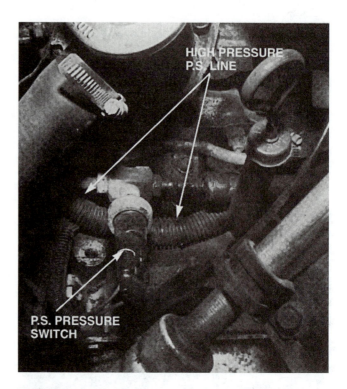

FIGURE 13–49   Typical power steering pressure switch installed in the high-pressure line. This switch closes and signals the engine computer whenever the pressure exceeds 300 psi (2000 kPa), indicating a moderate turn or a parking situation. The engine computer then raises the engine speed enough to compensate for the power steering load.

  6. **Engine idle speed below specifications.**
  7. **A defective power steering pressure switch.** If this switch fails, the computer will not increase engine idle speed while turning (see Figure 13–49).
  8. **Internal steering gear mechanical binding.**

As part of a complete steering system inspection and diagnosis, a steering wheel turning effort test should be performed as shown in Figure 13–50. The power steering force, as measured by a spring scale during turning, should be less than 5 lb (22 N).

POWER STEERING FLUID

The correct power steering fluid is *critical* to the operation and service life of the power steering system! The *exact* power steering fluid to use varies as to vehicle manufacturer and sometimes differences within vehicle

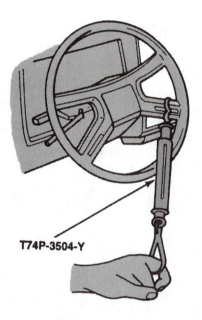

T74P-3504-Y

FIGURE 13–50 When diagnosing a hard steering complaint, use a spring scale to measure the amount of force needed to turn the steering wheel. The results should then be compared with specifications or compared to the effort of a similar vehicle. (Courtesy of Ford Motor Company)

manufacturers because of various steering component manufacturers.

NOTE: Remember, multiple-purpose power steering fluid does not mean *all*-purpose power steering fluid. Always consult the power steering reservoir cap, service manual, or owner's manual for the exact fluid to be used in the vehicle being serviced.

POWER STEERING FLUID FLUSHING PROCEDURE

Whenever there is any power steering service performed such as replacement of a defective pump or steering gear or rack-and-pinion unit, the entire system should be flushed. If all the old fluid is not flushed from the system, small pieces of a failed bearing or rotor could be circulated through the system. These metal particles can block paths in the control valve and cause failure of the new power steering pump or gear assembly.

◀ TECH TIP ▶

"PINKY TEST"

Whenever diagnosing any power steering complaint, always check the level *and* condition of the power steering fluid. Often this is best accomplished by putting your finger ("pinky") down into the power steering fluid reservoir and pulling it out to observe the texture and color of the fluid (see Figure 13–51).

A common problem with some power rack-and-pinion units is the wearing of grooves in the housing by the Teflon sealing rings of the spool (control) valve. When this wear occurs, aluminum particles become suspended in the power steering fluid, giving it a grayish color and also thickening the fluid.

Normally, clear power steering fluid that is found to be grayish in color and steering that is difficult when cold are clear indications as to what has occurred and why the steering is not functioning correctly.

NOTE: Some vehicles use power steering reservoir caps with *left-hand threads.* Always clean the top of the cap and observe and follow all directions and cautions. Many power steering pump reservoirs and caps have been destroyed by technicians attempting to remove a cap in the wrong direction using large pliers.

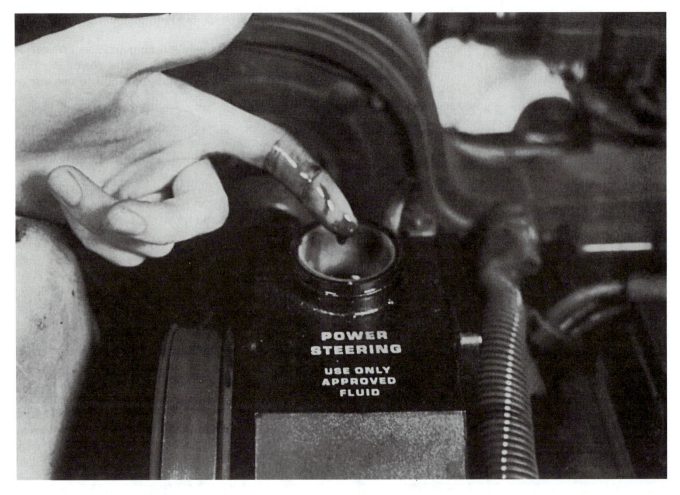

FIGURE 13–51

NOTE: Besides flushing the old power steering fluid from the system and replacing with new fluid, many technical experts recommend installing a filter in the low-pressure return line as an added precaution against serious damage from debris in the system. Power steering filters are commonly available through vehicle dealers parts departments as well as aftermarket sources from your local auto supply stores.

Two persons are needed to flush the system using the following steps:

Step 1. Remove the low-pressure return hose from the pump and plug the line fitting on the pump.

Step 2. Place the low-pressure return hose into an empty container.

Step 3. Fill the pump reservoir with fresh fluid and start the engine.

Step 4. As the dirty power steering fluid is being pumped into the container, keep the reservoir full of clean fluid as the assistant turns the steering wheel full lock one way to full lock the other way.

CAUTION: Never allow the pump reservoir to run dry of power steering fluid. Severe internal pump damage can result.

Step 5. When the fluid runs clean, stop the engine and reattach the low-pressure return hose to the pump reservoir.

Step 6. Restart the engine and fill the reservoir to the full mark. Turn the steering wheel lock to lock one or two times to bleed out any trapped air in the system.

BLEEDING AIR OUT OF THE POWER STEERING SYSTEM

If the power steering fluid is pink (if ATF is the power steering fluid) or tan (if clear power steering fluid is used), there may be air bubbles trapped in the fluid. Stop the engine and allow the air to "burp" out to the surface for several minutes. Lift the vehicle off the ground and then restart the engine and rotate the steering wheel. This method prevents the breakup of large air bubbles into thousands of smaller bubbles that are more difficult to bleed out of the system.

POWER STEERING PUMP SERVICE

Some power steering pump service can usually be performed without removing the pump, including:

1. Replacing the high-pressure and return hoses
2. Removing and cleaning the flow control valve assembly (see Figure 13–52).

Most power steering pump service requires the removal of the pump from the engine mounting and/or removal of the drive pulley.

NOTE: Most replacement pumps are not equipped with a pulley. The old pulley must be removed and installed on the new pump. The old pulley should be carefully inspected for dents, cracks, or warpage. If the pulley is damaged, it must be replaced.

The pulley must be removed and installed with a pulley removal and installation tool (see Figure 13–53).

CAUTION: Do not hammer the pump shaft or pulley in an attempt to install the pulley. The shock blows will damage the internal components of the pump.

ON-THE-VEHICLE STEERING GEAR SERVICE

Manual steering gears usually require SAE 80W-90 gear lube as the lubricant specified by most manufacturers. The fluid level should be one of the first items to check if hard steering is the problem. Power steering gears use

as a lubricant the power steering fluid that is circulated throughout the system.

Pitman Shaft Seal Replacement. A pitman shaft seal that is leaking is a major source of power steering leaks. Full system pressure is applied to these seals. (The Saginaw 800 series gear uses two seals: a single lip seal and a double lip seal toward the pitman arm end of the pitman shaft.) To replace the pitman seals on a typical steering gear, follow these steps:

Step 1. Hoist the vehicle safely and remove the pitman arm nut and pitman arm from the pitman shaft using an appropriate puller.

Step 2. Remove the snap ring and backup washer.

Step 3. Lower the vehicle and place the oil drain pan under the steering gear.

Step 4. Start the engine and rotate the steering wheel to the stops. (On vehicles equipped with a Saginaw 800 steering gear, turn to the left stop. On vehicles equipped with a Saginaw 605 steering gear, turn to the right stop. (If the type of gear is unknown, turn the wheel first to the left and then to the right.) The high pressure will blow the seals out of the steering gear and into the drain pan. Stop the engine immediately to prevent a complete loss of power steering fluid.

Step 5. Clean the seal pocket and pitman shaft. Use a crocus cloth, if necessary, to remove any surface rust or corrosion.

NOTE: If the pitman shaft has more severe rust than can be smoothed with crocus cloth, the pitman shaft should be replaced. Rust pitting can easily damage the new replacement seals and the lips of the seals will not be able to seal the high-pressure power steering fluid.

Step 6. Install both seals, backup washers, and snap ring using the appropriate seal installer (see Figure 13–54).

Step 7. Before installing the pitman arm, refill the power steering reservoir and start the engine. Turn the steering wheel to bleed air from the system and check for leaks.

Step 8. Install the pitman arm and double check for leaks. Turn the steering wheel until bleeding is completed.

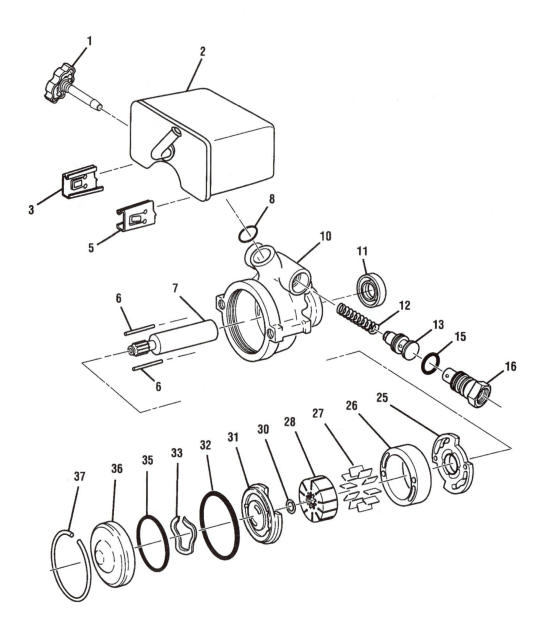

Key No.	Part Name
1 -	CAPSTICK ASM, RESERVOIR
2 -	RESERVOIR ASM, HYD PUMP (TYPICAL)
3 -	CLIP, RESERVOIR RETAINING (LH)
5 -	CLIP, RESERVOIR RETAINING (RH)
6 -	PIN, PUMP RING DOWEL
7 -	SHAFT, DRIVE
8 -	SEAL, O-RING
10 -	HOUSING ASM, HYD PUMP
11 -	SEAL, DRIVE SHAFT
12 -	SPRING, FLOW CONTROL
13 -	VALVE ASM, CONTROL
15 -	SEAL, O-RING

Key No.	Part Name
16 -	FITTING, O-RING UNION
25 -	PLATE, THRUST
26 -	RING, PUMP
27 -	VANE
28 -	ROTOR, PUMP
30 -	RING, SHAFT RETAINING
31 -	PLATE, PRESSURE
32 -	SEAL, O-RING
33 -	SPRING, PRESSURE PLATE
35 -	SEAL, O-RING
36 -	COVER, END
37 -	RING RETAINING

FIGURE 13–52 Typical power steering pump showing the order of assembly. The high-pressure (outlet) hose attaches to the fitting (number 16). The flow control valve can be removed from the pump by removing the fitting. (Courtesy of Oldsmobile)

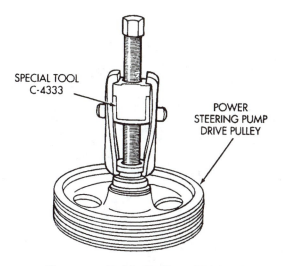

Remove Drive Pulley (Typical)

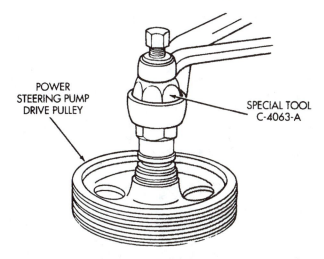

Install Drive Pulley (Typical)

FIGURE 13–53 (Courtesy of Chrysler Corporation)

Over-Center Adjustment. As mileage accumulates, some wear occurs in the steering gear. When this wear occurs, more and more clearance (lash) develops between the pitman shaft teeth (sector gear teeth) and the teeth on the rack piston. The result of this normal wear is excessive steering wheel free play. A common customer complaint is that the steering "feels loose" even though all steering linkage is okay.

NOTE: This "play in the steering" can also cause a steering wheel shimmy that an alignment or normal steering linkage parts replacement cannot solve.

An *over-center sector lash adjustment* may be necessary to correct this common problem.

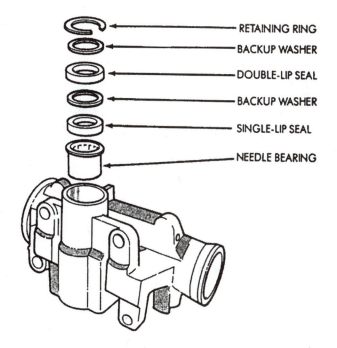

FIGURE 13–54 Pitman shaft seal and bearing locations in a typical Saginaw integral power steering gear. (Courtesy of Chrysler Corporation)

CAUTION: While many automotive manufacturers and experts recommend the following procedure, some vehicle manufacturers do not recommend that this procedure be performed with the steering gear installed in the vehicle.

With the front wheels in the straight-ahead position, visually check for lack of movement of the pitman shaft when the steering wheel is being moved *slightly* back and forth. [Move the steering wheel about 2½ in. (1 cm) total.] If the pitman arm does not move when the stub shaft (intermediate shaft) is being moved, there is clearance (lash) in the steering gear itself. If there is excessive clearance in the steering gear, carefully perform the following steps:

1. Drive the vehicle into the work area, keeping the front wheels straight ahead. Stop the engine.
2. Loosen the over-center adjustment locking nut. This usually requires a ⅝-in. socket or box end wrench.

NOTE: Most steering gears use conventional right-hand threads. The smaller Saginaw 605 steering gear uses left-hand threads. The 605 can be identified by the round side cover that is retained by a snap ring instead of four bolts as in the 800 series.

◄ TECH TIP ►

MORNING SICKNESS

Many technicians are asked to repair hard steering that occurs only when the vehicle is cold, usually first thing in the morning. After a couple of minutes, normal steering effort returns. As the vehicle gets older, the problem tends to get worse at higher and higher temperatures until the steering remains hard to turn even when warm. This condition occurs when the Teflon sealing rings cause wear grooves in the aluminum spool valve area of the rack-and-pinion steering unit (see Figure 13–55).

During cold weather, these Teflon seals are stiff and the power steering fluid leaks past these seals. This fluid leakage causes hydraulic pressure to be applied to both sides of the power piston on the rack. With hydraulic pressure on both sides instead of one side only, the steering wheel is *extremely* difficult to turn. The seals on the spool valve that are leaking determine which direction is hard to steer. Some vehicles are hard to steer in only one direction, while other vehicles may be difficult to steer in both directions. As the power steering fluids heat up, the Teflon seals become more pliable and seal correctly, thereby restoring proper steering effort.

NOTE: The power steering fluid leaking past the Teflon sealing rings of the spool valve will not cause or create an external power steering fluid leak. This leakage around the spool valve area is simply an internal leak that causes the power steering fluid, under pressure, to be applied to the wrong end of the rack piston. External power steering fluid leaks are commonly caused by leaking seals on the rack itself.

There are several methods that many technicians use to "cure" morning sickness, including:

Method 1. Replace the entire rack-and-pinion steering unit. While this is the most expensive method, it is also the most commonly used repair (see Figures 13–56 and 13–57).

Method 2. Replace the Teflon sealing rings with lap-joint Teflon seals. This procedure is usually performed when the vehicle is under warranty and involves the removal, disassembly, and replacement of the seals followed by reinstallation in the vehicle.

Method 3. Flush the old power steering fluid and refill with new fluid. This method is the lowest cost and is recommended for mild cases of cold weather reduction of power assist.

Method 4. Additive and power steering repair kits are also available from aftermarket manufacturers. One method changes the calibration of the flow control at the pump to help compensate for the internal leakage that occurs at the spool valve area of the rack. Flushing the system and using special additives is another commonly available method.

FIGURE 13–55   A section was cut from the housing of a power rack-and-pinion steering gear to show the wear grooves that have been cut into the aluminum.

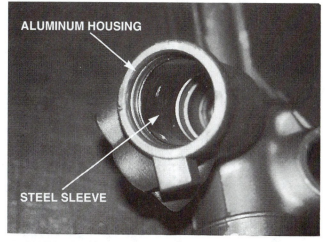

FIGURE 13–57   Completed repair using a steel sleeve.

FIGURE 13–56   Remanufacturers purchase these precut steel sleeves to be pressed into bored-out control valve housings to correct for the grooves created by the Teflon sealing rings.

3. After loosening the lock nut, turn the adjusting screw clockwise until resistance is felt; then turn the screw back (counterclockwise) ¼ turn (90°). (With the Saginaw 605 gear, turn the adjusting screw counterclockwise to remove the excessive clearance.)

NOTE: It is very important that the adjusting screw be loosened ¼ turn to assure proper clearance between the sector gear and the rack piston teeth.

---

4. While holding the adjusting screw to keep it from moving, tighten the locking nut.
5. The adjustment procedure is complete, but the vehicle should be driven before returning it to the customer. This step is very important! For example, if the adjusting screw did move when the lock nut was tightened, the steering may feel too tight and may not return properly after turning a corner. *Carefully test drive the vehicle, being especially careful when turning the first time.*

If the steering is too tight, repeat the adjustment procedure or remove the steering gear for repair.

POWER RACK-AND-PINION SERVICE

The service work on most power rack-and-pinion steering units is limited to replacement of outer and inner tie-rod ends, boots (bellows), and rack mounting bushings. Whenever there is an internal leak or excessive wear is determined, the most economical repair is usually replacement of the entire unit as an assembly. There are certain advantages and disadvantages to any repair procedure, such as the one summarized here.

STEERING SYSTEM TROUBLESHOOTING GUIDE

Problem	Possible causes
Excessive play or looseness in the steering	Worn idler arm or other steering linkage, loose or defective front wheel bearings; perform a dry park test
Hard steering	Low tire pressure, loose or defective power steering belt, defective power rack (spool valve wear), binding steering gear, incorrect wheel alignment, lack of lubrication of the steering linkage, sticking control value in the pump, steering column binding
Steering wheel fails to return after a turn	Steering gear adjusted too tightly, steering linkage bent or binding, lack of lubrication of the steering linkage, incorrect wheel alignment, kinked power steering return hose
Hissing noise in the rack and pinion	May be normal valve noise; compare with similar vehicles
Growl noise in the power steering pump	Restriction in the power steering hoses or steering gear, defective or worn pump.
Groan noise in the power steering pump	Air in the system; bleed or use a vacuum pump to remove
Temporary reduction of power steering assist when cold	Worn rack housing

SUMMARY

1. Steering wheel movement from the driver is transmitted through the steering column to the intermediate shaft and flexible coupling to the steering gear or rack-and-pinion unit.

2. Most steering system components use greasable ball-and-socket joints to allow for suspension travel and steering. Some manufacturers use rubber-bonded sockets (RBS) that are not to be greased.

3. Most conventional steering gears use a recirculating ball nut design.

4. Power steering pumps supply hydraulic fluid to the steering gear, power piston assembly, or power and pinion unit. The spool (rotary) valve controls and directs the high-pressure fluid to the power piston to provide power-assisted steering.

5. In a rack-and-pinion steering unit, the stub shaft connects to the small pinion gear. The pinion gear meshes with gear teeth cut into a long shaft called a rack.

6. The dry park test is a very important test to detect worn or damaged steering parts. While the vehicle is on the ground, have an assistant move the steering wheel back and forth while the technician feels for any looseness in every steering system part.

7. The idler arm usually is the first steering system component to wear out in a conventional parallelogram-type steering system. Following the idler arm in wear are the tie-rods, center link, and pitman arm.

8. The steering system must be level side to side to prevent unwanted bump steer, when the vehicle's direction is changed when traveling over bumps or dips in the road.

9. Always use a belt tension gauge when checking, replacing, or tightening a power steering drive belt. The proper power steering fluid should always be used to prevent possible seal or power steering hose failure.

REVIEW QUESTIONS

1. List all the parts that move when the steering wheel is turned with a conventional steering unit and a rack-and-pinion steering unit.

2. Describe the hydraulic fluid flow from the pump through the flow control valve and to the steering gear.

3. Describe how to perform a dry park test.

4. Explain the procedure for flushing a power steering system.

MULTIPLE-CHOICE QUESTIONS

1. Which type of steering linkage component must not be lubricated?
   a. Zerk fittings
   b. Alamite fittings
   c. RBS tie-rods
   d. Ball guides

2. What steering component is between the intermediate shaft from the steering column and the stub shaft of the steering gear on a rack-and-pinion unit?
   a. Pitman shaft
   b. Flexible coupling
   c. Sector shaft
   d. Tie-rod

3. A dry park test to determine the condition of the steering components and joints should be performed with the vehicle
   a. on level ground.
   b. on turn plates that allow the front wheels to move.
   c. on a frame contact lift with the wheels off the ground.
   d. lifted off the ground about 2 in. (5 cm).

4. Two technicians are discussing bump steer. Technician A says that an unlevel steering linkage can be the cause. Technician B says that if the steering wheel moves when the vehicle is bounced up and down, the steering linkage may be bent. Which technician is correct?
   a. A only
   b. B only
   c. Both A and B
   d. Neither A nor B

5. Two technicians are discussing the proper procedure for bleeding air from a power steering system. Technician A says that the front wheels of the vehicle should be lifted off the ground before bleeding. Technician B says that the steering wheel should be turned left and right while cranking or running the engine during the procedure. Which technician is correct?
   a. A only
   b. B only
   c. Both A and B
   d. Neither A nor B

6. Excessive steering wheel free-play is being discussed. Technician A says that a worn steering flexible coupler could be the cause. Technician B says that a worn groove in the housing around the control valve could be the cause. Which technician is correct?
   a. A only
   b. B only
   c. Both A and B
   d. Neither A nor B

7. A vehicle with a power rack-and-pinion steering is hard to steer when cold (temporary loss of power assist when cold). The most likely cause is
   a. leaking rack seals.
   b. a defective or worn power steering pump.
   c. worn grooves in the housing by the spool valve seals.
   d. using the incorrect power steering fluid.

8. Integral power steering gears use _____ for lubrication of the unit.
   a. SAE 80W-90 gear lube
   b. chassis grease (NLGI No.2)
   c. power steering fluid in the system
   d. molybdenum disulfide

9. Two technicians are discussing the replacement of the pitman shaft seal on an integral power steering gear. Technician A says that the pitman arm must be removed before the old seal can be removed. Technician B says that the steering gear unit should be removed from the vehicle before the seal can be removed. Which technician is correct?
   a. A only
   b. B only
   c. Both A and B
   d. Neither A nor B

10. Two technicians are discussing replacement of a tie-rod end to the steering knuckle. Technician A says to tighten the nut to specifications and loosen slightly, if necessary, to align the cotter key. Technician B says to tighten the nut tighter if necessary if the cotter key hole does not line up. Which technician is correct?
    a. A only
    b. B only
    c. Both A and B
    d. Neither A nor B

◀ Chapter 14 ▶

SUSPENSION SYSTEM OPERATION, DIAGNOSIS, AND SERVICE

OBJECTIVES

After studying Chapter 14, the reader will be able to:

1. List various types of suspensions and their component parts.
2. Describe how the front suspension components function to allow wheel movement up and down and provide for turning.
3. Discuss rear suspension function for both front-wheel-drive and rear-wheel-drive as well as four-wheel-drive vehicles.
4. Explain how to perform diagnosis of the suspension system.
5. Discuss the procedures for testing load-carrying and follower-type ball joints.
6. Describe ball-joint replacement procedures.

Street-driven cars and trucks use a suspension system to keep the tires on the road and to provide acceptable riding comfort. A vehicle with a solid suspension or no suspension would bounce off the ground when the tires hit a bump. If the tires are off the ground, even for a fraction of a second, loss of control is possible.

TYPES OF SUSPENSIONS

Early suspension systems on old horse wagons, buggies, and older vehicles used a solid axle for front and rear wheels (see Figure 14–1). If one wheel hit a bump, the other wheel was affected as shown in Figure 14–2.

Most vehicles today use a separate control arm type of suspension for each front wheel which allows for movement of one front wheel without affecting the other front wheel. This type of suspension is called **independent front suspension,** as shown in Figure 14–3. Many rear suspensions also use independent-type suspension systems. Regardless of the design type of suspension, all suspensions use springs in one form or another.

COIL SPRINGS

Coil springs are made of special round spring steel wrapped in a helix shape. Coil springs are used in front and/or rear suspensions.

Variable-Rate Coil Springs. Many coil springs are designed to provide a variable spring rate. This means that as the spring is being compressed, the spring be-

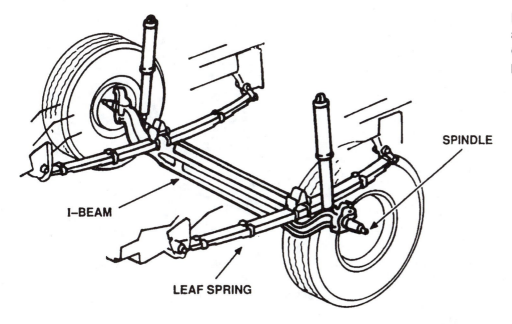

FIGURE 14–1 Solid I-beam axle with leaf springs. (Courtesy of Hunter Engineering Company)

SPINDLE

I–BEAM

LEAF SPRING

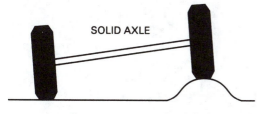

SOLID AXLE

FIGURE 14–2 When one wheel hits a bump or drops into a hole, both left and right wheels are moved. Because both wheels are affected, the ride is often harsh and stiff feeling.

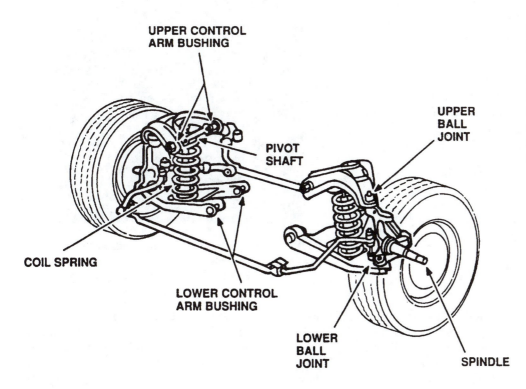

FIGURE 14–3 A typical independent front suspension used on a rear-wheel-drive vehicle. Each wheel can hit a bump or hole in the road *independently* without affecting the opposite wheel. (Courtesy of Hunter Engineering Company)

UPPER CONTROL ARM BUSHING

PIVOT SHAFT

UPPER BALL JOINT

COIL SPRING

LOWER CONTROL ARM BUSHING

LOWER BALL JOINT

SPINDLE

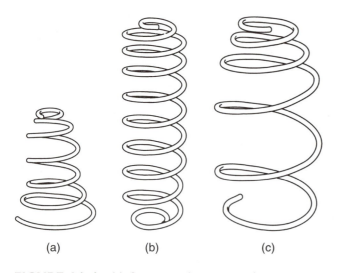

(a)   (b)   (c)

FIGURE 14–4 (a) Semiconical spring used in rear suspension of passenger cars. It provides a variable rate because it is designed so that the coils bottom out on the spring seat. Because coils can compress into themselves (nest), they conserve space. (b) Typical application on the rear coil suspensions of light trucks. This cylindrical spring is manufactured from tapered wire to achieve a variable rate. (c) A shaped linear-rate spring conserves space by compressing into itself. This type of design permits body stylists to achieve an aerodynamic wedge-shaped vehicle.

comes stiffer. This allows for a smooth ride when bumps and dips in the road are small and provides load-carrying capacity and resistance to "bottoming out" when traveling over rough roads (see Figure 14–4).

Coil Spring Mounting. Coil springs are usually installed in a **spring pocket** or **spring seat**. Hard rubber or plastic **cushions** or **insulators** are usually mounted between the coil spring and the spring seat (see Figure 14–5).

Spring Coatings. All springs are painted or coated with epoxy to help prevent breakage. A scratch, nick, or pit caused by corrosion can cause a **stress riser** that can lead to spring failure. Always use a tool that will not scratch or nick the surface of the spring.

LEAF SPRINGS

Leaf springs are constructed of one or more strips or leaves of long, narrow spring steel. These metal strips, called *leaves,* are assembled with plastic or synthetic rubber insulators between the leaves, allowing freedom of movement during spring operation (see Figure 14–6).

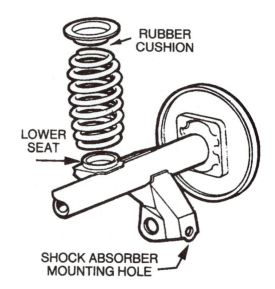

FIGURE 14–5 (Courtesy of Moog)

FIGURE 14–6 Typical leaf spring used on the rear of a rear-wheel-drive vehicle.

The ends of the spring are rolled or looped to form eyes. Rubber bushings are installed in the eyes of the spring and act as noise and vibration insulators, as shown in Figure 14–7. The leaves are held together by a **center bolt**, also called a **centering pin** (see Figure 14–8).

One end of a leaf spring is mounted to a hanger with a bolt and rubber bushings directly to the frame and the other end of the leaf spring is attached to the frame with movable mounting hangers called **shackles,** as shown in Figure 14–9. **Rebound clips** or **spring alignment clips** help prevent the leaves from separating whenever the leaf spring is rebounding from hitting a bump or rise in the roadway (see Figure 14–10). Single leaf steel springs are used on some vehicles and are called **mono leaf.**

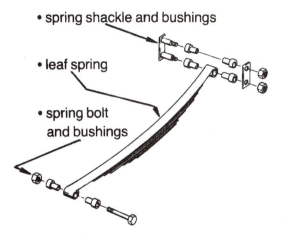

- spring shackle and bushings

- leaf spring

- spring bolt and bushings

FIGURE 14–7   (Courtesy of Dana Corporation)

COMPOSITE LEAF SPRINGS

Since the early 1980s, fiberglass-reinforced epoxy plastic leaf springs have been used on production vehicles. They save weight since an 8-lb spring can replace a conventional 40-lb steel leaf spring. The single leaf composite spring helps isolate road noise and vibrations (see Figure 14–11).

◀ **TECH TIP** ▶

DON'T CUT THOSE COIL SPRINGS!

Chassis service technicians are often asked to lower a vehicle. One method is to remove the coil springs and cut off one-half or more "coils" from the spring. Although this *will* lower the vehicle, this method is generally not recommended because:

1. A coil spring could be damaged during the cutting-off procedure, especially if a torch is used to do the cutting.
2. The spring will get stiffer when shortened, often resulting in a very harsh ride.
3. The amount the vehicle is lowered is *less* than the amount cut off from the spring. This is because as the spring is shortened, it becomes stiffer. The stiffer spring will compress less than the original.

Instead of cutting springs to lower a vehicle, there are several methods available that are preferred if the vehicle *must* be lowered:

1. Use shorter replacement springs that are designed specifically to lower the exact vehicle. A change in shock absorbers may be necessary because the shorter springs change the operating height of the stock (original) shock absorbers. Consult the spring manufacturers for exact installation instructions and recommendations.
2. Replacement spindles are designed to *raise* the location of the wheel spindle, thereby lowering the body in relation to the ground. Except for ground clearance problems, this method is the method recommended by many chassis service technicians. Replacement spindles keep the same springs, shock absorbers, and ride while lowering the vehicle without serious problems.

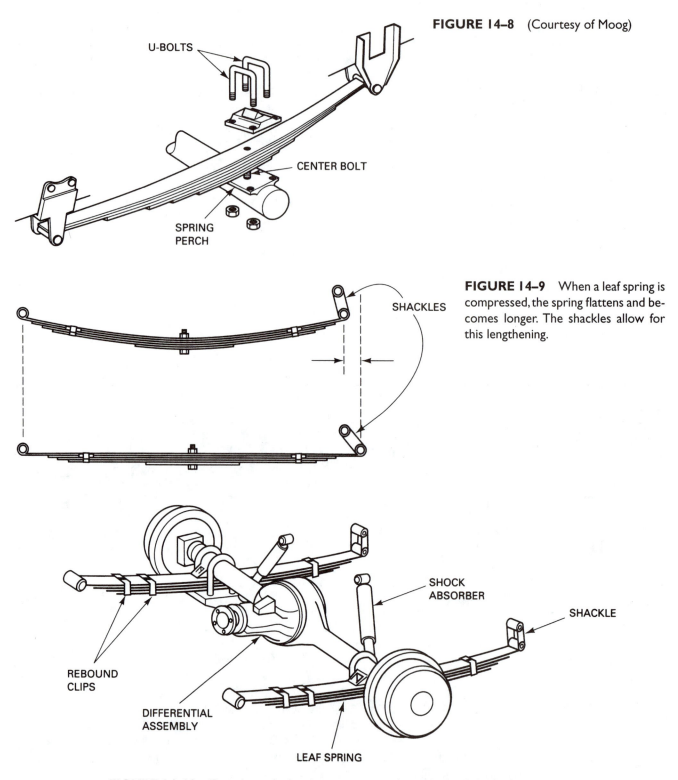

FIGURE 14–8 (Courtesy of Moog)

U-BOLTS

CENTER BOLT

SPRING
PERCH

FIGURE 14–9 When a leaf spring is compressed, the spring flattens and becomes longer. The shackles allow for this lengthening.

SHACKLES

SHOCK
ABSORBER

SHACKLE

REBOUND
CLIPS

DIFFERENTIAL
ASSEMBLY

LEAF SPRING

FIGURE 14–10 Typical rear leaf-spring suspension of a rear-wheel-drive vehicle.

TORSION BARS

A torsion bar is a spring that is a long, *round* hardened steel bar similar to a coil spring except a *straight* bar, as

shown in Figure 14–12. One end is attached to the lower control arm of a front suspension and the other end to the frame. When the wheels hit a bump, the bar twists and then untwists. Chrysler Corporation cars used tor-

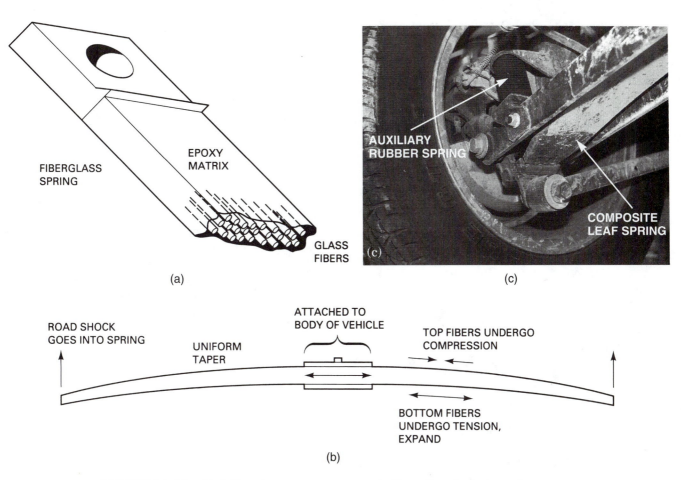

FIGURE 14–11 (a) A fiberglass spring is composed of long fibers locked together in an epoxy (resin) matrix. (b) When the spring compresses, the bottom of the spring expands and the top compresses. (c) Composite transverse leaf spring used on the rear suspension of a General Motors front-wheel-drive car. The stiffness of the auxiliary rubber spring is used to fine tune the rear suspension.

FIGURE 14–12 (Courtesy of FMC)

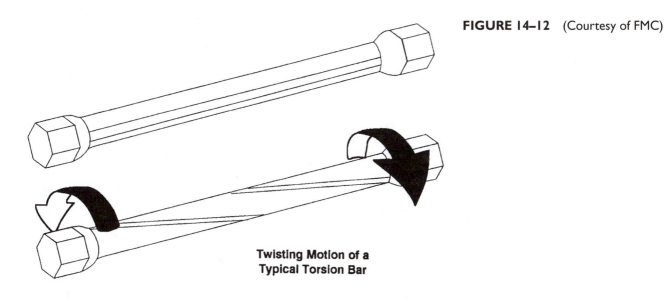

**Twisting Motion of a
Typical Torsion Bar**

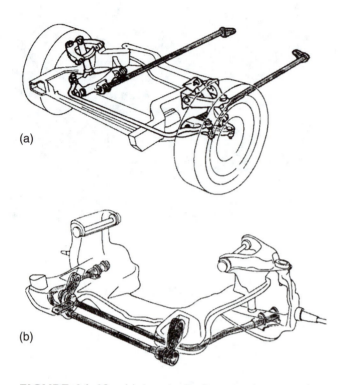

(a)

(b)

FIGURE 14–13 (a) Longitudinal torsion bars attach at the lower control arm at the front and an adjustable frame mount at the rear of the bar. (b) Transverse-mounted torsion bars attach to the lower control arm and the front of the frame. (Courtesy of Dana Corporation)

sion bar front suspension both longitudinally and transversely (see Figure 14–13).

Many manufacturers of pickup trucks currently use torsion bar suspensions, especially on their four-wheel-drive models. Torsion bars allow room for the front drive axle and constant-velocity joint and still provide for a strong suspension. Unlike other types of springs, torsion bars are *adjustable* for correct ride height (see Figure 14–14).

SUSPENSION PRINCIPLES

Suspensions use various links, arms, and joints to allow the wheels to move up and down freely. Front suspensions also have to allow the front wheels to turn. All suspensions must provide for all of the following supports:

1. *Transverse (or side-to-side) wheel support.* As the wheels of the vehicle move up and down, the suspension must allow for this movement and still hold the wheel from moving away from the vehicle or inward toward the center of the vehicle (see Figure 14–15).

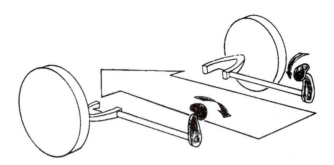

FIGURE 14–14 Most torsion bars are adjustable to raise or lower the vehicle height. The arm that the torsion bar slides into is retained by a large bolt. As the bolt is turned, more or less preload is applied to the torsion bar and this raises or lowers the vehicle. (Courtesy of Dana Corporation)

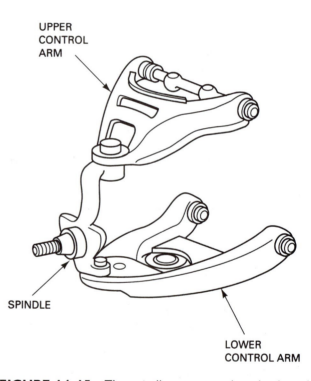

UPPER CONTROL ARM

SPINDLE

LOWER CONTROL ARM

FIGURE 14–15 The spindle supports the wheels and attaches to the control arm with ball-and-socket joints called ball joints. The control arm attaches to the frame of the vehicle through rubber bushings to help isolate noise and vibration between the road and the body. (Courtesy of Moog)

2. *Longitudinal (front-to-back) wheel support.* As the wheels of the vehicle move up and down, the suspension must allow for this movement and still hold the wheels from moving backward whenever a bump is hit.

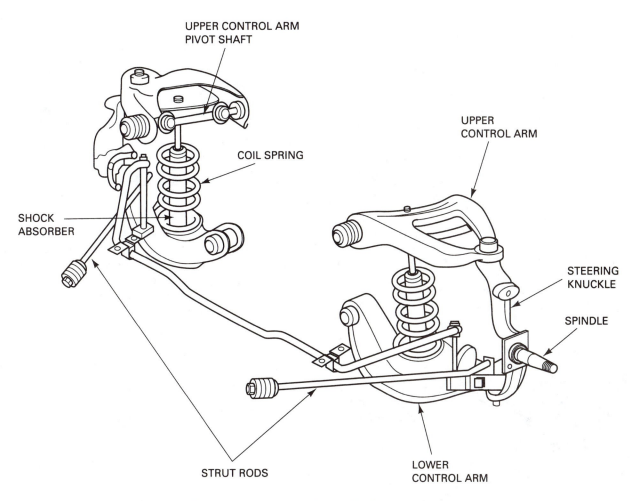

UPPER CONTROL ARM
PIVOT SHAFT

COIL SPRING

UPPER
CONTROL ARM

SHOCK
ABSORBER

STEERING
KNUCKLE

SPINDLE

STRUT RODS

LOWER
CONTROL ARM

FIGURE 14–16 The strut rods provide longitudinal support to the suspension to prevent forward or rearward movement of the control arms.

At least two suspension links or arms are required, to provide for freedom of movement up or down and *prevent* any in–out or forward–back movement (see Figure 14–16).

TWIN I-BEAMS

A twin I-beam front suspension was used for many years on Ford pickup trucks and vans since the mid-1960s. Strong steel twin beams that cross provide independent front suspension operation with the strength of a solid front axle (see Figure 14–17).

To control longitudinal (front-to-back) support, a **radius rod** is attached to each beam and anchors to the frame of the truck using rubber bushings. These bushings allow the front axle to move up and down while insulating road noise and vibration from the frame and body.

◀ **TECH TIP** ▶

RADIUS ROD BUSHING NOISE

When the radius rod bushing (see Figure 14–18) on a Ford truck or van deteriorates, the most common complaint from the driver is noise. Besides causing tire wear, worn or defective radius rod bushing deterioration can cause:

1. A clicking sound when braking (sounds as if the brake caliper may be loose)
2. A clunking noise when hitting bumps

When the bushing deteriorates, the axles can move forward and backward with less control. The noise is the first sign that something is wrong. Without proper axle support, handling and cornering can also be affected.

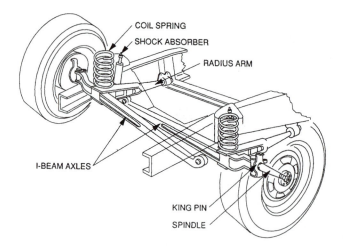

FIGURE 14-17 Twin-I-beam front suspension. (Courtesy of Dana Corporation)

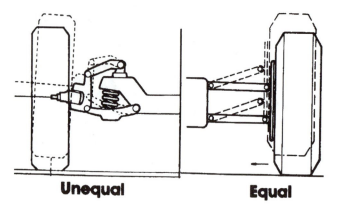

FIGURE 14-19 If the control arms were both the same length, the tire would move side to side as the wheels moved up and down as illustrated on the right. To keep the wheel moving straight up and down as the suspension moves, the upper control arm is shorter, as shown on the left. (Courtesy of Moog)

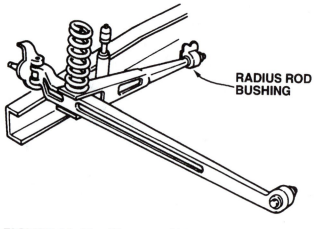

FIGURE 14-18 (Courtesy of Moog)

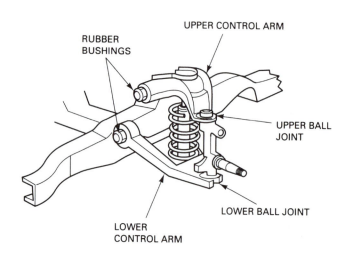

FIGURE 14-20 (Courtesy of Moog)

SHORT/LONG ARM SUSPENSIONS

This type of suspension uses a short upper control arm and a longer lower control arm, and is usually referred to as the SLA (short/long arm) type (see Figure 14–19). SLA-type suspension can be used with either coil springs or torsion bars. Most vehicles use two A-shaped steel arms: at the bottom of the A, connected to the frame by rubber bushings, and the other end, at the top of the A, connected to the steering knuckle by **ball joints** (see Figure 14–20). These **A arms** are usually called **control arms** because they "control" the location of the front wheels and allow for up-and-down movement of the front wheels as well as turning of the front wheels for steering.

NOTE: SLA-type suspension is also called a double-wishbone suspension because both control arms often resemble the shape of a chicken or turkey wishbone.

The top control arm is called the **upper control arm**. The bottom control arm is called the **lower control arm**. The same terms also apply to the other parts of the control arms such as *lower ball joint, upper ball joint, upper control arm bushings,* and *lower control arm bushings.*

The upper control arm is shorter than the lower control arm. This permits the tires to remain as vertical as possible during suspension travel. Short/long

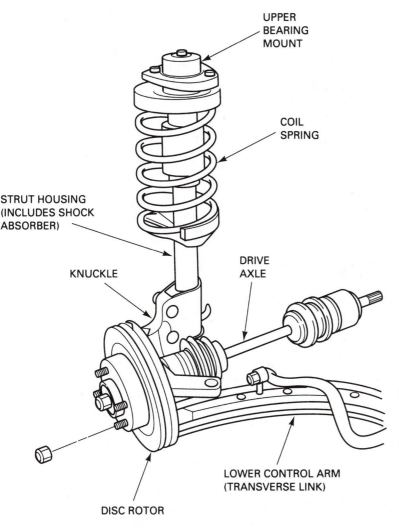

UPPER
BEARING
MOUNT

COIL
SPRING

STRUT HOUSING
(INCLUDES SHOCK
ABSORBER)

KNUCKLE

DRIVE
AXLE

LOWER CONTROL ARM
(TRANSVERSE LINK)

DISC ROTOR

FIGURE 14–21 Typical MacPherson strut-type front suspension on a front-wheel-drive vehicle. Note the drive axle shaft and CV joints.

arm suspensions are used in the front as well as the rear of many rear-wheel-drive and front-wheel-drive vehicles.

MACPHERSON STRUTS

The **MacPherson strut** suspension was patented in 1958 by Earle S. MacPherson, a vice president of engineering at Ford Motor Company. The most commonly used strut suspension combines the coil spring and the shock absorber into one structural suspension component, as shown in Figure 14–21.

A MacPherson strut suspension is light weight and saves space in a vehicle because only one control arm is needed and the top of the strut simply attaches to the body of the vehicle. *The entire strut rotates when the front wheels are turned.* The pivot points of a strut are at the lower ball joint and a bearing assembly at the top

of the strut. Some vehicles use a modified strut suspension, as shown in Figure 14–22.

BALL-JOINTS

Ball joints are actually ball-and-socket joints, similar to the joints in a person's shoulder. Ball joints allow the front wheels to move up and down as well as side to side for steering. A vehicle can be equipped with coil springs, mounted either above the upper control arm *or* mounted on the lower control arm (see Figure 14–23).

If the coil spring is attached to the top of the upper control arm, the upper ball joint is carrying the weight of the vehicle and is called the **load-carrying ball joint**. The lower ball joint is called the **follower ball joint** (see Figure 14–24).

If the coil spring is attached to the lower control arm, the lower ball joint is the load-carrying ball joint

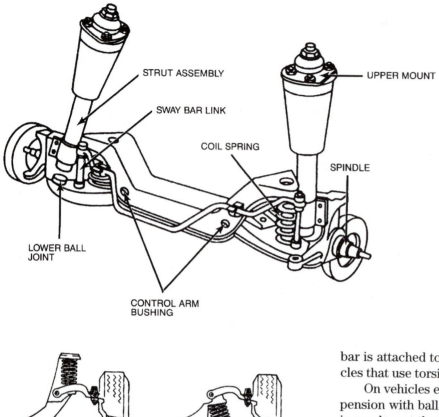

FIGURE 14–22 Modified strut suspension. The strut provides the structural support to the body with the coil spring acting directly on the lower control arm. (Courtesy of Dana Corporation)

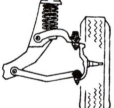

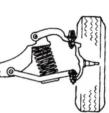

COIL SPRING OR TORSION BAR MOUNTED ON UPPER CONTROL ARM

COIL SPRING OR TORSION BAR MOUNTED ON LOWER CONTROL ARM

MACPHERSON STRUT

COIL SPRING MOUNTED ON LOWER CONTROL ARM WITH MODIFIED STRUT

FIGURE 14–23 Ball joints provide the freedom of movement necessary for steering and suspension movements. (Courtesy of Dana Corporation)

and the upper ball joint is the follower ball joint (see Figure 14–25).

If a torsion bar-type spring is used, the lower ball joint is the load-carrying ball joint because the torsion

bar is attached to the lower control arm on most vehicles that use torsion bars (see Figure 14–26).

On vehicles equipped with a twin-I-beam front suspension with ball joints, *both* ball joints are load-carrying and must therefore be replaced together if worn or defective (see Figure 14–27). MacPherson struts use a ball joint on the lower control arm. Since the weight of the vehicle is applied to the upper strut mount, the ball joint is non-load carrying (see Figure 14–28).

A specific amount of stud turning resistance is built into each ball joint to stabilize steering. A ball joint that does not support the weight of the vehicle and the outer suspension pivot is often called a **follower ball joint** or a **friction ball joint.** The load-carrying (weight-carrying) ball joint is subjected to the greatest amount of wear and is the most frequently replaced (see Figure 14–29).

STRUT RODS

Some vehicles are equipped with round steel rods that are attached between the lower control arm at one end and the frame of the vehicle with rubber bushings called **strut rod bushings** at the other end. The purpose of these strut rods is to provide forward/backward support to the control arms (see Figure 14–30).

Strut rods are also called **tension** or **compression** rods or simply **TC rods.** If a strut rod has a nut on *both* sides of the bushings, the strut rod is used to adjust *caster.* See Chapter 15 for information on caster and other alignment angles.

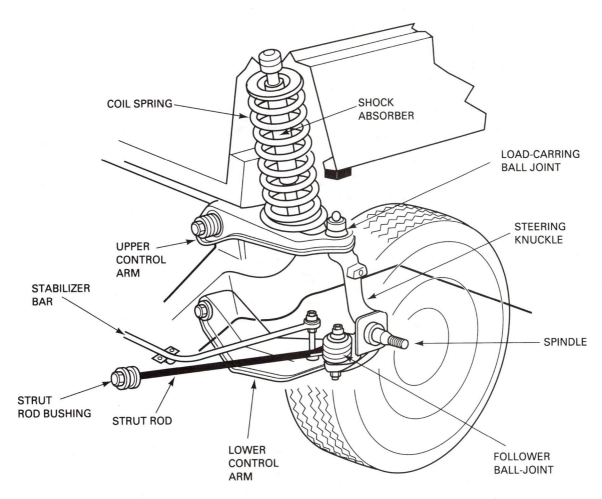

COIL SPRING

SHOCK ABSORBER

LOAD-CARRING BALL JOINT

STEERING KNUCKLE

UPPER CONTROL ARM

STABILIZER BAR

SPINDLE

STRUT ROD BUSHING

STRUT ROD

LOWER CONTROL ARM

FOLLOWER BALL-JOINT

FIGURE 14–24 The upper ball joint is load carrying in this type of suspension because the weight of the vehicle is applied through the spring, upper control arm, and ball joint to the wheel. The lower control arm is a lateral link and the lower ball joint is called a follower ball joint.

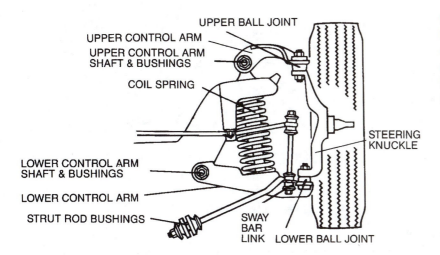

UPPER BALL JOINT

UPPER CONTROL ARM

UPPER CONTROL ARM SHAFT & BUSHINGS

COIL SPRING

LOWER CONTROL ARM SHAFT & BUSHINGS

LOWER CONTROL ARM

STRUT ROD BUSHINGS

SWAY BAR LINK

LOWER BALL JOINT

STEERING KNUCKLE

FIGURE 14–25 The lower ball joint is load carrying in this type of suspension because the weight of the vehicle is applied through the spring, lower control arm, and ball joint to the wheel. (Courtesy of Dana Corporation)

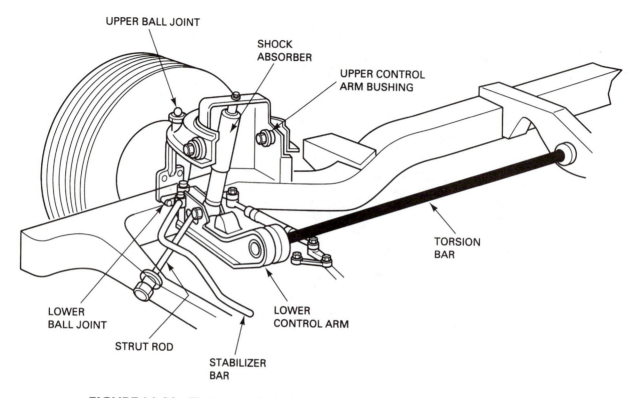

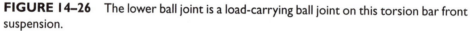

FIGURE 14–26 The lower ball joint is a load-carrying ball joint on this torsion bar front suspension.

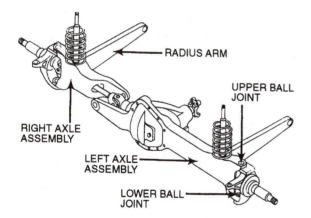

FIGURE 14–27 Both ball joints are load carrying on twin-I-beam suspension. This front suspension is on a four-wheel-drive Ford truck and is called a twin traction beam suspension. (Courtesy of Dana Corporation)

STABILIZER BARS

Most cars and trucks are equipped with a stabilizer bar on the front suspension, which is a round hardened steel bar which is attached to both lower control arms with bolts and rubber bushing washers called **stabilizer bar bushings** (see Figure 14–31). A stabilizer bar

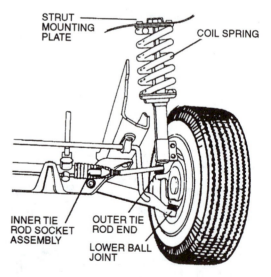

FIGURE 14–28 The ball joint used in a MacPherson suspension is non-load carrying. The vehicle weight is transferred through the spring and upper strut mount plate. (Courtesy of Dana Corporation)

is also called an **antisway bar (sway bar)** or **anti-roll bar (roll bar)**. The purpose of the stabilizer bar is to prevent excessive body roll while cornering and to add to stability while driving over rough road surfaces (see Figure 14–32). The stabilizer bar is also used at a longitudinal (front/back) support to the lower control

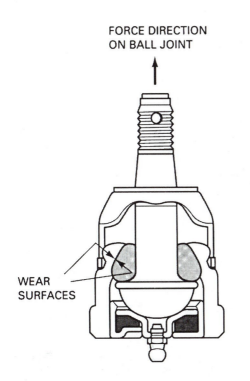

FIGURE 14–29 In this tension design ball joint, the wear surface is above the pivot ball and socket as shown.

arm on many vehicles equipped with MacPherson struts.

Stabilizer links connect the ends of the stabilizer bar to the lower control arm (see Figure 14–33). Stabilizer bar links are commonly found to be defective (cracked rubber washers or broken spacer bolts) because of the great amount of force that is transmitted through the links and the bushings.

SHOCK ABSORBERS

Shock absorbers are used on all conventional suspension systems to dampen and control the motion of the vehicle's springs. Without shock absorbers (dampers), the vehicle would continue to bounce after hitting bumps (see Figure 14–34). Struts are shock absorbers that are part of a MacPherson strut assembly. *The major purpose of any shock or strut is to control ride and handling.* Standard shock absorbers *do not* support the weight of a vehicle. *The springs support the weight of the vehicle and the shock absorbers control the actions and reactions of the springs.* Most shock absorbers are "direct acting" because they are connected directly between the vehicle frame or body and the axles (see Figure 14–35).

The shock absorber helps dampen the rapid up-and-down movement of the vehicle springs by converting energy of movement into heat by forcing hydraulic

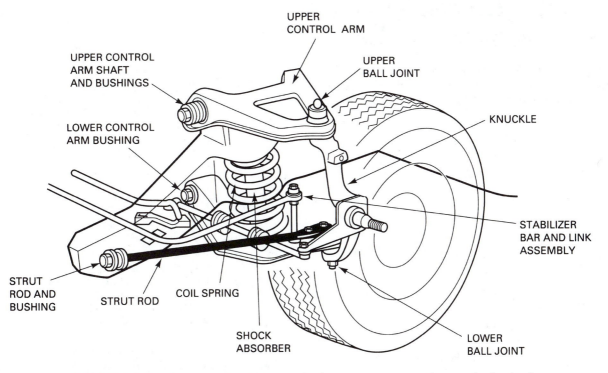

FIGURE 14-30 A strut rod is the longitudinal support to prevent front-to-back wheel movement.

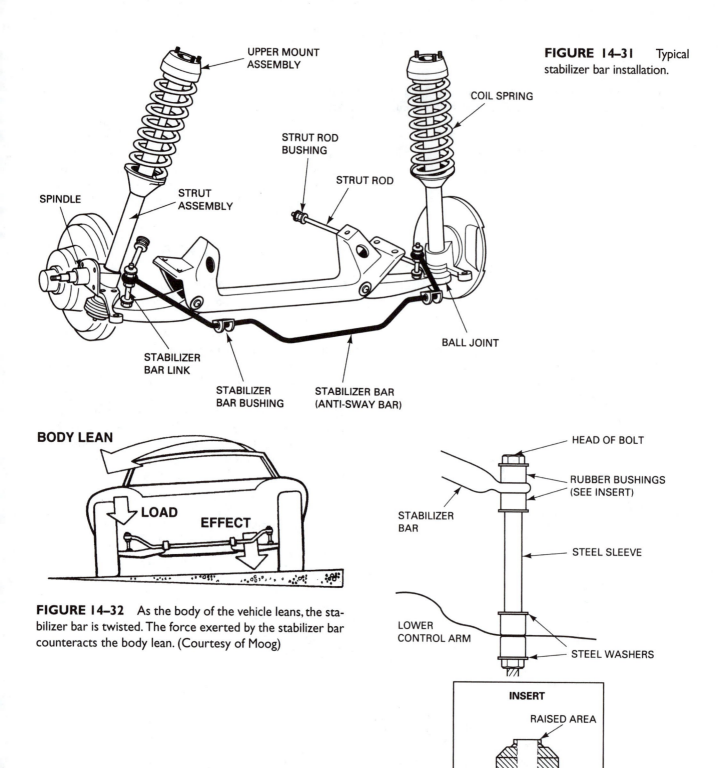

UPPER MOUNT
ASSEMBLY

FIGURE 14–31 Typical stabilizer bar installation.

COIL SPRING

STRUT ROD
BUSHING

STRUT ROD

SPINDLE

STRUT
ASSEMBLY

BALL JOINT

STABILIZER
BAR LINK

STABILIZER
BAR BUSHING

STABILIZER BAR
(ANTI-SWAY BAR)

BODY LEAN

LOAD

EFFECT

FIGURE 14–32 As the body of the vehicle leans, the stabilizer bar is twisted. The force exerted by the stabilizer bar counteracts the body lean. (Courtesy of Moog)

HEAD OF BOLT

RUBBER BUSHINGS
(SEE INSERT)

STABILIZER
BAR

STEEL SLEEVE

LOWER
CONTROL ARM

STEEL WASHERS

INSERT

RAISED AREA

RAISED AREA OF BUSHING
SHOULD BE PLACED
TOWARD STABILIZER
BAR OR CONTROL ARM

FIGURE 14–33 Stabilizer bar links are sold as a kit consisting of the long bolt with steel sleeve and rubber bushings. Steel washers are used on both sides of the rubber bushings as shown.

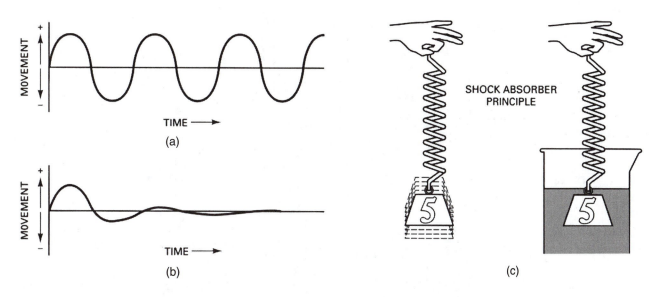

FIGURE 14–34 (a) Movement of vehicle is supported by springs without a dampening device. (b) Spring action dampened with a shock absorber. (c) The function of any shock absorber is to dampen the movement or action of a spring similar to using a liquid to control the movement of a weight on a spring. (Courtesy of FMC)

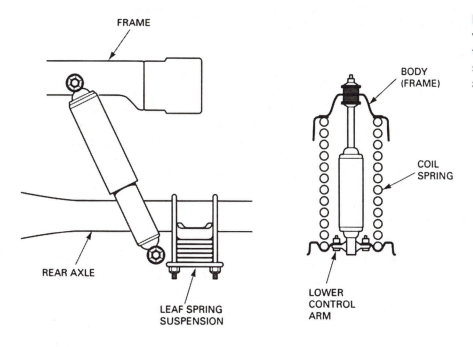

FIGURE 14–35 Shock absorbers work best when mounted as close to the spring as possible. Shock absorbers that are mounted straight up and down offer the most dampening.

fluid through small holes inside the shock absorber (see Figure 14–36).

GAS-CHARGED SHOCKS

Most shock absorbers on new vehicles are gas charged. Pressurizing the oil inside the shock absorber helps smooth the ride over rough roads. This pressure helps prevent air pockets from forming in the shock absorber oil as it passes through the small passages in the shock. Typical gas-charged shocks are pressurized with nitrogen with 130 to 150 psi (900 to 1030 kPa) to aid in both handling and ride control.

Some gas-charged shock absorbers use a single tube that contains two pistons that separate the high-pressure gas from the working fluid. Single-tube shocks are also called **monotube** or **DeCarbon** after the French inventor of the principle and manufacturer of suspension components (see Figure 14–37).

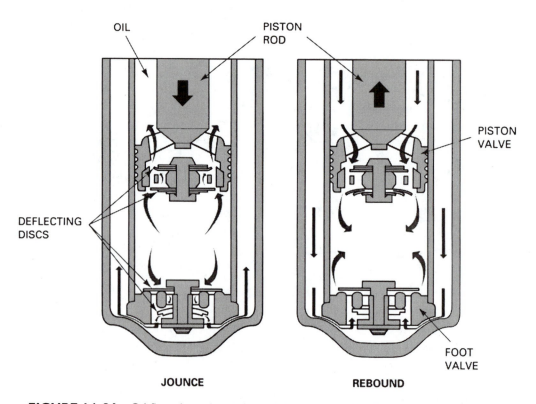

OIL

PISTON ROD

PISTON VALVE

DEFLECTING DISCS

FOOT VALVE

JOUNCE   REBOUND

FIGURE 14–36 Oil flow through a deflected disc piston valve. The deflecting disc can react rapidly to suspension movement. For example, if a large bump is hit at high speed, the disc can deflect completely and allow the suspension to reach its maximum jounce distance while maintaining a controlled rate of movement.

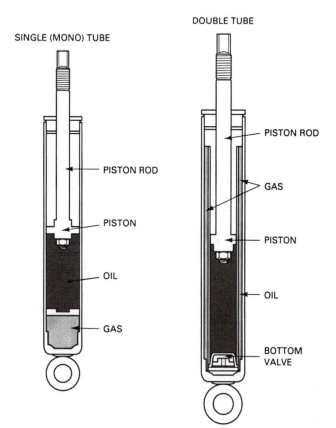

SINGLE (MONO) TUBE

DOUBLE TUBE

PISTON ROD

PISTON

OIL

GAS

PISTON ROD

GAS

PISTON

OIL

BOTTOM VALVE

FIGURE 14–37 Gas-charged shock absorbers are manufactured with a double tube design similar to conventional shock absorbers and with a single or mono-tube design.

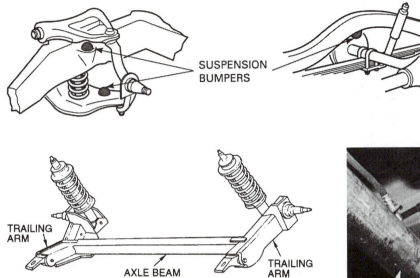

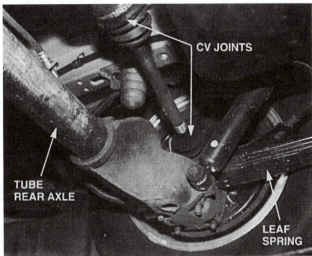

FIGURE 14-38 Suspension bumpers are used on all suspension systems to prevent metal-to-metal contact between the suspension and the frame or body of the vehicle when the suspension "bottoms out" over large bumps or dips in the road. (Courtesy of Moog)

FIGURE 14-39 A semi-independent axle beam rear suspension. It is called semi-independent because as a wheel hits a bump in the road, the trailing arm moves upward and twists the axle beam. This twisting of the axle beam helps isolate one side of the vehicle from the other. (Courtesy of Dana Corporation)

FIGURE 14-40 A solid rear axle tube suspension with leaf springs.

JOUNCE BUMPERS

All suspension systems have a limit of travel. If the vehicle hits a large bump in the road, the wheels are forced upward toward the vehicle with tremendous force. This force is absorbed by the springs of the suspension system. If the bump is large enough, the suspension is compressed to its mechanical limit. Instead of allowing the metal components of the suspension to hit the frame or body of the vehicle, a rubber or foam bumper is used to absorb and isolate the suspension from the frame or body (see Figure 14-38). These bumpers are called **suspension bumpers, strike-out bumpers, bump stops,** or **jounce bumpers.** Most suspensions also use a rubber or foam stop to limit the downward travel of the suspension during rebound. Some stops are built into the shock absorber or strut.

SOLID AXLE REAR SUSPENSIONS

Solid or straight axle rear suspension can use coil springs or leaf springs (see Figures 14-39 and 14-40). Leaf springs function as a load-carrying member as well as providing side-to-side support and stability. Coil springs, however, can function only as a load-carrying

member and must depend on various other suspension members to provide side-to-side (lateral) as well as front-to-back rear axle support.

Longitudinal (back-and-forth) support is provided by rear **control arms,** also called rear **trailing arms,** because the rear axle trails behind the control arm frame mounts. These rear control arms are usually angled to provide transverse (side-to-side) support as well as longitudinal support. Some manufacturers add an additional rear support member to ensure that the center of the body is kept directly over the center of the rear axle. These horizontal rear bars are called **track rods** or **panhard rods** and are bolted to the rear axle at one end and the vehicle frame at the other end (see Figure 14-41).

INDEPENDENT REAR SUSPENSIONS

Most newer front-wheel-drive vehicles use an independent rear suspension (**IRS**). An independent rear suspension provides a smoother ride than a solid axle

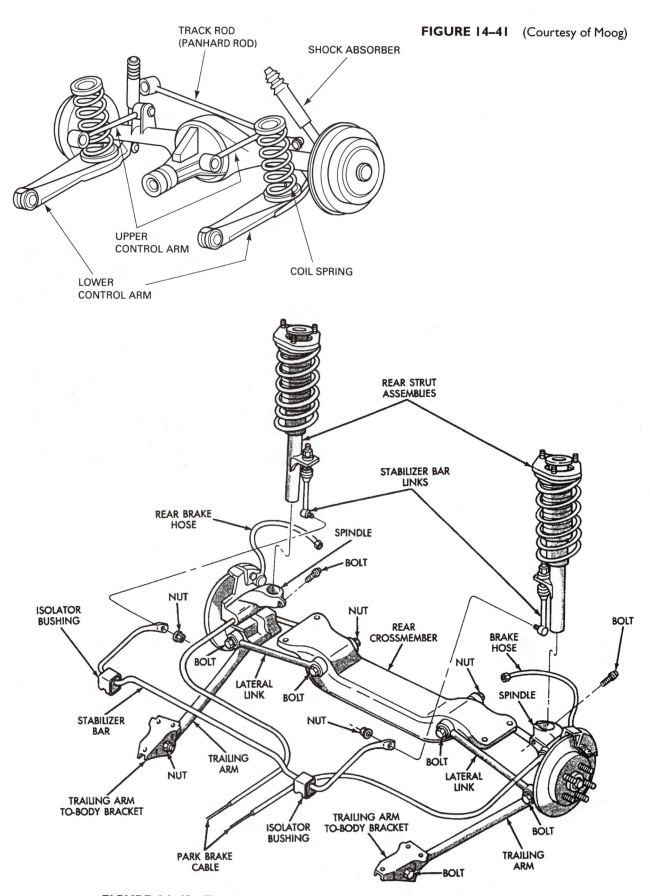

FIGURE 14-41 (Courtesy of Moog)

TRACK ROD
(PANHARD ROD)

SHOCK ABSORBER

UPPER
CONTROL ARM

LOWER
CONTROL ARM

COIL SPRING

REAR STRUT
ASSEMBLIES

STABILIZER BAR
LINKS

REAR BRAKE
HOSE

SPINDLE

BOLT

ISOLATOR
BUSHING

NUT

NUT

REAR
CROSSMEMBER

BRAKE
HOSE

BOLT

NUT

BOLT

LATERAL
LINK

BOLT

SPINDLE

STABILIZER
BAR

NUT

TRAILING
ARM

NUT

BOLT

LATERAL
LINK

TRAILING ARM
TO-BODY BRACKET

ISOLATOR
BUSHING

TRAILING ARM
TO-BODY BRACKET

BOLT

TRAILING
ARM

PARK BRAKE
CABLE

BOLT

FIGURE 14-42 Typical independent rear suspension on a front-wheel-drive vehicle.
(Courtesy of Chrysler Corporation)

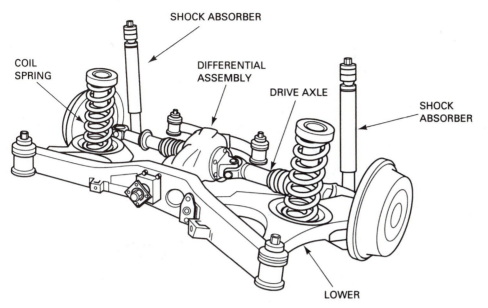

FIGURE 14–43 The differential is attached to the frame of the vehicle in this example of an IRS. (Courtesy of Moog)

SHOCK ABSORBER

COIL SPRING

DIFFERENTIAL ASSEMBLY

DRIVE AXLE

SHOCK ABSORBER

LOWER

suspension because each rear wheel can react to bumps and dips in the road without moving the entire rear axle. Many front-wheel-drive vehicles with IRS also provide for some alignment adjustments (see Figures 14–42 and 14–43).

AIR SHOCKS/STRUTS

Air-inflatable shock absorbers or struts are used in the rear of vehicles to provide proper vehicle ride height while carrying heavy loads. Many air shock/strut units are original equipment and are often combined with a built-in air compressor and ride height sensor(s) to provide automatic ride height control.

Air-inflatable shocks are standard shock absorbers with an air chamber with a rubber bag built into the dust cover (top) of the shock (see Figure 14–44). Air pressure is used to inflate the bag, which raises the installed height of the shock. *It is important that the load capacity of the vehicle not be exceeded or serious damage can occur to the vehicle's springs, axles, bearings, and shock support mounts.*

ELECTRONICALLY CONTROLLED SUSPENSIONS

Many vehicle manufacturers offer some type of electronically controlled suspensions. Most use conventional springs or air springs. Few production vehicles use true **active** or **reactive** suspensions.

Passive electronically controlled suspension systems are more common and use conventional or air bag

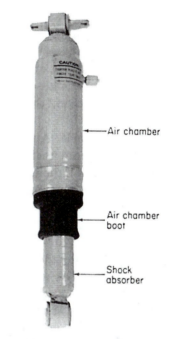

Air chamber

Air chamber boot

Shock absorber

FIGURE 14–44 Typical air shock.

springs to support the weight of the vehicle instead of hydraulic cylinders used on active systems. Most nonactive electronically controlled suspension systems can be grouped into four basic categories:

1. **Electronically controlled rear air-inflatable shock absorbers.** The main purpose of this system is to maintain controlled rear ride (trim) height under all vehicle load conditions. Some vehicles are equipped with rear air shocks that can be controlled electronically to adjust the

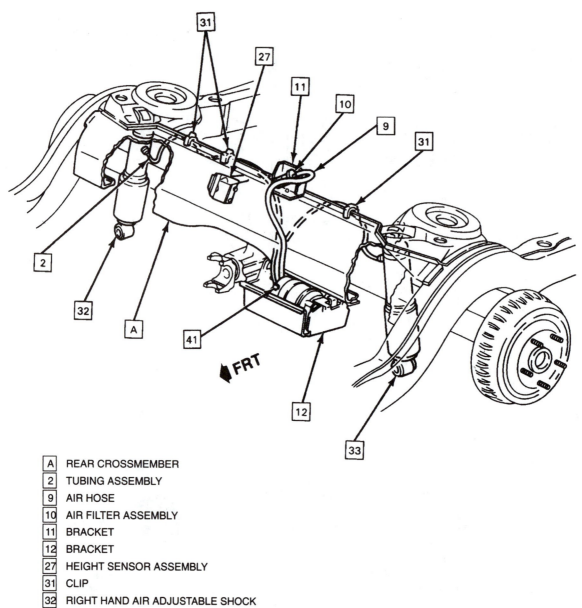

A	REAR CROSSMEMBER
2	TUBING ASSEMBLY
9	AIR HOSE
10	AIR FILTER ASSEMBLY
11	BRACKET
12	BRACKET
27	HEIGHT SENSOR ASSEMBLY
31	CLIP
32	RIGHT HAND AIR ADJUSTABLE SHOCK
33	LEFT HAND AIR ADJUSTABLE SHOCK

FIGURE 14–45 Typical electronic level control system showing the location of the height sensor and the compressor assembly. (Courtesy of Oldsmobile)

ride height of the vehicle regardless of vehicle load. A typical electronic level-control system includes the following components:

a. Air-adjustable rear shocks (or struts in some cases)
b. Small air compressor (mounted under the hood or under the vehicle at the rear)
c. Electronic height sensor
d. Air dryer
e. Exhaust solenoid
f. Relay, wiring, and tubing

The compressor is usually a small single-piston air pump powered by a 12-volt permanent-magnet (PM) electric motor (see Figure 14–45). An air dryer is usually attached to the pump to remove moisture from the air before being sent to the shocks and through the dryer (to dry the chemical dryer) during the release of air from the shocks (see Figure 14–46). The height sensor operates the compressor or exhaust solenoid, based on the height of the rear of the vehicle.

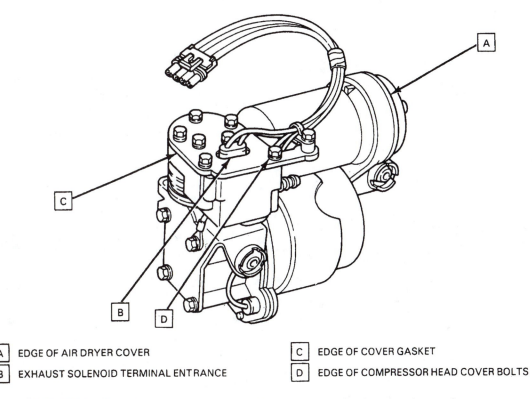

| A | EDGE OF AIR DRYER COVER | C | EDGE OF COVER GASKET |
| B | EXHAUST SOLENOID TERMINAL ENTRANCE | D | EDGE OF COMPRESSOR HEAD COVER BOLTS |

FIGURE 14–46   All air compressors used for suspension height control use a dryer to remove moisture from the air before it is used to pressurize the air chambers of the shock absorbers, struts, or air springs. (Courtesy of Oldsmobile)

2. **Four-wheel ride height control.** This type of system uses air-inflatable springs to support vehicle weight with the inflation pressure varied by an electronic controller. The main purpose of this type of system is to maintain the same ride height both front and rear as well as side to side under all driving conditions (see Figures 14–47 and 14–48).

CAUTION: Many vehicles equipped with air suspension, such as many Lincoln automobiles, have a trunk-mounted on/off switch. Before servicing, hoisting, jacking, or towing, this switch must be in the "off" position.

---

3. **Computer-controlled shock absorbers.** These are used with conventional metal or fiberglass composite springs. The main purpose of this type of system is to permit a smoother ride over rough road surfaces, yet still permit a stiffer ride at higher speeds or to stiffen the shocks during braking, accelerating, or cornering.

4. **Active suspension system.** These systems were first developed by Lotus of England. An active sus-

pension system uses hydraulic cylinders at each wheel to keep the tires on the road and the body of the vehicle level under all driving conditions. The purpose of an active suspension system is to provide a smooth ride and superior handling in one vehicle. The active suspension system can provide the smooth ride of a soft suspension and still provide the control and handling of a stiff suspension. The typical active suspension system consists of the following components:

a. **Controller.** The on-board computer (controller) uses information from sensors to activate hydraulically operated cylinders located at each wheel.

b. **Sensors.** Each wheel has a height sensor and an acceleration sensor. Other sensors used by the controller include vehicle speed sensor, steering wheel movement sensor, and brake switch sensor.

c. **Hydraulic pump and actuators.** An engine-driven hydraulic pump provides the power necessary to raise and lower the vehicle and maintain proper vehicle height. These high-speed actuators must be capable of raising or lowering the vehicle in as little as 3 milliseconds ($\frac{3}{1000}$ of 1 second).

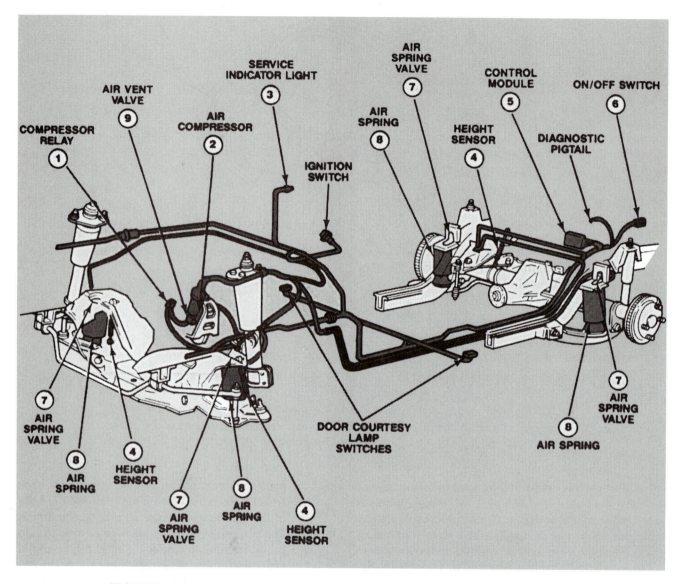

FIGURE 14–47 Air suspension components. Note on/off switch at the rear of the vehicle. This switch *must* be turned "off" before hoisting the vehicle. (Courtesy of Ford Motor Company)

SUSPENSION SYSTEM DIAGNOSIS

Suspension systems are designed and manufactured to provide years of trouble-free service with a minimum amount of maintenance. The smart technician should always road test any vehicle before and after servicing. (See the Tech Tip "Road Test—Before and After" for details.) *The purpose of any diagnosis is to eliminate known good components.* See the suspension diagnostic chart, Figure 14–49, for a list of components that can cause the problem or customer complaint.

Road Test Diagnosis. If possible, perform a road test of the vehicle with the owner of the vehicle. It is also helpful to have the owner drive the vehicle. A proper road test for any suspension system problem should include:

1. **Drive beside parked vehicles**. Any noise generated by the vehicle suspension or tires is reflected off solid objects such as a row of parked vehicles along a street. Repeat the drive for the right side. Defective wheel bearings or power steering pumps usually make noise and can be heard during this test.

DESIGN

- Air spring is at normal trim height
- Air pressure contained in rubber membrane maintains vehicle height and acts like coil spring
- Air spring valve mounted in end cap opens to allow air to enter and exit spring
- When air is added, vehicle will rise
- When air is removed, vehicle will lower

JOUNCE

- When control arm moves upward, piston moves upward into rubber membrane
- As the arm moves upward toward jounce the rate of the air spring increases

REBOUND

- When control arm moves downward, piston extends outward from rubber membrane
- Rubber membrane unfolds from around piston to allow downward suspension movement

FIGURE 14–48   Air spring operation. (Courtesy of Ford Motor Company)

2. **Drive into driveways**. Suspension problems often occur when turning at the same time the suspension hits a bump. The curb causes the suspension to compress while the wheels are turned. Defective stabilizer bar bushings, control arm bushings, and ball joints will usually make noise during this test procedure.

3. **Drive in reverse while turning.** This technique is usually used to find possible defective outer CV joints used on the drive axle shaft of front-wheel-drive vehicles.

4. **Drive over a bumpy road.** Worn or defective suspension (and steering) components can cause the vehicle to bounce or dart side to side while traveling over bumps and dips in the road. Worn or defective ball joints, control arm bushings, stabilizer bar bushings, stabilizer bar links, and worn shock absorbers can be the cause.

Once the problem has been confirmed, a further inspection can be performed in the service bay.

VISUAL INSPECTION

All suspension components should be carefully inspected for signs of wear or damage. While an assistant bounces the vehicle up and down, check if there is any free play in any of the suspension components (see Figure 14–50). Many load-carrying ball joints have wear indicators with a raised area around the grease fitting (see Figures 14–51 and 14–52). Always check wear-indicator ball joints with the wheels of the vehicle on the ground. If the raised area around the grease fitting is flush or recessed with the surrounding area, the ball joint is worn more than 0.050 in. and must be replaced.

SUSPENSION PROBLEM DIAGNOSIS CHART

CHECK	PROBLEM					
	Noise	**Instability**	**Pull to One Side**	**Excessive Steering Play**	**Hard Steering**	**Shimmy**
Tires/Wheels	Road/tire noise	Low/uneven air pressure; radials mixed with bias-belted ply tires	Low/uneven air pressure; mismatched tire sizes	Low/uneven air pressure	Low/uneven air pressure	Wheel out of balance/uneven tire wear/ over worn tires; radials mixed with belted bias ply tires
Shock Dampners (Struts/ Absorbers)	Loose/worn mounts/ bushings	Loose/worn mounts/ bushings; worn/ damaged struts/ shock absorbers	Loose/worn mounts/ bushings		Loose/worn mounts/ bushings on strut assemblies	Worn/damaged struts/shock absorbers
Strut Rods	Loose/worn mounts/ bushings	Loose/worn mounts/ bushings	Loose/worn mounts/ bushings			Loose/worn mounts/ bushings
Springs	Worn/damaged	Worn/damaged	Worn/damaged, especially rear		Worn/damaged	
Control Arms	Steering knuckle contacting control arm stop; worn/damaged mounts/bushings	Worn/damaged mounts/ bushings	Worn/damaged mounts/ bushings		Worn/damaged mounts/ bushings	Worn/damaged mounts/ bushings
Steering System	Component wear/damage	Component wear/damage	Component wear/damage	Component wear/damage	Component wear/damage	Component wear/damage
Alignment		Front and rear, especially caster	Front, camber and caster	Front	Front, especially caster	Front, especially caster
Wheel Bearings	On turns/speed changes: front-wheel bearings	Loose/worn (front and rear)	Loose/worn (front and rear)	Loose/worn (front and rear)		Loose/worn (front and rear)
Brake System			On braking		On braking	
Other	Clunk on speed changes: trans-axle: click on turn: CV joints; ball joint lubrication				Ball joint lubrication	Loose/worn friction ball joints

CAUTION: More than one factor may be the cause of a problem. Be sure to inspect all suspension components, and repair all parts that are worn or damaged. Failure to do so may allow the problem to reoccur and cause premature failure of other suspension components.

FIGURE 14–49   (Courtesy of Dana Corporation)

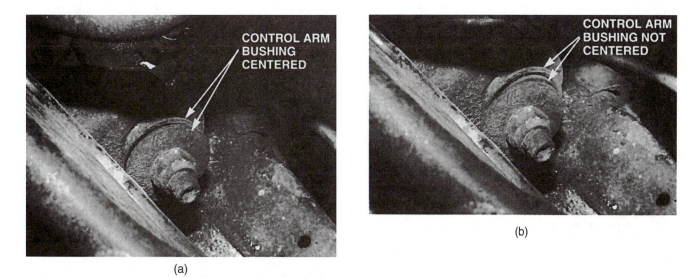

(a)

(b)

FIGURE 14–50   (a) This upper control arm bushing did not look worn or defective until the vehicle was pushed down. (b) As the vehicle was pushed down, the upper control arm moved off center, indicating a defective bushing.

FIGURE 14–51   Grease fitting projecting down from surrounding area of ball joint. The ball joint should be replaced when the area around the grease fitting is flush or recessed.

HINT: Most ball joints must be replaced if the joint has more than 0.050 in. of axial (up-and-down) movement. To help visualize this distance, consider that the thickness of an American nickel is about 0.060 in. (see Figure 14–53).

◄ **TECH TIP** ►

ROAD TEST—BEFORE AND AFTER

Often, technicians will start to work on a vehicle based on the description of the problem from the driver or owner. A typical conversation was overheard where the vehicle owner complained that the vehicle handled "funny," especially when turning. The owner wanted a wheel alignment and the technician and shopowner wanted the business. The vehicle was aligned, yet the problem was still present. The real problem was a defective tire. The service technician should have road tested the vehicle *before* any service work was done to confirm the problem and try to determine its cause. Every technician should test drive the vehicle *after* any service work is performed to confirm that the service work was performed correctly and that the customer complaint has been solved. This is especially true for any service work involving the steering, suspension, or braking systems.

To perform a proper ball-joint inspection, the force of the vehicle's springs *must* be *unloaded* from the ball joint. If this force is not relieved, the force of the spring forces the ball-and-socket joint tightly together, and any wear caused by movement will not be detected.

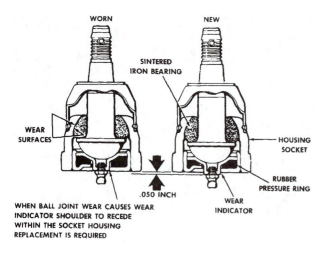

FIGURE 14–52 (Courtesy of Moog)

HINT: The location of the load-carrying ball joint is closest to the seat of the spring or torsion bar.

If the coil spring or torsion bar is attached to the *lower* control arm, the *lower* ball joint is the load-carrying ball joint (see Figure 14–54). This includes vehicles equipped with modified MacPherson strut suspension (see Figure 14–55). If the coil spring is attached to the upper control arm, a special tool or a block of wood is needed, as shown in Figure 14–56.

BALL-JOINT REMOVAL

Take care to avoid damaging grease seals when separating ball joints from their mounts. *The preferred method to separate tapered parts is to use a puller-type tool that applies pressure to the tapered joint as the bolt is tightened on the puller* (see Figure 14–57). Sometimes the shock of a hammer can be used to separate the ball joint from the steering knuckle. For best results, another hammer should be used as a backup while striking the joint to be separated on the side with a heavy hammer.

CAUTION: Using tapered "pickle forks" should be avoided, unless the part is to be replaced, because they often damage the grease seal of the part being separated.

Some ball-joint studs have a slot or groove where a **pinch bolt** is used to hold the ball joint to the steering knuckle (see Figure 14–58).

When removing ball joints that are riveted in place, always cut off or drill rivet heads before separating the

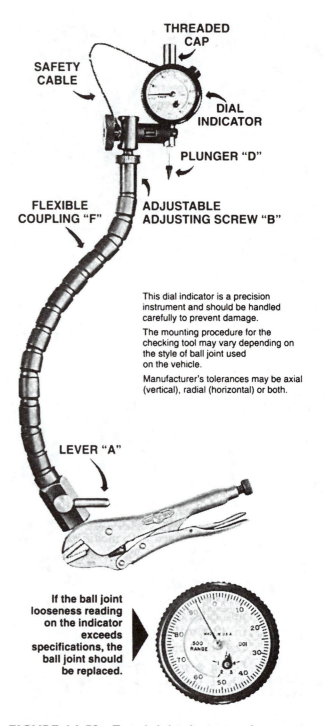

This dial indicator is a precision instrument and should be handled carefully to prevent damage.

The mounting procedure for the checking tool may vary depending on the style of ball joint used on the vehicle.

Manufacturer's tolerances may be axial (vertical), radial (horizontal) or both.

If the ball joint looseness reading on the indicator exceeds specifications, the ball joint should be replaced.

FIGURE 14–53 Typical dial indicator used to measure suspension component movement. The locking pliers attach the gauge to a stationary part of the vehicle and the flexible coupling allows the dial indicator to be positioned at any angle. (Courtesy of Moog)

ball joint from the spindle. This provides a more solid base to assist in removing rivets. *The preferred method to remove rivets from ball joints is to center punch and drill out the center of the rivet before using a drill or*

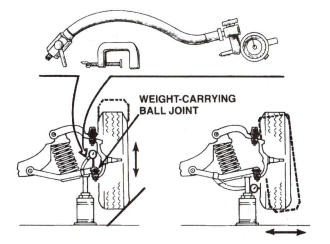

FIGURE 14–54 If the spring is attached to the lower control arm as in this SLA suspension, the jack should be placed under the lower control arm as shown. A dial indicator should be used to measure the amount of free play in the ball joints. Be sure that the looseness being measured is not due to normal wheel bearing end play. (Courtesy of Dana Corporation)

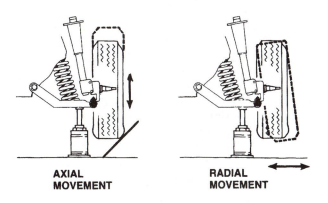

FIGURE 14–55 The jack should be placed under the lower control arm of this modified MacPherson suspension. (Courtesy of Dana Corporation)

an air-powered chisel to remove the rivet heads. Be careful not to drill or chisel into the control arms (see Figures 14–59 through 14–62).

Press-in ball joints are removed and installed using a special C-clamp tool, (see Figure 14–63).

NOTE: Many replacement press-in ball joints are slightly larger in diameter [about 0.050 in. (1.3 mm)] than the original ball joint to provide the same press fit. If the ball joints have been replaced before, the control arm must be replaced.

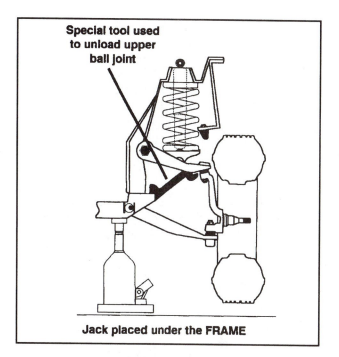

FIGURE 14–56 A special tool or a block of wood should be inserted between the frame and the upper control arm before lifting the vehicle off the ground. This tool stops the force of the spring against the upper ball joint so that a true test can be performed on the condition of the ball joint. (Courtesy of FMC)

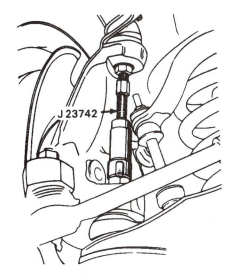

FIGURE 14–57 Taper breaker tool being used to separate the upper ball joint from the steering knuckle. (Courtesy of Oldsmobile)

Avoid using heat to remove suspension or steering components. Many chassis parts use rubber and plastic that can be damaged if heated with a torch. *If heat is used to remove a part, the part must be replaced.* Many

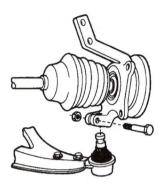

FIGURE 14–58 A pinch bolt attaches the steering knuckle to the ball joint. Remove the pinch bolt by turning the nut, not the bolt. (Courtesy of Moog)

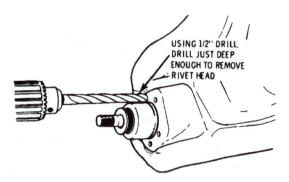

FIGURE 14–60 The head of the rivet can be removed by using a larger-diameter drill bit as shown. (Courtesy of Oldsmobile)

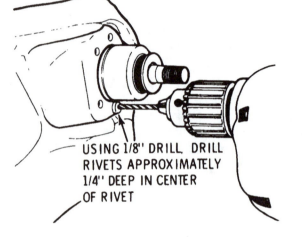

FIGURE 14–59 By drilling into the rivet, the holding force is released. (Courtesy of Oldsmobile)

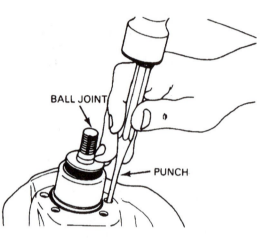

FIGURE 14–61 Using a punch and a hammer to remove the rivet after drilling down through the center and removing the head of the rivet. (Courtesy of Oldsmobile)

vehicles are equipped with nonreplaceable ball joints, and the entire control arm must be replaced if the ball joint is worn or defective.

CAUTION: Always follow the manufacturer's recommended installation instructions whenever replacing any suspension or other chassis component part. Tie-rod ends and ball joints use a taper to provide the attachment to other components. Whenever a nut is used to tighten a tapered part, it is important not to back off (loosen) the nut after tightening. As the nut is being tightened, the taper is being pulled into the taper of the adjoining part. The specified torque on the nut assures that the two pieces of the taper are joined properly. If the cotter key does not line up with the hole in the tapered stud when the nut has been torqued properly, tighten it more to line up a hole—never loosen the nut.

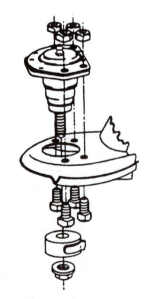

FIGURE 14–62 The replacement ball joint is bolted in using the same holes as the rivets. (Courtesy of Oldsmobile)

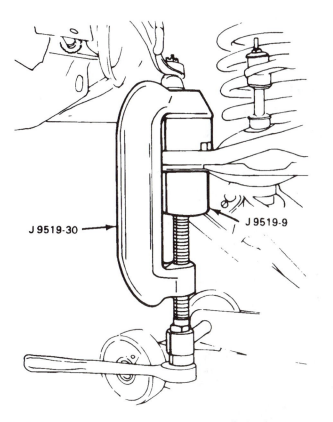

J 9519-30 →

J 9519-9

FIGURE 14–63 Press-in ball joints are best removed using a large C-clamp press as shown. The same tool is used to remove the old and install the new ball joint by using different adapters and sleeves for the tool. (Courtesy of Oldsmobile)

SHOCK ABSORBER AND STRUT DIAGNOSIS

Replacement shock absorbers may be required when any or all of the following symptoms appear:

1. **Ride harshness.** Worn shocks can cause ride harshness and yet not cause the vehicle to bounce after hitting a bump.
2. **Frequent "bottoming out" on rough roads.** Shock absorbers provide controlled movement of the axle whenever the vehicle hits a bump or dip in the road.

NOTE: Frequent bottoming out is also a symptom of reduced ride height due to sagging springs. Before replacing the shock absorbers, always check for proper ride height as specified in the vehicle service manual or any alignment specification booklet available from suppliers or companies of alignment or chassis parts and equipment. See the appendix for names and addresses of chassis and alignment equipment manufacturers.

◄ DIAGNOSTIC STORY ►

RATTLE STORY

A customer complained that rattling was heard every time the vehicle hit a bump. The noise sounded as if it came from the rear. All parts of the exhaust system and suspension system were checked. Everything seemed okay until the vehicle was raised with a frame-type hoist instead of a drive-on type. Then, whenever the right rear wheel was lifted, the noise occurred. The problem was a worn (elongated) shock absorber mounting hole. A washer with the proper-size hole was welded over the worn lower frame mount and the shock absorber was bolted back into place.

3. **Extended vehicle movement after driving on dips or a rise in the road.** The most common shock absorber test is the **bounce test**. Push down on the body of the vehicle and let go. The vehicle should return to its normal ride height and stop. If the vehicle continues to bounce two or three times, the shocks or struts are worn and must be replaced.
4. **Cuppy-type tire wear.** Defective shock absorbers can cause cuppy-type tire wear. This type of tire wear is caused by the tire bouncing up and down as it rotates.
5. **Leaking hydraulic oil.** When a shock or strut leaks oil externally, this indicates a defective seal. The shock absorber or strut cannot function correctly when low on oil inside.

Shock absorbers should be replaced in pairs. Both front or both rear shocks should be replaced together to provide the best handling and control.

NOTE: Shock absorbers do not affect ride height except where special air shocks or coil overload carrying shocks are used. If a vehicle is sagging on one side or in the front or the rear, the springs should be checked and replaced if necessary.

Shock absorbers are filled with fluid and sealed during production. A slight amount of fluid may bleed by the rod seal in cold weather and deposit a light film on the upper area of the shock absorber. This condition will not hurt the operation of the shock and should be considered normal.

FIGURE 14–64 Using correction fluid to mark just one strut mounting stud. This helps assure that the assembly is correctly positioned back in the vehicle.

MACPHERSON STRUT REPLACEMENT

On most vehicles equipped with MacPherson strut suspensions, strut replacement involves the following basic steps:

CAUTION: Always follow the manufacturer's recommended methods and procedures whenever replacing a MacPherson strut assembly or component.

---

1. Hoist the vehicle, remove the wheels, and mark the attaching bolts/nuts as shown in Figure 14–64.
2. Remove the upper strut mounting bolts except for one to hold the strut until ready to remove the strut assembly (see Figure 14–65).
3. Remove the brake caliper or brake hose from the strut housing (see Figures 14–66 and 14–67).
4. After removing all lower attaching bolts, remove the final upper strut bolt and remove the strut assembly from the vehicle (see Figure 14–68). Place the strut assembly into a strut spring compressor fixture as shown in Figures 14–69 and

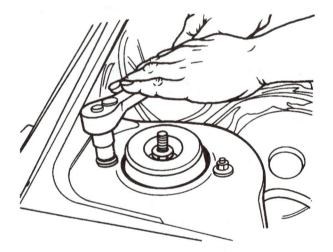

FIGURE 14–65 Removing the upper strut mounting bolts. (Courtesy of Dana Corporation)

14–70. Manual spring compressors can also be used as shown in Figures 14–71 and 14–72.

5. Compress the coil spring enough to relieve the tension on the strut rod nut. Remove the strut rod nut as shown in Figure 14–73.
6. After removing the strut rod nut, remove the upper strut bearing assembly and the spring (see Figure 14–74).

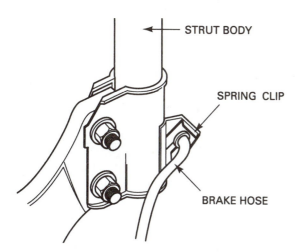

FIGURE 14–66 A brake hydraulic hose is often attached to the strut housing. Sometimes all that is required to separate the line from the strut is to remove a spring clip.

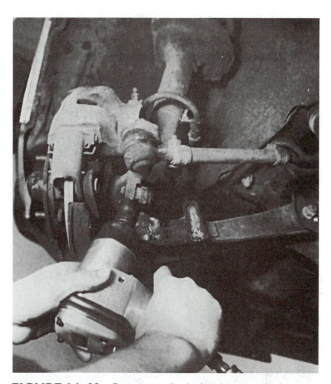

FIGURE 14–68 Removing the bolts that attach the strut to the knuckle.

FIGURE 14–67 A commonly used procedure of cutting through the strut clamp eliminates the need to disconnect the brake hose and open the hydraulic system. This operation saves time because the technician does not have to bleed the air out of the brake system after installing new struts.

FIGURE 14–69 Typical spring compressor designed for General Motors MacPherson strut assemblies. The bottom of the strut is attached to the fixture with pins and secured to the top with the top bearing nuts. The unit is held to a workbench with a conventional holding fixture that can also hold transmissions or transaxles.

285

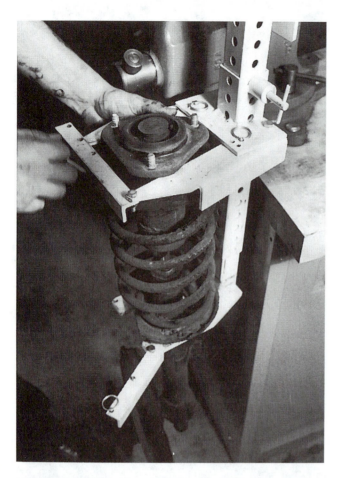

FIGURE 14–70 Universal-type MacPherson strut spring compressor. Note where the compressor is mounted—under the bearing and on top of the spring. This spring compressor is clamped in a vise attached to a workbench.

NOTE: The bearing assembly should be carefully inspected and replaced if necessary. Some automotive experts recommend replacing the bearing assembly whenever the strut is replaced.

7. Many MacPherson struts are replaced as an entire unit assembly. Other struts can have the cartridge inside the housing replaced. The cartridge is installed after removing the **gland nut** at the top of the strut tube and removing the original strut rod, valves, and hydraulic oil. Always replace the cartridge assembly as per the manufacturer's recommended procedure.

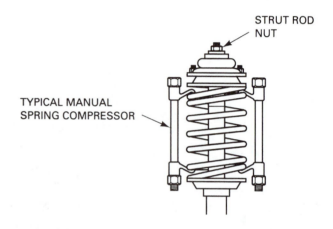

STRUT ROD NUT

TYPICAL MANUAL SPRING COMPRESSOR

FIGURE 14–71 Two individual spring compressors can also be used to compress the spring on a MacPherson strut before removing the strut retaining nut.

FIGURE 14–72 These individual spring compressors have a rubberized coating on the jaws that touch the spring. This is important because the use of metal jaws can remove the rust preventative coating on the spring. If this coating is removed, even in a small area, rust can start and can weaken or cause the spring to break.

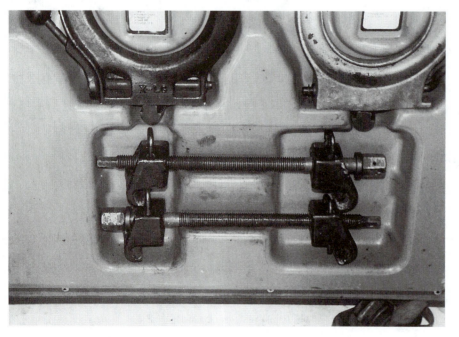

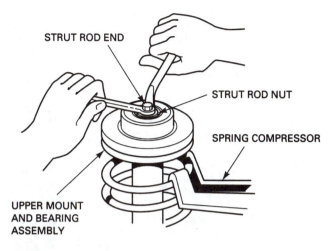

FIGURE 14–73   Removing the strut rod nut. The strut shaft is being held with one wrench while the nut is being removed with the other wrench. Notice that the spring is compressed before the nut is removed.

8. Reinstall the strut in the vehicle.

CAUTION: Many GM front-wheel-drive vehicles use a MacPherson strut with a large-diameter spring seat area. GM chassis engineers call this a "cow catcher" design. If the coil spring breaks, the extra large cow catcher will prevent the end of the broken spring from moving outward where it could puncture a tire and possibly cause an accident. Always use a replacement strut that has this same feature (see Figure 14–75).

FRONT COIL SPRING REPLACEMENT

Coil springs should be replaced in pairs if the vehicle ride height is lower than specifications. Sagging springs can cause the tires to slide laterally (side to side) across the pavement, causing excessive tire wear. Sagging springs can also cause the vehicle to "bottom out" against the suspension bumpers when traveling over normal bumps and dips in the road.

If a vehicle is overloaded, the springs of the vehicle can "take a set" and not recover to the proper ride height. This commonly occurs with all types of vehicles whenever a heavy load is carried even on a short trip. Most control arms use two holes for the purpose of coil spring seating. The end of the spring should cover one hole completely and partially cover the second hole (see Figures 14–76 through 14–78).

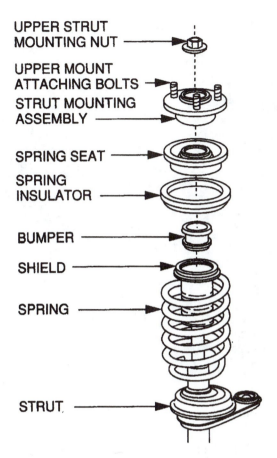

FIGURE 14–74   Typical MacPherson strut showing the various parts and components. (Courtesy of Dana Corporation)

FIGURE 14–75

FIGURE 14–76   Spring compressing tool in place to hold the spring as the ball joint is separated.

TORSION BAR ADJUSTMENT/REPLACEMENT

Most torsion bar suspensions are designed with an adjustable bolt to permit the tension on the torsion bar to be increased or decreased to change the ride height. Torsion bar adjustment should be made if the difference in ride height from one side to another exceeds ⅛ in. (0.125 in. or 3 mm). If the ride height difference side to side is greater than ⅛ in., the vehicle will tend to wander or be unstable with constant steering wheel movements required to maintain straight-ahead direction while driving on a straight, level road. See Figure 14–79 for an example of a clamp used to hold the adjusting lever if the torsion bar is being removed from the vehicle. Otherwise, the adjustment is performed by turning the adjusting bolt.

CONTROL ARM BUSHING REPLACEMENT

Defective control arm bushings are a common source of vehicle handling and suspension noise problems. Most suspension control arm bushings are constructed of three parts: an inner metal sleeve, the rub-

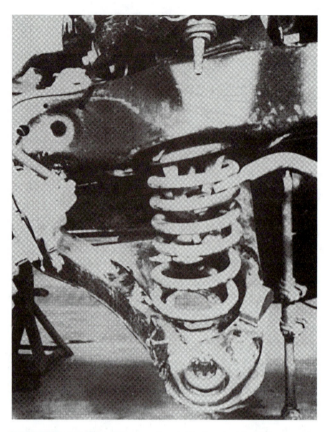

FIGURE 14–77   The spring is being held with a spring compressor as the lower control arm is being pushed down to release the spring.

Spring to be installed with tape at lowest position. Bottom of spring is coiled helical, and the top is coiled flat with a gripper notch near end of wire.

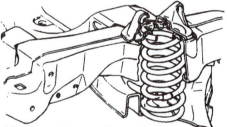

After assembly, end of spring coil must cover all or part of one inspection drain hole. The other hole must be partly exposed or completely uncovered.

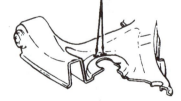

FIGURE 14–78   (Courtesy of Oldsmobile)

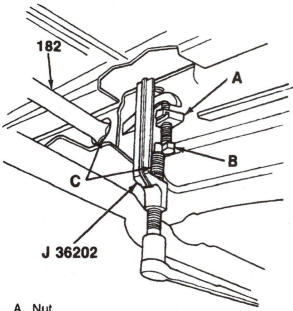

A. Nut
B. Torsion Bar Adjusting Bolt
C. Apply Lubricant at Points to Ease Installation
182. Torsion Bar

FIGURE 14–79 (Courtesy of Oldsmobile)

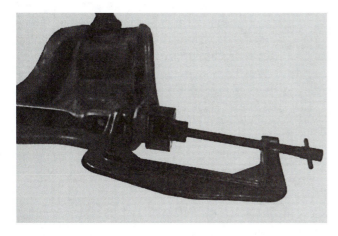

FIGURE 14–80 A C clamp type tool being used to remove and install control arm bushings.

ber bushing itself, and an outer steel sleeve. (Some vehicles use a two-piece bushing that does not use an outer sleeve.) To remove an old bushing from a control arm, the control arm must first be separated from the suspension and/or frame of the vehicle. While an air chisel is frequently used to force the steel sleeve out of the suspension member, a puller tool such as shown in Figure 14–80 is the method most recommended by vehicle manufacturers. All bushings should be tightened

◀ **DIAGNOSTIC STORY** ▶

IT'S NOT FAR—IT CAN TAKE IT

An automotive instructor needed to transport several V-8 engines just a couple of miles. A truck was not available, so the instructor carried the three engines in a station wagon. The rear of the station wagon sagged under the load. After the engines were unloaded, the rear of the station wagon remained lower than normal. The steel of the coil spring had exceeded its *yield point* and did not return to its original position. As a result, the rear coil springs are ruined and took a *set* due to the excessive load. The rear coil springs had to be replaced to restore the proper ride height.

NOTE: Leaf springs too can be easily overloaded and take a set or break! Overloading *any* vehicle can also damage the wheel bearings.

Never carry a load that exceeds the design capacity of the vehicle.

with the vehicle on the ground and the wheels in a straight-ahead position.

TROUBLESHOOTING ELECTRONIC LEVELING SYSTEMS

The first step of any troubleshooting procedure is to check for normal operation. Some leveling systems require that the ignition key be "on," while other systems operate all the time. Begin troubleshooting by placing approximately 300 lb (135 kg) on the rear of the vehicle. If the compressor does not operate, check to see if the sensor is connected to a rear suspension member and that the electrical connections are not corroded.

If the ride-height compressor runs excessively, check the air compressor, the air lines, and the air shocks (or struts) with soapy water for leaks. Most air shocks or air struts are not repairable and must be replaced. Most electronic level-control systems provide some adjustments of the rear ride height by adjusting

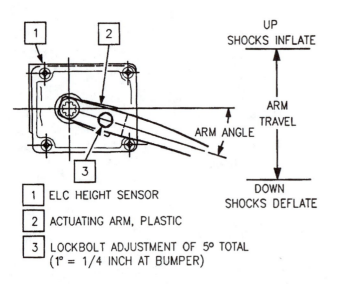

1 ELC HEIGHT SENSOR

2 ACTUATING ARM, PLASTIC

3 LOCKBOLT ADJUSTMENT OF 5° TOTAL
(1° = 1/4 INCH AT BUMPER)

FIGURE 14–81 Most electronic level control sensors can be adjusted such as this General Motors unit. (Courtesy of Oldsmobile)

the linkage between the height sensor and the rear suspension (see Figure 14–81).

SUMMARY

1. Spring types include coil, leaf, and torsion bar.
2. Suspension designs include a straight or solid axle, a two-control-arm type called an SLA, or a MacPherson strut.
3. All shock absorbers dampen the motion of the suspension to control ride and handling.
4. Ball joints attach to control arms and allow the front wheels to move up and down as well as turn.
5. Active (or reactive) suspension systems use sensors and a hydraulic pump to maintain ride height under all vehicle maneuvers.
6. A thorough road test of a suspension problem should include driving beside parked vehicles and into driveways in an attempt to determine when and where the noise occurs.
7. Defective shock absorbers can cause ride harshness as well as frequent "bottoming out" on rough roads.
8. Ball joints must be unloaded before testing. The ball joints used on vehicles with a MacPherson strut suspension are *not* load carrying. Wear indicator ball joints are observed with the wheels on the ground.
9. Always use a taper breaker puller or two hammers to loosen tapered parts to remove them. Never use heat unless you are replacing the part because heat from a torch can damage rubber and plastic parts.
10. When installing a tapered part, always tighten the attaching nut to specifications and never loosen the nut to install a cotter key. If the cotter key will not line up with a hole in the tapered part, tighten the nut more until the cotter key hole lines up with the nut and stud.
11. Always follow the manufacturer's recommended procedures whenever replacing springs or MacPherson struts. Never remove the strut rod nut until the coil spring is compressed and the spring force is removed from the upper bearing assembly.

REVIEW QUESTIONS

1. List the types of suspensions and name their component parts.
2. Describe the purpose and function of a stabilizer bar.
3. Describe how to perform a proper road test for the diagnosis of suspension-related problems.
4. List four symptoms of worn or defective shock absorbers.
5. Describe the testing procedure for ball joints.
6. Describe the general procedure to correctly remove and replace tapered suspension components.

MUTIPLE-CHOICE QUESTIONS

1. Technician A says that torsion bars are adjustable. Techician B says some coil springs are variable rate. Which technician is correct?
   a. A only
   b. B only
   c. Both A and B
   d. Neither A nor B

2. A vehicle makes a loud noise while traveling over bumpy sections of road. Technician A says that worn or deteriorated control arm bushings could be the cause. Technician B says that worn or deteriorated strut rod bushings could be the cause. Which technician is correct?
   a. A only
   b. B only
   c. Both A and B
   d. Neither A nor B

3. Two technicians are discussing MacPherson struts. Technician A says that the entire strut assembly rotates when the front wheels are turned. Technician B says that a typical MacPherson strut suspension system uses only one control arm and one ball joint per side. Which technician is correct?
   a. A only
   b. B only
   c. Both A and B
   d. Neither A nor B

4. Technician A says that replacement regular shock absorbers will raise the rear of a vehicle that is sagging down. Technician B says that replacement springs will be required to restore the proper ride height. Which technician is correct?
   a. A only
   b. B only
   c. Both A and B
   d. Neither A nor B

5. A vehicle used for 24-hour-a-day police work was found to have damaged suspension limiting rubber jounce bumpers. Technician A says that sagging springs could be the cause. Technician B says that defective or worn shock absorbers could be the cause. Which technician is correct?
   a. A only
   b. B only
   c. Both A and B
   d. Neither A nor B

6. Two technicians are discussing air shocks. Technician A says that air is forced through small holes to dampen the ride. Technician B says that air shocks are conventional hydraulic shock absorbers with an air bag to control vehicle ride height. Which technician is correct?
   a. A only
   b. B only
   c. Both A and B
   d. Neither A nor B

7. Two technicians are discussing ball-joint inspection. Technician A says that the vehicle should be on the ground with the ball joints *loaded*, then checked for free-play movement. Technician B says that the ball joints should be *unloaded* before checking for free play in the ball joints. Which technician is correct?
   a. A only
   b. B only
   c. Both A and B
   d. Neither A nor B

8. The preferred method to use to separate tapered chassis parts is to use
   a. a "pickle fork" tool.
   b. a torch to heat the joint until it separates.
   c. two hammers to shock and deform the taper or a puller tool.
   d. a drill to drill out the tapered part.

9. A light film of oil is observed on the upper area of a shock absorber. Technician A says that this condition should be considered normal. Technician B says that a rod seal may bleed fluid during cold weather causing the oil film. Which technician is correct?
   a. A only
   b. B only
   c. Both A and B
   d. Neither A nor B

10. Technician A says that the spring must be compressed before removing the strut rod nut. Technician B says that all MacPherson struts use a replaceable cartridge. Which technician is correct?
    a. A only
    b. B only
    c. Both A and B
    d. Neither A nor B

◀ Chapter 15 ▶

ALIGNMENT DIAGNOSIS AND SERVICE

OBJECTIVES

After studying Chapter 15, the reader will be able to:

1. Define camber, caster, toe, SAI, included angle, turning radius, thrust line, and setback.
2. Explain how camber, caster, and toe affect the handling and tire wear of the vehicle.
3. List the many checks that should be performed before aligning a vehicle.
4. Describe the proper alignment setup procedure.

A wheel alignment is the adjustment of the suspension and steering to ensure proper vehicle handling with minimum tire wear. The change in alignment angles may result from one or more of the following factors:

1. Wear of the steering and the suspension components
2. Bent or damaged steering and suspension parts
3. Sagging springs that can change the ride height of the vehicle and, therefore, the alignment angles

By adjusting the suspension and steering components, the proper alignment angles can be restored. An alignment includes checking and adjusting, if necessary, both front and rear wheels.

ALIGNMENT-RELATED PROBLEMS

Most alignment diagnosis is symptom based. The definitions of alignment symptom terms used in this book include:

- **Pull:** generally defined as a definite "tug" on the steering wheel toward the left or the right while driving straight on a level road (see Figure 15–1).
- **Lead or drift:** a mild pull that does not cause a force on the steering wheel that the driver must counteract. When the vehicle moves toward one side or the other, this is called a *lead* or a *drift*. A lead or drift could be caused by the crown of the road as shown in Figure 15–2.

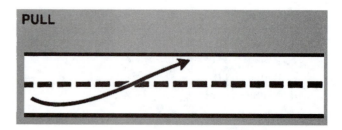

FIGURE 15–1 (Courtesy of Ford Motor Company)

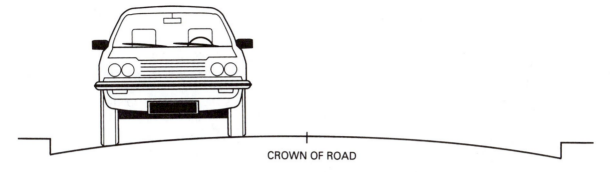

FIGURE 15–2 The *crown* of the road is the angle or slope of the roadway needed to drain water off the pavement.

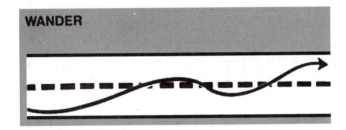

FIGURE 15–3 Wander is an unstable condition requiring constant driver corrections. (Courtesy of Ford Motor Company)

CAUTION: When test driving a vehicle for a lead or a drift, make sure that the road is free of traffic and that your hands remain close to the steering wheel. Your hands should simply be held away from the steering wheel for just a second or two, just long enough to check for a lead or drift condition.

- **Wander:** a condition where almost constant steering wheel corrections by the driver are necessary to maintain a straight-ahead direction on a straight, level road (see Figure 15–3).

CAMBER

Camber is the inward or outward tilt of the wheels from true vertical as viewed from the front or rear of the vehicle (see Figure 15–4).

1. If the top of the tire is tilted out, camber is positive (+).
2. If the top of the tire is tilted in, camber is negative (−).

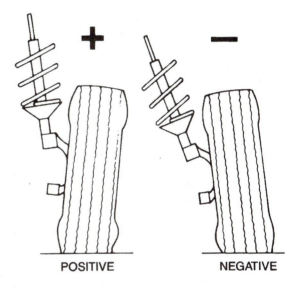

FIGURE 15–4 Positive and negative camber. (Courtesy of Dana Corporation)

3. Camber is zero (0°) if the tilt of the wheel is true vertical, as shown in Figure 15–5.
4. Camber is measured in degrees or fractions of degrees.
5. Camber can cause tire wear if not correct.
   a. **Excessive positive camber:** causes scuffing and wear on the outside edge of the tire as shown in Figure 15–6.
   b. **Excessive negative camber:** causes scuffing and wear on the inside edge of the tire as shown in Figure 15–7.
6. Camber can cause pull if it is unequal side to side. The vehicle will pull toward the side with the *most* camber. A difference of more than ½° from one side to the other will cause the vehicle to pull (see Figure 15–8).

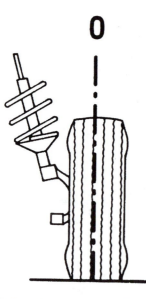

FIGURE 15–5 Zero camber. (Courtesy of Hunter Engineering Company)

7. Incorrect camber can cause excessive wear on wheel bearings as shown in Figures 15–9 and 15–10.

   Many vehicle manufacturers specify positive camber so that the vehicle's weight is applied to the larger inner wheel bearing and spindle. As the vehicle is loaded or when the springs sag, camber usually decreases. If camber is kept positive, the running camber is kept near zero degrees for best tire life.

NOTE: Many front-wheel-drive vehicles that use sealed wheel bearings often specify negative camber.

---

8. Camber is *not* adjustable on many vehicles.
9. If camber is adjustable, the change is made by moving the upper or the lower control arm or

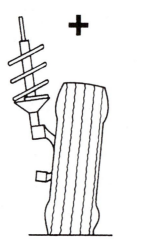

SCUFFING

OUTER SHOULDER WEAR

FIGURE 15–6 Excessive positive camber. (Courtesy of Hunter Engineering Company)

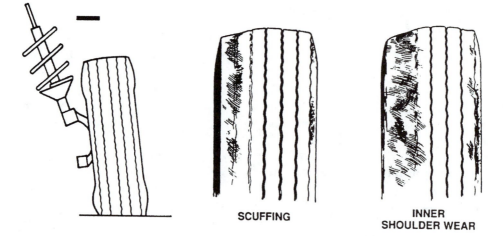

SCUFFING

INNER SHOULDER WEAR

FIGURE 15–7 Excessive negative camber. (Courtesy of Hunter Engineering Company)

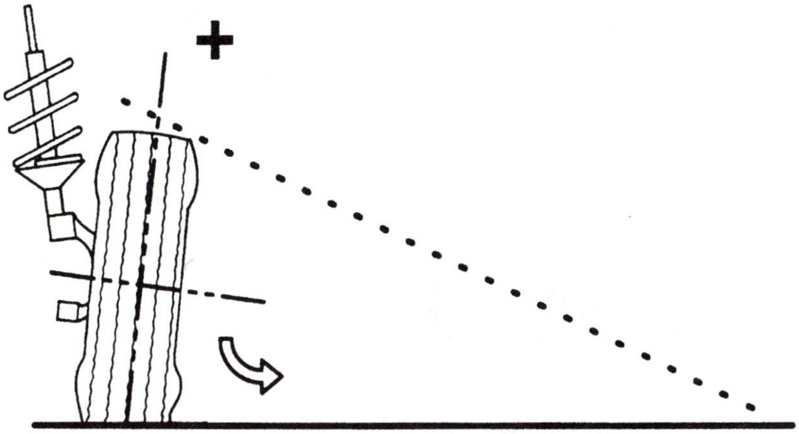

FIGURE 15–8 Camber tilts the tire and forms a cone shape that causes the wheel to roll away or pull outward toward the point of the cone. (Courtesy of Hunter Engineering Company)

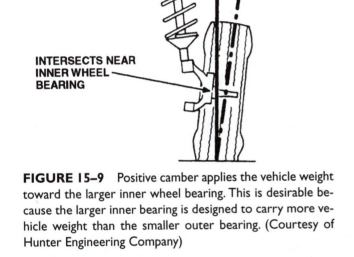

FIGURE 15–9 Positive camber applies the vehicle weight toward the larger inner wheel bearing. This is desirable because the larger inner bearing is designed to carry more vehicle weight than the smaller outer bearing. (Courtesy of Hunter Engineering Company)

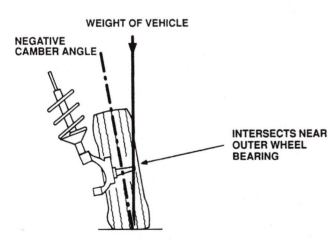

FIGURE 15–10 Negative camber applies the vehicle weight to the smaller outer wheel bearing. Excessive negative camber, therefore, may contribute to outer wheel bearing failure. (Courtesy of Hunter Engineering Company)

strut assembly by means of one of the following methods:

    a. Shims

    b. Eccentric cams

    c. Slots

  10. Camber should be equal on both sides; however, if camber cannot be adjusted exactly equal, make certain that there is more camber on the front of the left side to help compensate for the road crown ($\frac{1}{2}°$ maximum difference).

TOE

Toe is the difference in distance between the front and rear of the tires. As viewed from the top of the vehicle (bird's-eye view), zero toe means that both wheels on the same axle are parallel as shown in Figure 15–11. If the front of the tires are closer than the rear of the same tires, the toe is called *toe-in* or positive (+) toe (see Figure 15–12). If the front of the tires are farther apart than the rear of the same tires, the wheels are *toed-out* or have negative (−) toe (see Figure 15–13).

The purpose of the correct toe setting is to provide maximum stability with the minimum of tire wear when the vehicle is being driven.

  1. Toe is measured in fractions of degrees or fractions of an inch (usually $\frac{1}{16}$'s), millimeters (mm), or decimals of an inch (such as 0.06 in.).

  2. Incorrect toe is the major cause of excessive tire wear (see Figure 15–14).

NOTE: If the toe is improper by just $\frac{1}{8}$ in. (3 mm) the resulting tire wear is equivalent to dragging a tire sideways 28 ft (8.5 m) for every mile traveled (1.6 km).

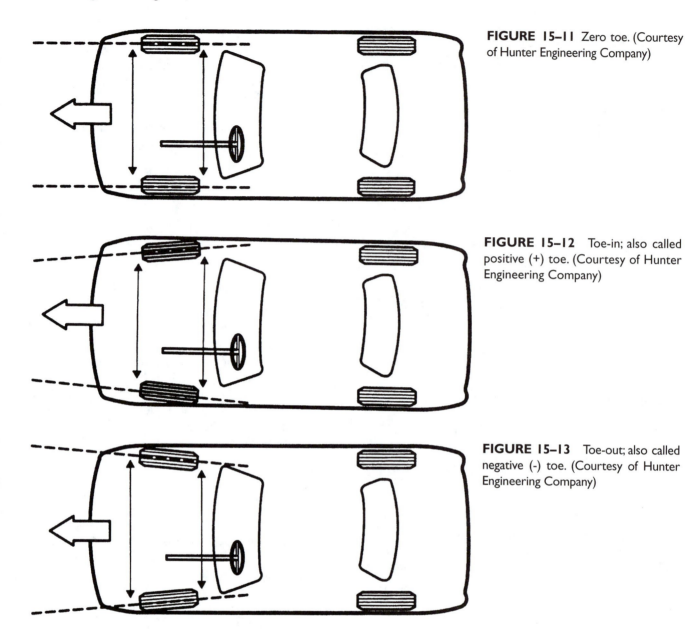

FIGURE 15–11 Zero toe. (Courtesy of Hunter Engineering Company)

FIGURE 15–12 Toe-in; also called positive (+) toe. (Courtesy of Hunter Engineering Company)

FIGURE 15–13 Toe-out; also called negative (-) toe. (Courtesy of Hunter Engineering Company)

Toe causes camber wear on one side of the tire if not correct, as shown in Figure 15–15.

3. Incorrect *front* toe does *not* cause a pull condition. Incorrect toe on the front wheels is split equally as the vehicle is driven because the forces acting on the tires are exerted through the tie rod and steering linkage to both wheels.

4. Incorrect (unequal) *rear* toe can cause tire wear (see Figures 15–16, 15–17, and 15–18). If the toe of the rear wheels is not equal, the steering wheel will not be straight and pull toward the side with the most toe-in.

5. Front toe adjustment must be made correctly by adjusting the tie rod sleeves (see Figure 15–19).

6. Most vehicle manufacturers specify a slight amount of toe-in to compensate for the natural tendency of the front wheels to spread apart (become toed-out) due to centrifugal force of the rolling wheels acting on the steering linkage.

NOTE: Some manufacturers of front-wheel-drive vehicles specify a toe-out setting to compensate for the toe-in forces created by the engine drive forces on the front wheels.

---

7. Normal wear to the tie rod ends and other steering linkage parts usually causes toe-out.

FIGURE 15-14   This tire is just one month old! It was new and installed on the front of a vehicle that had about ¼ in. (6 mm) of toe-out. By the time the customer returned to the tire store for an alignment, the tire was completely bald on the inside. Note the almost new tread on the outside.

CASTER

Caster is the forward or rearward tilt of the steering axis in reference to a vertical line as viewed from the side of

◀ TECH TIP ▶

SMOOTH IN, TOED IN—SMOOTH OUT, TOED OUT

Whenever the toe setting is not zero, a rubbing action occurs to the tire tread that causes a feather-edge wear (see Figure 15-20). A quick and easy method to determine if incorrect toe could be the cause for excessive tire wear or other problems is simply to rub your hand across the tread of the tire. If it feels smoother moving your hand toward the center of the vehicle than when you move your hand toward the outside, the cause is excessive toe-in. The opposite effect is caused by toe-out. This may be felt on all types of tires, including radial ply tires, where the wear may not be seen as feather edged. Just remember this simple saying, "smooth in, toed in—smooth out, toed out."

FIGURE 15-15   Excessive toe-out and the type of wear that can occur to the inside of the left front tire. (Courtesy of Hunter Engineering Company)

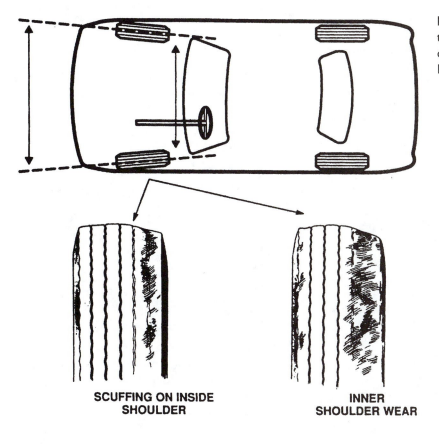

SCUFFING ON INSIDE
SHOULDER

INNER
SHOULDER WEAR

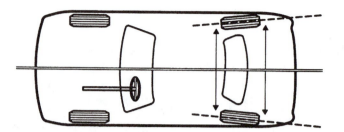

FIGURE 15–16 Rear toe-in (+). (Courtesy of Hunter Engineering Company)

FIGURE 15–17 Incorrect toe can cause the tire to run sideways as it rolls, resulting in a diagonal wipe. (Courtesy of Hunter Engineering Company)

FIGURE 15–18 Diagonal wear such as shown here is usually caused by incorrect toe on the rear of a front-wheel-drive vehicle.

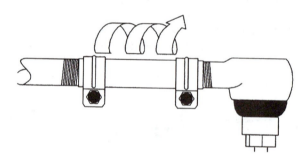

FIGURE 15–19 Toe on the front of most vehicles is adjusted by turning the tie rod sleeve as shown. (Courtesy of FMC)

the vehicle. The steering axis is defined as the line drawn through the upper and lower steering pivot points. On a SLA suspension system, the upper pivot is the upper ball joint and the lower pivot is the lower ball joint. On a MacPherson strut system, the upper pivot is the center of the upper bearing mount and the lower pivot point is the lower ball joint. Zero caster means that the steering axis is straight up and down, also called zero degrees or perfectly vertical as shown in Figure 15–21.

1. Positive (+) caster is present when the upper suspension pivot point is behind the lower pivot point (ball joint) as viewed from the side (see Figure 15–22).
2. Negative (−) caster is present when the upper suspension pivot point is ahead of the lower pivot point (ball joint) as viewed from the side (see Figure 15–23).
3. Caster is measured in degrees or fractions of degrees.
4. Caster is not a tire wearing angle, but positive caster does cause changes in camber during a

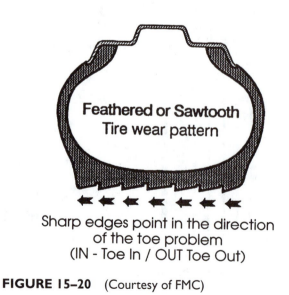

Feathered or Sawtooth Tire wear pattern

Sharp edges point in the direction of the toe problem
(IN - Toe In / OUT Toe Out)

FIGURE 15–20 (Courtesy of FMC)

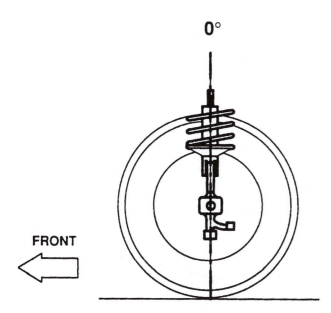

FIGURE 15–21   Zero caster. (Courtesy of Hunter Engineering Company)

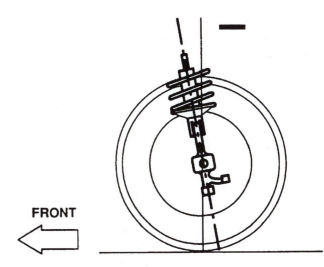

FIGURE 15–23   Negative (−) caster. (Courtesy of Hunter Engineering Company)

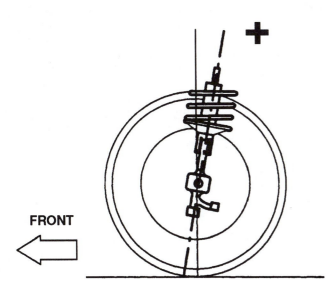

FIGURE 15–22   Positive (+) caster. (Courtesy of Hunter Engineering Company)

FIGURE 15–24   As the spindle rotates, it lifts the weight of the vehicle due to the angle of the steering axis. (Courtesy of Hunter Engineering Company)

turn (see Figure 15–24). This condition is called **camber roll**. (See the Tech Tip "Caster Angle Tire Wear.")

5. Caster is a stability angle.

   a. If the caster is excessively positive, the vehicle steering will be very stable (the vehicle will tend to go straight with little steering wheel correction needed) and help with steering wheel **returnability** after a turn.

   b. If the caster is positive, the steering effort will increase with increasing positive caster. Greater road shocks will be felt by the driver when driving over rough road surfaces (see Figure 15–25).

   Vehicles with as much as 11° positive caster usually have a steering dampener to control possible shimmy at high speeds and to dampen the snapback of the spindle after a turn (see Figure 15–26).

   c. If the caster is negative or excessively unequal, the vehicle will not be as stable and will tend to wander. If a vehicle is heavily loaded in the rear, the caster increases as shown in Figure 15–27.

6. Caster could cause pull if unequal. The vehicle will pull toward the side with the least caster.
7. Caster is *not* adjustable on many vehicles.
8. If caster is adjustable, the change is made by moving either the lower or the upper pivot point

forward or backward by means of one of the following methods:
  a. Shims
  b. Eccentric cams
  c. Slots
  d. Strut rods

NOTE: Caster is measured only on the front turning wheels of the vehicle. While some caster is built into the rear suspension of many vehicles, rear caster is not measured as part of a four-wheel alignment.

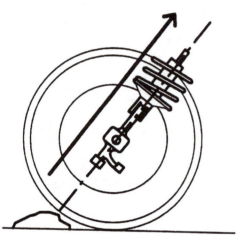

FIGURE 15–25 High caster provides a road shock path to the vehicle. (Courtesy of Hunter Engineering Company)

STEERING AXIS INCLINATION (SAI)

The steering axis is the angle formed between true vertical and an imaginary line drawn between the upper and lower pivot points of the spindle (see Figure 15–29). Steering axis inclination (SAI) is the inward tilt of the steering axis. SAI, also known as king pin inclination (KPI), is the imaginary line drawn through the king pin as viewed from the front.

FIGURE 15–26 Steering damper. (Courtesy of Hunter Engineering Company)

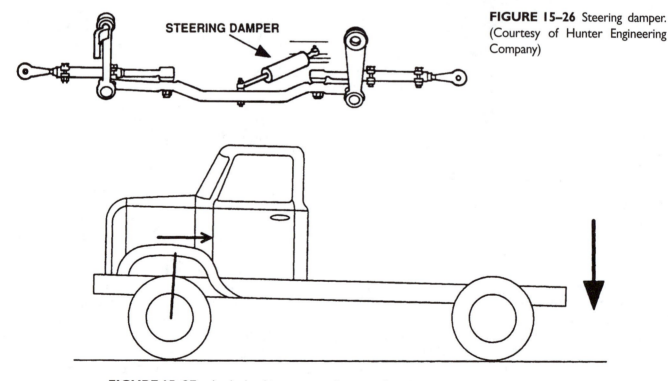

FIGURE 15–27 As the load increases in the rear of a vehicle, the top steering axis pivot point moves rearward, increasing positive (+) caster. (Courtesy of Hunter Engineering Company)

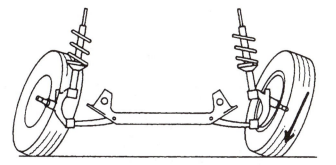

**OUTSIDE TURN
SPINDLE MOVES DOWN**

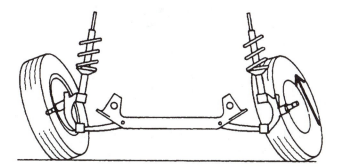

**INSIDE TURN
SPINDLE MOVES UP**

FIGURE 15-28 (Courtesy of Hunter Engineering Company)

◀ TECH TIP ▶

CASTER ANGLE TIRE WEAR

The caster angle is generally considered to be a *non*tire wearing angle. Although this statement is true, excessive or unequal caster can *indirectly* cause tire wear. When the front wheels are turned on a vehicle with a lot of positive caster, the front wheels become angled, called **camber roll**. (The caster angle is a measurement of the difference in camber angle from when the wheel is turned inward compared to when the wheel is turned outward.) Many vehicle manufacturers have positive caster designed into the suspension system. This positive caster has increased the directional stability of these vehicles. However, if the vehicle is used exclusively in city-type driving, the positive caster can cause tire wear to the outside shoulders of both front tires as seen in Figure 15–28.

INCLUDED ANGLE

Included angle is the SAI added to the camber reading of the front wheels only. The included angle is determined by the design of the steering knuckle, or strut

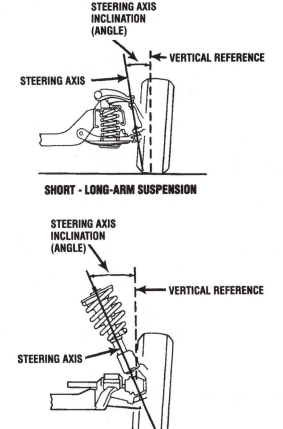

SHORT - LONG-ARM SUSPENSION

STRUT SUSPENSION

FIGURE 15–29 (Courtesy of Oldsmobile)

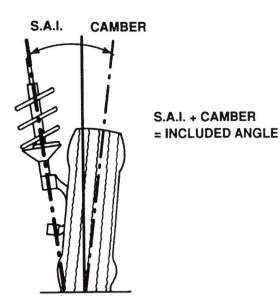

FIGURE 15–30 Included angle on a MacPherson strut suspension. (Courtesy of Hunter Engineering Company)

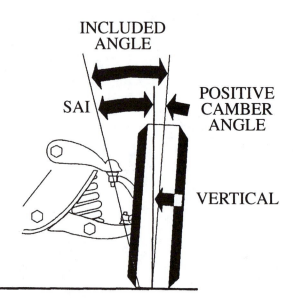

FIGURE 15–31 Included angle on an SLA suspension. (Courtesy of FMC)

construction (see Figures 15–30 and 15–31). Included angle is an important angle to measure for diagnosis of vehicle handling or tire wear problems.

TURNING RADIUS (TOE-OUT ON TURNS)

Whenever a vehicle turns a corner, the inside wheel has to turn at a sharper angle than the outside wheel because the inside wheel has a shorter distance to travel, as shown in Figure 15–32. **Turning radius,** also called **toe-out on turns** (abbreviated **TOT** or more commonly, **TOOT**), is determined by the angle of the steering knuckle arms. **Turning radius is a nonadjustable angle**. The turning radius can and should be measured as part of an alignment to check if the steering arms are bent or damaged. Symptoms of out-of-specification turning angle include tire squeal noise during normal cornering even at low speeds and/or scuffed tire wear. This angle is called the *Ackerman effect*, named for its promoter, an English publisher, Rudolph Ackerman, circa 1898.

SETBACK

Setback is the angle formed by a line drawn perpendicular (90°) to the front axles (see Figure 15–33). Setback is a nonadjustable measurement even though it may be corrected. Positive setback means that the right

front wheel is set back farther than the left. Negative setback means that the left front wheel is set back farther than the right.

THRUST ANGLE

Thrust angle is the angle of the rear wheels as determined by the total rear toe. If both rear wheels have zero toe, the thrust angle is the same as the geometric centerline of the vehicle. The total of the rear toe setting determines the **thrust line,** the direction in which the rear wheels are pointed (see Figure 15–34). On vehicles with an independent rear suspension, if both wheels do not have equal toe, the vehicle will pull in the direction of the side with the most toe-in.

TRACKING

Tracking is the term used to describe the fact that the rear wheels should "track" directly behind the front wheels. If the vehicle has been involved in an accident, it is possible that the frame or rear axle mounting could cause dog tracking.

PREALIGNMENT CHECKS

Before checking or adjusting the front-end alignment, the following items should be checked and corrected if necessary:

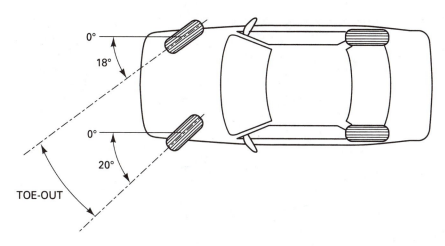

FIGURE 15–32 To provide handling, the inside wheel has to turn at a greater turning radius than the outside wheel.

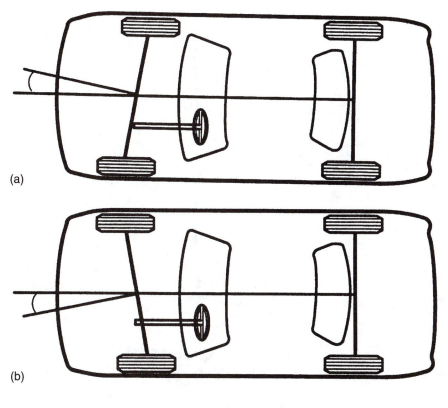

FIGURE 15–33 (a) Positive setback. (b) Negative setback. (Courtesy of Hunter Engineering Company)

1. Check all the tires for proper inflation pressures and approximately the same size and tread depth and that they are the recommended size for the vehicle (see Figure 15–35).
2. Check the front wheel bearings for proper adjustment.
3. Check for loose ball joints or torn ball-joint boots.
4. Check the tie rod ends for damage or looseness.
5. Check the center link or rack bushings for play.
6. Check the pitman arm for any movement.
7. Check for runout of the wheels and the tires.

8. Check for vehicle ride height (should be level front to back as well as side to side). Make sure that if there is a factory load-leveling system, it is functioning correctly. Check the height according to the manufacturer's specifications (see Figure 15–36).

NOTE: Manufacturers often have replacement springs or spring spacers that can be installed between the coil spring and the spring seat to restore proper ride level. Ride (trim) height is also called *chassis height*.

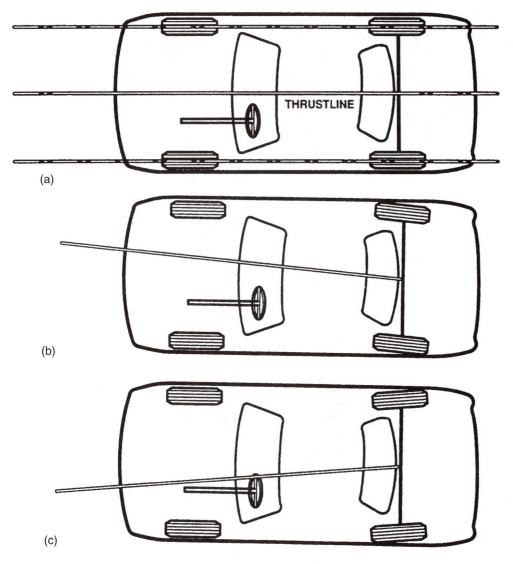

(a)

(b)

(c)

FIGURE 15–34 (a) Zero thrust angle. (b) Thrust line to the right. (c) Thrust line to the left. (Courtesy of Hunter Engineering Company)

9. Check for steering gear looseness at the frame.
10. Check for improperly operating shock absorbers.
11. Check for worn control arm bushings or ball joints.
12. Check for loose or missing stabilizer bar attachments.
13. Check the trunk for excess loads (see Figure 15–37).
14. Check for dragging brakes.

NOTE: Checking for dragging brakes is usually performed when installing alignment heads to the wheels prior to taking an alignment reading. A dragging brake can cause the vehicle to pull or lead toward the side with the dragging brake.

READING ALIGNMENT SPECIFICATIONS

There are several methods used by vehicle manufacturers and alignment equipment manufacturers to specify alignment angles.

Maximum/Minimum/Preferred Method. This method indicates the preferred setting for each alignment angle and the minimum and maximum allowable value for each. The alignment technician should always attempt to align the vehicle to the preferred setting (see Figure 15–38).

Plus or Minus Method. This method indicates the preferred setting with the lowest and highest allowable value indicated by a negative (−) and positive (+) sign, as in the specifications shown in Figure 15–39. For example, if a camber reading is specified as +½° with a + and −

	A tire worn on the outside edges, like this, has been run underinflated,		A bad toe adjustment can also cause "feathering" of a tire, which you have to feel to detect, since the tire may look perfectly good, as this one does.
	and a tire with just the center worn down, like this, has been overinflated.		
	When a tire is worn on only one side, like this one, it's a pretty good indication of a camber or toe problem.		Bald spots or scalloped effects are usually caused by unbalanced wheels, tire defects or worn suspension components.

FIGURE 15–35 (Courtesy of Ford Motor Company)

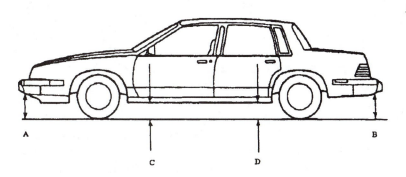

FIGURE 15–36 Measuring points for ride (trim) height vary by manufacturer. (Courtesy of Hunter Engineering Company)

FIGURE 15–37 Heavy or unequal loads can affect alignment angles.

value of ½°, it could be written as +½° ±½°. The minimum value would be 0° (½° - ½° = 0°) and the maximum value would be +1° (+½° + ½° = 1°). The range would be from 0 to 1°.

NOTE: The angle is assumed positive unless labeled with a negative (−) sign in front of the number.

◀ **DIAGNOSTIC STORY** ▶

FIVE-WHEEL ALIGNMENT

The steering wheel should always be straight when driving on a straight, level road. If the steering wheel is not straight, the customer will often think that the wheel alignment is not correct. One customer complained that the vehicle pulled to the right while driving on a straight road. The service manager test drove the vehicle and everything was perfect, except that the steering wheel was not perfectly straight, even though the toe setting was correct. Whenever driving on a straight road, the customer would "straighten" the steering wheel and, of course, the vehicle went to one side. After "correctly" adjusting the toe with the steering wheel straight, the customer and the service manager were both satisfied. The technician learned that regardless of how accurate the alignment, the steering wheel *must* be straight because it is this "fifth wheel" that the customer notices most.

***

NOTE: Many vehicle manufacturers now include the maximum allowable steering wheel angle variation from straight. This specification is commonly ± 3° (plus or minus three degrees) or less.

◀ **TECH TIP** ▶

SET EVERYTHING TO ZERO?

An apprentice service technician observed that the experienced alignment technician seldom looked at the specifications for the vehicle being aligned. When questioned, the technician said that for best tire life, the tires should rotate perpendicular to the road. After studying alignment specifications, the technician noticed that almost every camber and toe specification for both front and rear included zero within the range of the specifications. Caster, of course, varies from one vehicle to another and should be checked and adjusted to specifications. The beginning technician learned that zero camber and zero toe will be acceptable and "within specifications" on almost all vehicles and is easy to remember!

Specifications are often published in fractional or decimal degrees or in degrees and minutes. There are 60 minutes (written as 60′) in 1 degree.

Angle–Unit Conversions

Units	Conversions		
Fractional degrees	¼°	½°	¾°
Decimal degrees	0.25°	0.50°	0.75°
Degrees and minutes	0°15′	0°30′	0°45′

ALIGNMENT SETUP PROCEDURES

After confirming that the tires and all steering and suspension components are serviceable, the vehicle is ready for an alignment. The exact setup procedures for the equipment being used must always be followed. The typical alignment procedures include the following steps:

1. Drive onto the alignment rack straight and adjust the ramps and/or turn plates so that they are centered under the tires of the vehicle (see Figure 15–40).
2. Use chocks for the wheels to prevent the vehicle from accidentally rolling off the alignment rack.
3. Attach and calibrate the wheel sensors to each wheel as specified by the alignment equipment manufacturer as shown in Figure 15–41.
4. Unlock all rack or turn plates.
5. Lower the vehicle and "jounce" the vehicle by pushing down on the front, then the rear bumper of the vehicle. This motion allows the suspension to become centered.
6. Following the procedures for the alignment equipment, determine all alignment angles.

MEASURING CAMBER, CASTER, SAI, TOE, AND TOOT

Camber. Camber is measured with the wheels in the straight-ahead position on a level platform. Since camber is a vertical reference angle, alignment equipment reads camber directly.

ALIGNMENT SPECIFICATIONS AT CURB HEIGHT

FRONT WHEEL ALIGNMENT	ACCEPTABLE ALIGNMENT RANGE AT CURB HEIGHT	PREFERRED SETTING
CAMBER . . All*.................................	-0.6° to +0.6°	+0.0°
*Side To Side Differential	0.7° or less	0.0°
TOTAL TOE All Vehicles (See Note) Specified In Degrees	0.4° In to 0.0°	0.2° In
CASTER*	REFERENCE ANGLE	
All Models..	+2.0° to +4.0°	+3.0°
*Side To Side Caster Differential Not to Exceed..	1.0° or less	0.0°
REAR WHEEL ALIGNMENT	ACCEPTABLE ALIGNMENT RANGE AT CURB HEIGHT	PREFERRED SETTING
CAMBER . . All Models................................	-0.6° to +0.4°	-0.1°
TOTAL TOE* All Vehicles (See Note) Specified In Degrees	0.2° Out to 0.4° In	0.1° In
THRUST ANGLE............................... *TOE OUT When Backed On Alignment Rack Is TOE IN When Driving.	-0.15° to +0.15°	

NOTE: Total toe is the arithmetic sum of the left and right wheel toe settings. Positive is Toe-in, negative is Toe-out. Total Toe must be equally split between each front wheel to ensure a centered steering wheel. Left and Right toe must be equal to within 0.02 degrees.

FIGURE 15–38 Curb height is ride height or trim height as measured at the curb weight. Curb weight is when the vehicle has a full tank of fuel and all other fluids filled. (Courtesy of Chrysler Corporation)

WHEEL ALIGNMENT SPECIFICATIONS

	CASTER	CROSS CASTER (LH-RH)	CAMBER	CROSS CAMBER (LH-RH)	TOE (TOTAL IN) DEGREES	STEERING WHEEL ANGLE	THRUST ANGLE
FRONT	+3°±.5°	0°±.75°	+.2° ±.5°	0°±.75°	0°±.3°	0°±3°	- -
REAR	- -	- -	-.3° ±.5°	0°±.75°	+.1° ±.2°	- -	0°±.1°

NT076

FIGURE 15–39 (Courtesy of Oldsmobile)

FIGURE 15–40 Drive onto the alignment rack as straight as possible with the turn plates positioned so that the center of the wheel is directly over the center of the turn plates.

FIGURE 15–41 This wheel sensor uses a safety wheel that screws to the valve stem to keep it from falling onto the ground if the clamps slip on the wheel lip.

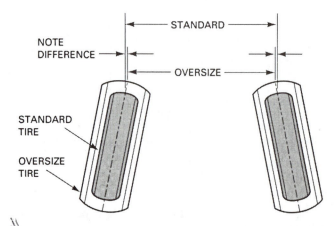

FIGURE 15–42 If toe for an oversize tire is set by distance, the toe angle will be too small. Toe angle is the same regardless of tire size.

Caster. Caster is measured by moving the front wheels through an arc both inward, then outward, from straight ahead. This necessary movement of the front wheels to measure caster is called **caster sweep**. What the alignment measuring equipment is actually doing is measuring the camber at one wheel sweep and measuring the camber again at the other extreme of the caster sweep. The caster angle itself is the difference between the two camber readings.

◄ **TECH TIP** ►

USING A TWO-WHEEL ALIGNMENT RACK FOR A FOUR-WHEEL ALIGNMENT

Alignment racks designed for total four-wheel alignment are equipped with movable plates under the rear wheels. Older racks used for front-end-only alignment do not move and therefore will not allow the rear suspension to become settled. This freedom of movement is necessary to perform a four-wheel alignment correctly, especially on a vehicle with independent rear suspension. One commonly used "trick of the trade" is to place a plastic garbage bag under each rear wheel before lowering the vehicle onto the rack. As the vehicle is lowered, the rear wheels will easily *slide* over the plastic-on-plastic surface. The rear wheels will resume the normal position, just as if the vehicle were lowered onto movable turn plates. Another method that is often used is to roll the vehicle back about 4 ft (1.2 m) and then forward to allow the rear independent suspension to settle.

NOTE: Some alignment machines do not have long enough cables to allow this method to be used.

SAI. Steering axis inclination (SAI) is also measured by performing a caster sweep of the front wheels. When measuring SAI separately, the usual procedure involves raising the front wheels off the ground and leveling and locking the wheel sensors before performing a caster sweep. When the suspension is extended, the SAI is more accurately determined because the angle itself is expanded.

Toe. Toe is determined by measuring the angle of both front and/or both rear wheels from the straight-ahead (0°) position. Most alignment equipment reads the toe angle for each wheel *and* the combined toe angle of both wheels on the same axle. This combined toe is called **total toe**. Toe angle is more accurate than the center-to-center distance, especially if oversized tires are installed on the vehicle (see Figure 15–42).

TOOT. Toe-out on turns (TOOT) is a diagnostic angle and is normally not measured as part of a regular alignment, but it is recommended to be performed as a part of

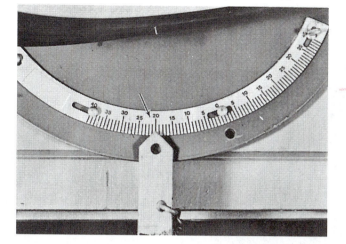

FIGURE 15–43 The protractor scale on the front turn plates allow the technician to test the turning radius by turning one wheel to an angle specified by the manufacturer and observing the angle of the other front wheel.

a total alignment check. TOOT is measured by recording the angle of the front wheels as indicated on the front turn plates (see Figure 15–43). If, for example, the inside wheel is turned 20°, the outside wheel should indicate about 18° on the turn plate. The exact angles are usually specified by the vehicle manufacturer. The turning angle should be checked only after the toe is correctly set. *The turning angle for the wheel on the outside of the turn should not vary more than 1½° from specifications.*

CHECKING FOR BENT STRUTS, SPINDLES OR CONTROL ARMS

Even a minor bump against a curb can bend a spindle or a strut housing, as shown in Figure 15–44. Before attempting to correct an alignment, check all the angles

◀ TECH TIP ▶

ASK YOURSELF THESE THREE QUESTIONS

An older technician told a beginning technician that the key to success in doing a proper alignment is to ask yourself three questions about the alignment angles.

Question 1. *"Is it within specifications?"* For example, if the specification reads 1 ±½°, any reading between +½ and +1½° is within specifications. All vehicles should be aligned to within this range. Individual opinions and experience can assist the technician whether the actual setting should be at one extreme or the other or held to the center of the specification range.

Question 2. *"Is it within ½° of the other side of the vehicle?"* Not only should the alignment be within specifications, it should also be as equal as possible from one side to the other. The difference between the camber from one side to the other side is called **cross camber**. **Cross caster** is the difference between the caster angle from one side to another. Some manufacturers and technicians recommend that this side-to-side difference be limited to ¼°!

Question 3. *"If the camber and caster cannot be exactly equal side to side in the front, is there more camber on the left and more caster on the right to help compensate for road crown?"* Seldom, if ever, are the alignment angles perfectly equal. Sometimes one side of the vehicle is more difficult to adjust than the other side. Regardless of the reasons, if there *has* to be a difference in front camber and/or caster angle, follow this advice to avoid a possible lead or drift problem even if the answers to the first two questions are yes.

FIGURE 15–44 Notice the difference in angle of the strut and spindle between the original on the right and the new replacement strut housing on the left.

and use the appropriate diagnostic chart to check for hidden damage that a visual inspection may miss.

The following charts can be used to determine what is wrong if the alignment angles are known. Simply use the chart that correctly identifies the type of suspension on the problem vehicle.

at spec = the alignment angle is within specifications.

over spec = the alignment angle is greater or higher than specified by the manufacturer.

under spec = the alignment angle is less than or lower than specified by the manufacturer.

Strut Suspension Diagnostic Chart

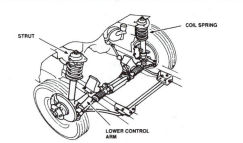

FIGURE 15–45 (Courtesy of Hunter Engineering Company)

SAI	Camber	Included angle	Possible cause
At spec	Under spec	Under spec	Bent spindle or strut

SAI	Camber	Included angle	Possible cause
At spec	Over spec	Over spec	Bent spindle or strut
Under spec	Over spec	At spec	Bent transverse link or tower out at top
Under spec	Over spec	Over spec	Bent transverse link or strut tower out at top as well as bent spindle or strut
Under spec	Over spec	Under spec	Bent transverse link or strut tower out at top as well as bent spindle or strut
Over spec	Under spec	At spec	Strut tower in at top
Over spec	Over spec	Over spec	Strut tower in at top and bent spindle or strut

SLA-Type Suspension Diagnostic Chart

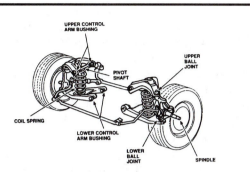

FIGURE 15–46 (Courtesy of Hunter Engineering Company)

SAI	Camber	Included angle	Possible cause
At spec	Under spec	Under spec	Bent spindle
Under spec	Over spec	At spec	Bent lower transverse link
Under spec	Over spec	Over spec	Bent lower transverse link and spindle
Over spec	Under spec	At spec	Bent upper transverse link

Multilink Suspension Diagnostic Chart

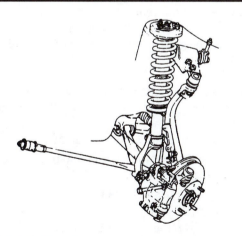

FIGURE 15–47 (Courtesy of Hunter Engineering Company)

SAI	Camber	Included angle	Possible cause
At spec	Under spec	Under spec	Steering knuckle

SAI	Camber	Included angle	Possible cause
At spec	Over spec	Over spec	Steering knuckle
Under spec	Over spec	At spec	Bent transverse link, upper link out, or third link out at bottom
Under spec	Under spec	Under spec	Bent steering knuckle and either bent transverse link or upper link out
Under spec	At spec	Under spec	Bent steering knuckle and either bent transverse link or upper link out
Over spec	Under spec	At spec	Upper link in
Over spec	Over spec	Over spec	Upper link in and bent steering knuckle

Straight (Mono) Axle Diagnostic Chart

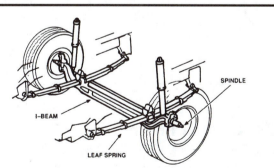

FIGURE 15–48 (Courtesy of Hunter Engineering Company)

SAI	Camber	Included angle	Possible cause
At spec	Over spec	Over spec	Bent spindle/ assembly

(continues)

SAI	Camber	Included angle	Possible cause
Over spec	Under spec	At spec	Bent axle housing
Under spec	Over spec	At spec	Bent axle housing
Under spec	Over spec	Over spec	Bent spindle and axle housing

Twin-I-Beam Diagnostic Chart

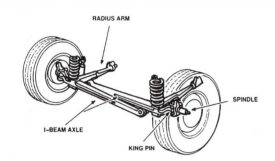

FIGURE 15–49   (Courtesy of Hunter Engineering Company)

SAI	Camber	Included angle	Possible cause
At spec	Over spec	Over spec	Bent spindle assembly
Over spec	Under spec	At spec	Bent I-beam
Under spec	Over spec	At spec	Bent I-beam
Under spec	Over spec	Over spec	Bent I-beam and spindle assembly

CHECKING FRAME ALIGNMENT OF FRONT-WHEEL-DRIVE VEHICLES

Many front-wheel-drive vehicles mount the drivetrain (engine and transaxle) and lower suspension arms to a subframe or cradle. The frame being shifted either left or right can cause differences in SAI, included angle, and camber (see Figures 15–50 and 15–51). Adjust the frame if the SAI and camber angles are different on the left and right sides, yet the included angles are equal.

TYPES OF ALIGNMENTS

There are three types of alignment: geometric centerline, thrust line, and total four-wheel alignment.

Geometric Centerline.   This type of alignment was simply an alignment that uses the geometric centerline of the vehicle as the basis for all measurements of toe (front or rear) (see Figure 15–52). This method is now considered to be obsolete.

Thrust Line.   A thrust line alignment uses the thrust angle of the rear wheels and sets the front wheels parallel to the thrust line (see Figure 15–53).

HINT: It has often been said that while the front wheels steer the vehicle, the rear wheels determine the direction the vehicle will travel. Just think of the rear wheels as being like a rudder of a boat. As the rudder turns, the front of the boat turns.

Thrust line alignment is *required* for any vehicle with a nonadjustable rear suspension. If a vehicle has an adjustable rear suspension, a total four-wheel alignment is necessary to ensure proper tracking.

Total Four-Wheel Alignment.   A total four-wheel alignment is the most accurate method and is necessary to ensure maximum tire wear and vehicle handling. The major difference between a thrust line alignment and a total four-wheel alignment is that the rear toe is adjusted to bring the thrust line to zero. In other words, the rear toe on both rear wheels is adjusted equally so that the actual direction the rear wheels are pointed is the same as the geometric centerline of the vehicle (see Figure 15–54).

Four-Wheel Alignment Procedure.   The procedure for a total four-wheel alignment includes these steps:

1. Adjust the rear camber (if applicable).
2. Adjust the rear toe (this should reduce the thrust angle to near zero).
3. Adjust the front camber and caster.
4. Adjust the front toe, being sure that the steering wheel is in the straight-ahead position.

ADJUSTING REAR CAMBER/TOE

Adjusting rear camber and/or toe is the first step in the total four-wheel alignment process. Rear camber is

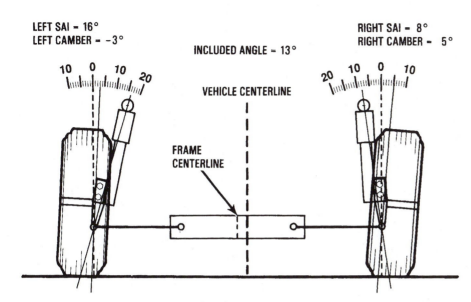

LEFT SAI = 16°
LEFT CAMBER = -3°

INCLUDED ANGLE = 13°

VEHICLE CENTERLINE

RIGHT SAI = 8°
RIGHT CAMBER = 5°

FRAME CENTERLINE

FIGURE 15–50 In this example both SAI and camber are way off from being equal side to side. However, both sides have the same included angle, indicating that the frame may be out of alignment. An attempt to align this vehicle by adjusting the camber on both sides either with factory or aftermarket kits would result in a totally incorrect alignment. (Courtesy of Oldsmobile)

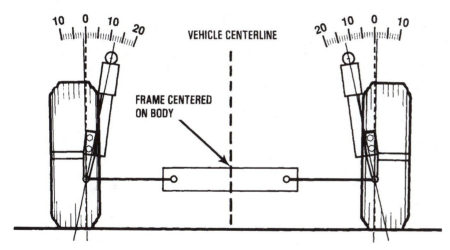

SAI = 12°
CAMBER = 1°
INCLUDED ANGLE = 13°

VEHICLE CENTERLINE

FRAME CENTERED ON BODY

FIGURE 15–51 This is the same vehicle as shown in Figure 15–50, but now the frame (cradle) has been shifted over and correctly positioned. Notice how both the SAI and camber became equal without any other adjustments necessary. (Courtesy of Oldsmobile)

rarely made adjustable but can be corrected by using aftermarket alignment kits or shims. See Figures 15–55 and 15–56 for examples of various methods used to adjust rear camber.

Many vehicle manufacturers provide adjustment for rear toe on many vehicles that use an independent rear suspension. Rear toe adjusts the thrust angle. A thrust angle that exceeds ½° (0.5°) on a vehicle with a solid axle is an indication that a component could be damaged or out of place in the rear of the vehicle. Rear toe is often adjusted using an adjustable tie rod

end or an eccentric cam on the lower control arm. Most solid rear axles do not have a method to adjust rear toe except for aftermarket shims or kits (see Figure 15–58).

NOTE: On vehicles equipped with four-wheel steering, refer to the service manual for the exact procedure to follow to lock or hold the rear wheels in position for a proper alignment check.

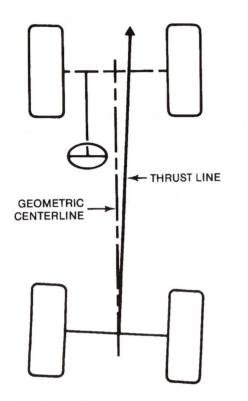

FIGURE 15–52 Geometric centerline alignment sets the front toe readings based on the geometric centerline of the vehicle and does not consider the thrust line of the rear wheel toe angles. (Courtesy of Hunter Engineering Company)

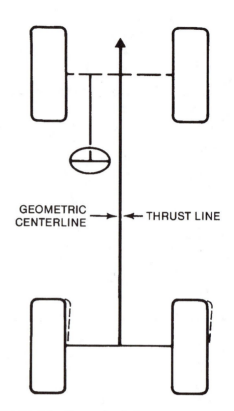

FIGURE 15–54 Four-wheel alignment corrects for any rear wheel toe to make the thrust line and the geometric centerline of the vehicle both the same. (Courtesy of Hunter Engineering Company)

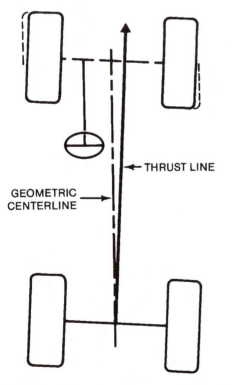

FIGURE 15–53 Thrust line alignment sets the front toe parallel with the rear wheel toe. (Courtesy of Hunter Engineering Company)

REAR CAMBER ADJUSTMENT

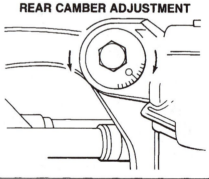

Rotate cam bolt to adjust.

FIGURE 15–55 Some rear suspensions use an eccentric cam to adjust camber. The bolt is offset from the center of the large washers. When the bolt is rotated, the control arm moves in and out, which changes the location of the suspension point. (Courtesy of FMC)

314

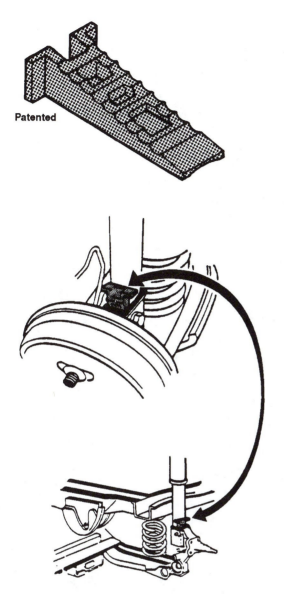

Patented

FIGURE 15–56 Aftermarket alignment parts or kits are available to change the rear camber. (Courtesy of Shimco International, Inc.)

◀ TECH TIP ▶

GRITTY SOLUTION

Often it is difficult to loosen a TORX bolt, especially those used to hold the backing plate onto the rear axle on many GM vehicles (see Figure 15–57).

A technique that always seems to work is to place some *valve grinding compound* on the fastener. The gritty compound keeps the TORX socket from slipping up and out of the fastener, and more force can be exerted to break loose a tight bolt. Valve grinding compound can also be used on Phillips head screws as well as other types of bolts, nuts, and sockets.

GUIDELINES FOR ADJUSTING FRONT CAMBER/SAI AND INCLUDED ANGLE

If the camber is adjusted at the base of the MacPherson strut, camber and included angle are changed and SAI remains the same (see Figure 15–59). If camber is adjusted by moving the upper strut mounting location, the included angle remains the same but the SAI and camber change (see Figure 15–60).

SAI and included angle should be within ½° (0.5°) side to side. If these angles differ, check the frame mount location before attempting to correct differences in camber. Cross camber/caster means the difference between the camber or caster on one side of the vehicle from the camber or caster on the other side of the vehicle (see Figure 15–61).

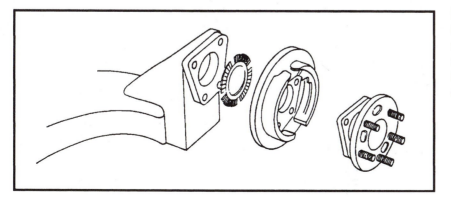

FIGURE 15–57 Full contact plastic or metal shims can be placed between the axle housing and the brake backing plate to change the rear camber or toe or both. (Courtesy of Northstar Manufacturing Company, Inc.)

HOW TO SELECT THE CORRECT ALIGNMENT SHIM

1. The fraction or whole number embossed on each shim is the approximate degree of correction.

2. The shim drawings illustrate the position in which the shims could be installed on the rear axle.

3. **TO REDUCE CAMBER,** the thick part of the shim should be at the bottom (six o'clock) of the axle.

4. **TO INCREASE CAMBER,** the thick part of the shim should be at the top (twelve o'clock) of the axle.

5. **TO REDUCE TOE-OUT,** the thick part of the shim should point towards the rear of the vehicle.

6. **TO INCREASE TOE-OUT,** the thick part of the shim should point towards the front of the vehicle.

7. Fine adjustment can be obtained by rotating the shim right or left with the slots provided.

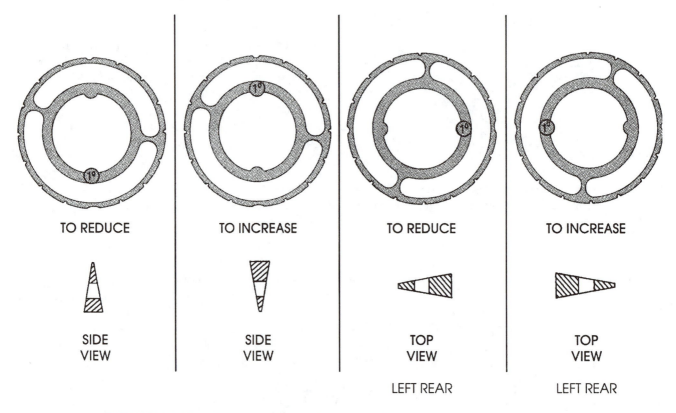

CAMBER   TOE-OUT

TO REDUCE	TO INCREASE	TO REDUCE	TO INCREASE
SIDE VIEW	SIDE VIEW	TOP VIEW	TOP VIEW
		LEFT REAR	LEFT REAR

FIGURE 15–58 (Courtesy of Shimco International, Inc.)

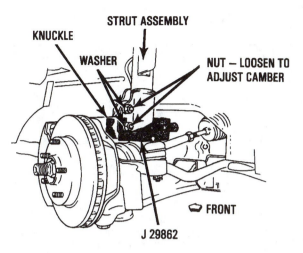

FIGURE 15–59 Many struts allow camber adjustment at the strut-to-knuckle fasteners. Here a special tool is being used to hold and move the strut into alignment with the fasteners loosened. Once the desired camber angle is achieved, the strut nuts are tightened and the tool removed. (Courtesy of Oldsmobile)

Alignment angle	Recommended maximum variation
SAI	Within ½° (0.5°) side to side
Included angle	Within ½° (0.5°) side to side
Cross camber	Within ½° (0.5°) side to side
Cross caster	Within ½° (0.5°) side to side

CAUTION: Do not attempt to correct a pull condition by increasing the cross camber or cross caster beyond the amount specified by the vehicle manufacturer.

ADJUSTING FRONT CAMBER/CASTER

Most SLA-type suspensions can be adjusted for caster and camber. Most manufacturers recommend adjusting caster, then camber, before adjusting the toe. As the caster is changed, the camber and toe also change. If the camber is then adjusted, the caster is unaffected. Many technicians adjust caster and camber at the same time using shims, slots, or eccentric cams.

Always follow the manufacturer's recommended alignment procedure. For example, many manufacturers include a **shim chart** in their service manual that gives the thickness and location of the shim changes

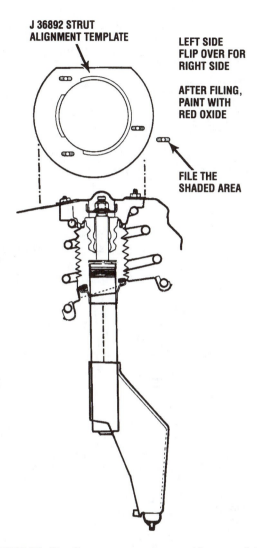

FIGURE 15–60 Some struts require modification of the upper mount for camber adjustment. (Courtesy of Oldsmobile)

based on the alignment reading. Shim charts are used to set camber *and* caster at the same time. Shim charts are designed for each specific model of vehicle (see Figure 15–62). Regardless of the methods or procedures used, toe is always adjusted after all the angles are set because caster and camber both affect the toe.

SETTING TOE

Front toe is the last angle that should be adjusted and is the most likely to need correction. Most newer alignment equipment displays in degrees of toe instead of inches of toe (see the toe unit conversion chart).

+ toe = toe-in

− toe = toe-out

METHODS OF ADJUSTMENT

Tools and adjustment devices may be available from aftermarket suppliers to perform adjustments in cases where manufacturers do not make such provisions.

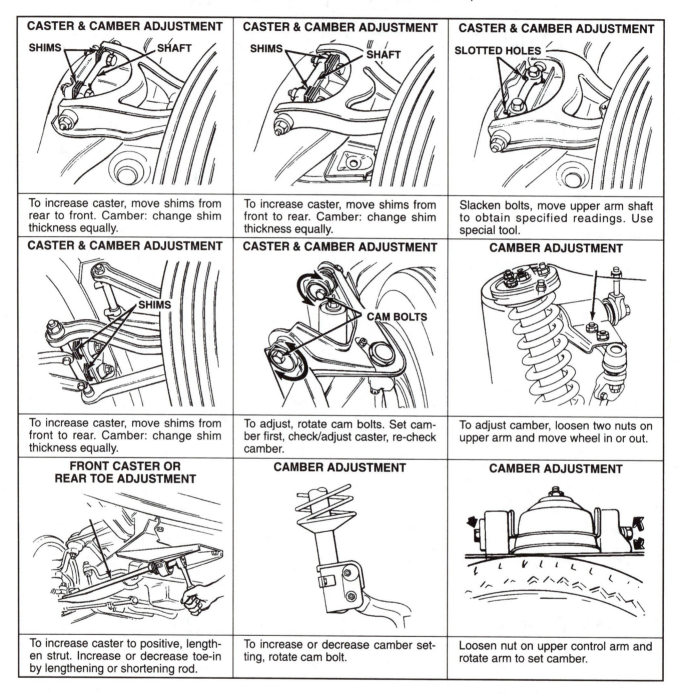

CASTER & CAMBER ADJUSTMENT	CASTER & CAMBER ADJUSTMENT	CASTER & CAMBER ADJUSTMENT
SHIMS / SHAFT	SHIMS / SHAFT	SLOTTED HOLES
To increase caster, move shims from rear to front. Camber: change shim thickness equally.	To increase caster, move shims from front to rear. Camber: change shim thickness equally.	Slacken bolts, move upper arm shaft to obtain specified readings. Use special tool.
CASTER & CAMBER ADJUSTMENT	CASTER & CAMBER ADJUSTMENT	CAMBER ADJUSTMENT
SHIMS	CAM BOLTS	
To increase caster, move shims from front to rear. Camber: change shim thickness equally.	To adjust, rotate cam bolts. Set camber first, check/adjust caster, re-check camber.	To adjust camber, loosen two nuts on upper arm and move wheel in or out.
FRONT CASTER OR REAR TOE ADJUSTMENT	CAMBER ADJUSTMENT	CAMBER ADJUSTMENT
To increase caster to positive, lengthen strut. Increase or decrease toe-in by lengthening or shortening rod.	To increase or decrease camber setting, rotate cam bolt.	Loosen nut on upper control arm and rotate arm to set camber.

FIGURE 15–61   (Courtesy of FMC)

DEGREES CASTER →

DEGREES CAMBER	BOLT	+4.9°	+4.7°	+4.5°	+4.3°	+4.1°	+3.9°	+3.7°	+3.5°	+3.3°	+3.1°	+2.9°	+2.7°	+2.5°	+2.3°	+2.1°
+2.2°	FRONT	+300	+211	+211	+210	+210	+201	+201	+201	+200	+200	+111	+111	+110	+110	+110
	REAR	+101	+101	+110	+110	+111	+200	+200	+201	+210	+210	+211	+211	+300	+301	+301
+2.0°	FRONT	+210	+210	+210	+201	+201	+200	+200	+111	+200	+110	+110	+110	+101	+101	+100
	REAR	+011	+100	+101	+101	+110	+110	+111	+200	+110	+201	+210	+210	+211	+300	+300
+1.8°	FRONT	+201	+201	+200	+200	+111	+111	+110	+110	+110	+101	+101	+100	+100	+100	+011
	REAR	+010	+011	+011	+100	+101	+101	+110	+110	+111	+200	+200	+201	+210	+210	+211
+1.6°	FRONT	+200	+200	+111	+111	+110	+110	+100	+101	+100	+100	+100	+011	+011	+010	+010
	REAR	+001	+001	+010	+011	+011	+100	+100	+101	+110	+110	+111	+200	+200	+201	+210
+1.4°	FRONT	+111	+110	+110	+101	+101	+100	+100	+011	+100	+011	+010	+010	+010	+001	+001
	REAR	-000	-000	+001	+001	+010	+011	+011	+100	+110	+101	+110	+200	+110	+200	+200
+1.2°	FRONT	+110	+101	+101	+100	+100	+011	+011	+010	+010	+010	+010	+001	+000	+111	-000
	REAR	-010	-001	-000	+000	+001	+001	+010	+011	+011	+100	+101	+101	+110	+111	+111
+1.0°	FRONT	+100	+100	+011	+011	+010	+010	+001	+001	+001	+000	-000	-000	-001	-001	-010
	REAR	-011	-010	-001	-001	-000	+000	+001	+001	+010	+011	+011	+100	+101	+101	+110
+0.8°	FRONT	+011	+011	+010	+010	+001	+001	+000	+000	-000	-001	-001	-010	-010	-010	-011
	REAR	-100	-100	-011	-010	-010	-001	-001	+000	+001	+001	+010	+011	+011	+100	+101
+0.6°	FRONT	+010	+001	+001	+000	+000	-000	-010	-001	-010	-010	-010	-011	-011	-100	-100
	REAR	-101	-101	-100	-100	-011	-010	-010	-001	-001	-001	+000	+001	+001	+011	+011
+0.4°	FRONT	+001	+000	+000	-000	-001	-001	-010	-010	-010	-011	-000	+000	+001	+010	+010
	REAR	-111	-110	-101	-101	-100	-100	-011	-100	-010	-001	-101	-101	-110	-110	-110
+0.2°	FRONT	-001	-001	-001	-010	-010	-011	-011	-100	-011	-101	-010	-001	-110	+000	+001
	REAR	-200	-111	-111	-110	-101	-101	-100	-100	-100	-101	-010	-001	-000	-111	-200
0.0°	FRONT	-010	-010	-011	-011	-100	-100	-100	-100	-011	-110	-010	-011	-111	-111	-000
	REAR	-201	-201	-200	-100	-111	-110	-101	-101	-101	-100	-111	-010	-010	-001	-201
-0.2°	FRONT	-011	-110	-010	-100	-101	-111	-110	-200	-101	-101	-100	-200	-200	-201	-010
	REAR	-210	-201	-201	-200	-200	-110	-111	-111	-200	-200	-201	-100	-011	-010	-210
-0.4°	FRONT	-100	-101	-101	-110	-110	-110	-200	-111	-111	-110	-101	-201	-201	-210	-210
	REAR	-300	-211	-210	-210	-201	-201	-200	-111	-111	-110	-101	-101	-100	-011	-011

INSTRUCTIONS FOR USING ALIGNMENT CHART

1. Determine vehicle's current caster and camber measurements.
2. Using the current caster reading, read down the appropriate column to the lines corresponding to the current camber reading.
3. Correction values will be given for the front and rear bolts.
   EXAMPLE: Current reading +1.6° caster +0.4° camber. By reading down the chart from +1.6° caster to +0.4° camber you will find that the front bolt requires an adjustment of −101 and the rear bolt requires an adjustment of +010.

+ = Shim Addition
− = Shim Removal

Correction Value Example: + 2 0 1

(as shown)
on current figure

└─ No. of 0.030 in. shims
└─── No. of 0.060 in. shims
└───── No. of 0.120 in. shims

FIGURE 15–62 Typical shim alignment chart. ⅛-in. (0.125-in.) shims can be substituted for the 0.120-in. shims. ¹⁄₁₆-in. (0.0625) shims can be substituted for the 0.060-in. shims. ¹⁄₃₂-in. (0.03125-in.) shims can be substituted for the 0.030-in. shims. (Courtesy of Oldsmobile)

319

Toe Unit Conversions

Units	Conversions			
Fractional inches	1/16 in.	1/8 in.	3/16 in.	1/4 in.
Decimal inches	0.062 in.	0.125 in.	0.188 in.	0.250 in.
Millimeters	1.60 mm	3.18 mm	4.76 mm	6.35 mm
Decimal degrees	0.125°	0.25°	0.375°	0.5°
Degrees and minutes	0°8'	0°15'	0°23'	0°30'
Fractional degrees	1/8°	1/4°	3/8°	1/2°

HELPFUL HINT: To convert from degrees to decimal inches, simply divide by 2. For example, if the total toe is 0.25°, one-half (divided by 2) is equal to 0.125 in. (1/8 in.). Toe is usually specified in degrees because it more accurately reflects the toe angle regardless of the size of the wheels/tires.

Toe is adjusted by turning the tie rod(s) or tie rod end sleeve(s) (see Figures 15–63 and 15–64). To make sure that the steering wheel is straight after setting toe, the steering wheel *must* be locked in the straight-ahead position while the toe is being adjusted. To lock the steering wheel, always use a steering wheel lock that presses against the seat and the outer rim of the steering wheel. *Do not* use the locking feature of the steering column to hold the steering wheel straight. Always "unlock" the steering column, straighten the steering wheel, and install the steering wheel lock.

NOTE: If the vehicle is equipped with power steering, the engine must be started and the steering wheel straightened with the engine running to be assured a straight steering wheel. Lock the steering wheel with the steering lock tool before stopping the engine.

After straightening the steering wheel, turn the tie rod adjustment until the toe for both wheels is within specifications. Many alignment machines include a screen that shows **straight-ahead steering.** Simply adjust the tie rod adjusters until the reading shows that the toe is correct and the front wheels will result in a straight steering wheel. This is often called **centerline steering** and is a centered steering wheel with the vehicle traveling a straight course. Test drive the vehicle for proper handling and centerline steering.

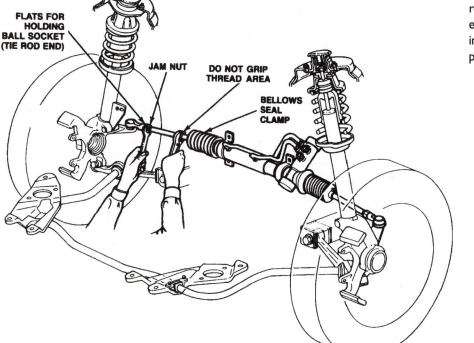

FIGURE 15–63 Adjusting toe by rotating the tie rod on a vehicle equipped with rack-and-pinion steering. (Courtesy of Ford Motor Company)

FLATS FOR HOLDING BALL SOCKET (TIE ROD END)

JAM NUT

DO NOT GRIP THREAD AREA

BELLOWS SEAL CLAMP

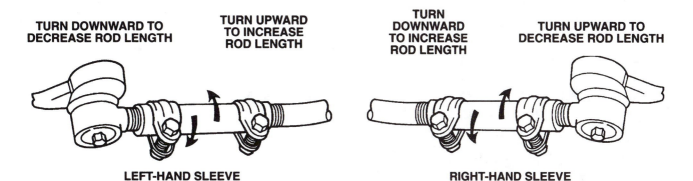

TURN DOWNWARD TO DECREASE ROD LENGTH

TURN UPWARD TO INCREASE ROD LENGTH

TURN DOWNWARD TO INCREASE ROD LENGTH

TURN UPWARD TO DECREASE ROD LENGTH

LEFT-HAND SLEEVE

RIGHT-HAND SLEEVE

FIGURE 15–64 Toe is adjusted on a parallelogram steering linkage by turning adjustable tie rod sleeves. Special tie rod sleeve adjusting tools should be used that grip the slot in the sleeve and will not crush the sleeve while it is being rotated. (Courtesy of Ford Motor Company)

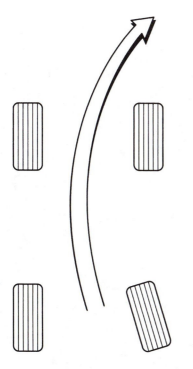

FIGURE 15–65 The toe-in on the right rear wheel creates a turning force toward the right.

AFTERMARKET ALIGNMENT METHODS

Accurate alignments are still possible on vehicles without factory methods of adjustment by using alignment kits or parts. Aftermarket alignment kits are available for most vehicles. Even when there are factory alignment methods, sometimes the range of adjustment is not enough to compensate for sagging frame members or other normal or accident-related faults (see Figures 15–66 and 15–67).

◀ **DIAGNOSTIC STORY** ▶

LEFT THRUST LINE BUT A PULL TO THE RIGHT!

A new four-door sport sedan had been aligned several times at the dealership in an attempt to solve a pull to the right. The car was front-wheel drive and had four-wheel independent suspension. The dealer rotated the tires without any difference. The alignment angles of all four wheels were in the center of specifications. The dealer even switched all four tires from another car in an attempt to solve the problem.

In frustration, the owner took the car to an alignment shop. Almost immediately the alignment technician discovered that the right rear wheel was slightly toed-in. The right rear being toed-in caused a thrust line to the left, causing a pull to the right (see Figure 15–65). The alignment technician adjusted the toe on the right rear wheel which was adjustable and reset the front toe. The car drove beautifully.

The owner was puzzled why the new car dealer was unable to correct the problem. It was later discovered that the alignment machine at the dealership was out of calibration by the exact amount that the right rear wheel was out of specification. The car pulled to the right because the independent suspension created a rear steering force toward the left that caused the front to pull to the right. Remember, the rear wheels can steer a vehicle just like a rudder on a boat. Alignment equipment manufacturers recommend that alignment equipment be calibrated regularly.

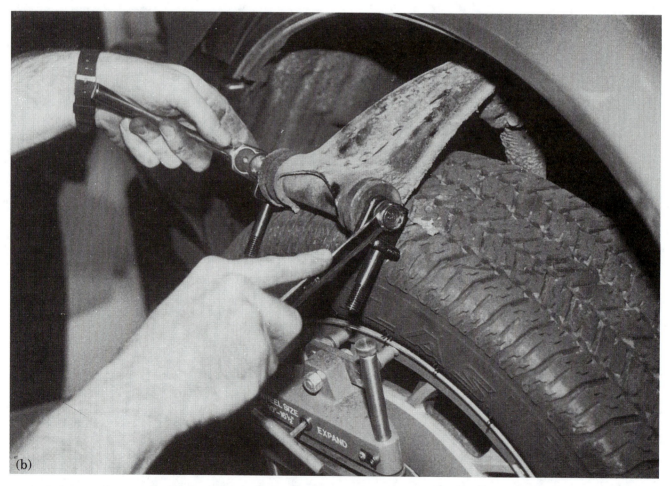

FIGURE 15–66 (a) Aftermarket camber kit designed to provide some camber adjustments for a vehicle that does not provide any adjustment. (b) Installation of this kit requires that the upper control arm shaft be removed. Note that the upper control arm was simply rotated out over the wheel pivoting on the upper ball joint.

ALIGNMENT TROUBLESHOOTING GUIDE

Problem	Possible Causes
Pull left/right	Uneven tire pressure, tire conicity, mismatched tires, unequal camber, unequal caster, brake drag, setback, suspension/frame sag, unbalanced power assist, bent spindle, bent strut, worn suspension components (front or rear), rear suspension misalignment
Incorrect steering wheel position	Incorrect individual or total toe, rear wheel misalignment, excessive suspension or steering component play, worn rack-and-pinion attachment bushings, individual toe adjusters not provided
Hard steering	Improper tire pressure, binding steering gear or steering linkage, low P/S fluid, excessive positive caster, lack of lubrication, upper strut mount(s), worn power steering pump, worn P/S belt
Loose steering	Loose wheel bearings, worn steering or suspension components, loose steering gear mount, excessive steering gear play, loose or worn steering coupler
Excessive road shock	Excessive positive caster, excessive negative camber, improper tire inflation, too wide wheel/tire combination for the vehicle, worn or loose shocks, worn springs
Poor returnability	Incorrect camber or caster, bent spindle or strut, binding suspension or steering components, improper tire inflation
Wander/instability	Incorrect alignment, defective or improperly inflated tires, worn steering or suspension parts, bent spindle or strut, worn or loose steering gear, loose wheel bearings

Problem	Possible Causes
Squeal/scuff on turns	Defective or improperly inflated tires, incorrect turning angle (TOOT), bent steering arms, excessive wheel setback, poor driving habits (too fast for conditions), worn suspension or steering parts
Excessive body sway	Loose or broken stabilizer bar links or bushings, worn shocks or mountings, broken or sagging springs, uneven vehicle load, uneven or improper tire pressure
Memory steer	Binding steering linkage, binding steering gear, binding upper strut mount, ball joint, or king pin
Bump steer	Misalignment of steering linkage, bent steering arm, frame, defective or sagged springs, uneven load, bent spindle or strut
Torque steer	Bent spindle or strut, bent steering arm, misaligned frame, worn torque strut, defective engine or transaxle mounts, drive axle misalignment, mismatched or unequally inflated tires

SUMMARY

1. Toe is the most important alignment angle because toe is usually the first alignment angle needed to be corrected and when incorrect causes severe tire wear.
2. Caster is the basic stability angle, yet does not cause tire wear (directly) if not correct or equal side to side.
3. Camber is both a pulling angle if not equal side to side as well as a tire wearing angle if not set to specifications.
4. SAI and included angle (SAI and camber added together) are an important diagnostic tool.
5. If the toe-out-on-turns (TOOT) reading is not within specifications, a bent steering spindle (steering knuckle) is the most likely cause.
6. A four-wheel alignment includes aligning all four wheels of the vehicle, whereas a thrust line alignment sets the front toe equal to the thrust line (total rear toe) of the rear wheels.

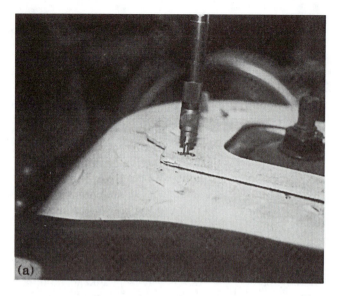

(a)

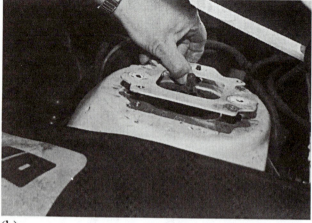

(b)

(c)

FIGURE 15–67 (a) Installation of some aftermarket alignment kits requires the use of special tools, such as this cutter being used to drill out spot welds on the original alignment plate on a strut tower. (b) Original plate being removed. (c) Note the amount of movement that the upper strut bearing mount has around the square openings in the strut tower. An aftermarket plate can now be installed to allow both camber and caster adjustment.

7. Before attempting to align any vehicle, it must be checked for proper ride height (trim height), tire conditions, and tire pressures as well as a thorough inspection of all steering and suspension components.

8. The proper sequence for a complete four-wheel alignment is:

   a. Rear camber
   b. Rear toe
   c. Front camber and caster
   d. Front toe

REVIEW QUESTIONS

1. Explain the three basic alignment angles of camber, caster, and toe.

2. Explain how knowing SAI, TOOT, and included angle can help in the correct diagnosis of an alignment problem.

3. List 10 prealignment checks that should be performed before the wheel alignment is checked and/or adjusted.

4. List the steps necessary to follow for a four-wheel alignment.

MULTIPLE-CHOICE QUESTIONS

1. Technician A says that upper control arms are not part of the steering system and therefore cannot cause play in the steering. Technician B says that a defective universal joint between the steering column and the steering gear box stub shaft can cause excessive steering wheel play. Which technician is correct?

   a. A only
   b. B only
   c. Both A and B
   d. Neither A nor B

2. Technician A says that a vehicle will pull (or lead) to the

side with the most camber (or least negative camber). Technician B says that a vehicle will pull (or lead) to the side with the most toe. Which technician is correct?

a. A only

b. B only

c. Both A and B

d. Neither A nor B

3. Technician A says that the vehicle will pull to the side with the most toe in the rear. Technician B says that the rear toe angle determines the thrust angle. Which technician is correct?

a. A only

b. B only

c. Both A and B

d. Neither A nor B

4. Strut rods (if adjustable) adjust

a. toe.

b. camber.

c. caster.

d. toe-out on turns.

5. If metal shims are used for alignment adjustment in the front, they adjust

a. camber.

b. caster.

c. toe.

d. camber and caster only.

6. Technician A says that as wear occurs, camber usually becomes negative. Technician B says that as steering linkage wear occurs, toe usually becomes toe-out from an original toe-in specification. Which technician is correct?

a. A only

b. B only

c. Both A and B

d. Neither A nor B

Use the following information to answer Question 7.

Specifications	Min.	Preferred	Max.
Camber (degree)	0	1.0	1.4
Caster (degree)	0.8	1.5	2.1
Toe (inch)	-0.10	0.06	0.15

Results	L	R
Camber (degree)	-0.1	0.6
Caster (degree)	1.8	1.6
Toe (inch)	1.12	+0.12

7. The vehicle above will

a. pull toward the right and feather edge both tires.

b. pull toward the left.

c. wear the outside of the left tire and the inside of the right tire.

d. none of the above.

Use the following information to answer Questions 8 and 9.

Specifications	Min.	Preferred	Max.
Camber (degree)	-¼	+½	1
Caster (degree)	0	+2	+4
Toe (inch)	-1⁄16	1⁄16	3⁄16

Results	L	R
Camber (degree)	-0.3	-0.1
Caster (degree)	3.6	1.8
Toe (inch)	-0.16	+0.32

8. The vehicle above will

a. pull toward the left.

b. pull toward the right.

c. wander.

d. lead to the left slightly.

9. The vehicle above will

a. wander.

b. wear tires but will not pull.

c. will pull but will not wear tires.

d. pull toward the left and cause feather-edge tire wear.

10. Which alignment angle is most likely to need correction and cause the most tire wear?

a. Toe

b. Camber

c. Caster

d. SAI/KPI

ANSWERS TO EVEN-NUMBERED MULTIPLE-CHOICE QUESTIONS

Chapter 1 2-D, 4-B, 6-D, 8-B

Chapter 2 2-A, 4-C, 6-C, 8-A

Chapter 3 2-D, 4-A, 6-B, 8-A, 10-A

Chapter 4 2-C, 4-D, 6-B, 8-C

Chapter 5 2-D, 4-D, 6-C, 8-A, 10-C

Chapter 6 2-A, 4-C, 6-A, 8-A, 10-C

Chapter 7 2-C, 4-A, 6-B, 8-B

Chapter 8 2-B, 4-B, 6-C, 8-B, 10-A

Chapter 9 2-C, 4-A, 6-A, 8-D

Chapter 10 2-B, 4-C, 6-C, 8-B, 10-C

Chapter 11 2-C, 4-A, 6-B, 8-B, 10-B

Chapter 12 2-B, 4-D, 6-B, 8-A, 10-B

Chapter 13 2-B, 4-C, 6-A, 8-C, 10-B

Chapter 14 2-C, 4-B, 6-B, 8-C, 10-A

Chapter 15 2-A, 4-C, 6-C, 8-B, 10-A

◀ Glossary ▶

ABS   Antilock brakes.

Ackerman principle   The angle of the steering arms causes the inside wheel to turn more sharply than the outer wheel when making a turn. This produces toe-out on turns (TOOT).

Alamite fitting   *See* Zerk.

Align   To bring the parts of a unit into the correct position with respect to each other.

Alloy   A metal that contains one or more other elements usually added to increase strength or give the base metal important properties.

Anchor pin   A steel stud firmly attached to the backing plate. One end of the brake shoes is either attached to or rests against it.

Antidive   A term used to describe the geometry of the suspension that controls the movement of the vehicle during braking. It is normal for a vehicle to nose dive slightly during braking and is designed into most vehicles.

Antiroll bar   *See* Stabilizer bar.

Antisquat   A term used to describe the geometry of the suspension that controls the movement of the vehicle body during acceleration. One hundred percent antisquat means that the body remains level during acceleration. Less than 100 percent indicates that the body "squats down" or lowers in the rear during acceleration.

Antisway bar   *See* Stabilizer bar.

API   American Petroleum Institute.

APRA   Automotive Parts Rebuilders Association.

Aramid   Generic name for aromatic polyamide fibers developed in 1972. Kevlar is the Dupont brand name for aramid.

Articulation test   A test specified by some vehicle manufacturers that tests the amount of force necessary to move the inner tie rod end in the ball socket assembly. The usual specification for this test is greater than 1 lb (0.5 kg) and less than 6 lbs (2.7 kg) of force.

Asbestosis   A health condition where asbestos causes scar tissue to form in the lungs, causing shortness of breath.

ASE   Abbreviation for the National Institute for Automotive Service Excellence; a nonprofit organization for the testing and certification of vehicle service technicians.

Aspect ratio   The ratio of height to width of a tire. A tire with an aspect ratio of 60 (a 60 series tire) has a height (from rim to tread) of 60 percent of its cross-sectional width.

ASR   Acceleration slip regulation.

ASTM   American Society for Testing and Materials.

ATe   Alfred Teves Engineering, a manufacturer of brake system components and systems.

Atmospheric pressure   Pressure exerted by the atmosphere on all things (14.7 pounds per square inch).

Axial   In line along with the axis or centerline of a part or component. Axial play in a ball joint means looseness in the same axis as the ball-joint stud.

Back spacing   The distance between the back rim edge and the center section mounting pad of a wheel.

Backing plate   A steel plate upon which the brake shoes are attached. The backing plate is attached to the steering knuckle or axle housing.

Backside setting   *See* Back spacing.

Ball joint   A flexible joint having a ball-and-socket type of construction, used in suspension systems.

Ball socket assembly   An inner tie rod end assembly that contains a ball-and-socket joint at the point where the assembly is threaded on to the end of the steering rack.

Barrel shaped   A brake drum having a frictional surface that is larger in the center than at the open end or at the rear of the drum.

Base brake   *See* Service brake.

Bellmouthed A brake drum with a frictional surface larger at the open end of the drum than at any other point toward the rear of the drum.

Belt Fabric or woven steel material over the body plies of a tire, and just under the tread area, to help keep tread from squirming.

Bias-belted A bias ply tire with additional belt material just under the tread area.

Bias-ply The body plies cover the entire tire and are angled as they cross from bead to bead.

Bleeder screw A valve in wheel cylinders (and other locations) for bleeding air from the hydraulic system.

Bolt circle The diameter (in inches or millimeters) of a circle drawn through the center of the bolt holes in a wheel.

Boots Rubber dust protectors on the ends of wheel or caliper cylinders.

Bounce test A test used to check the condition of shock absorbers.

BPMV Brake pressure modulator valve.

Brake fade A result of heat buildup. It is the reduction in braking force due to loss of friction between the brake shoes and the drum.

Brake lining A friction material fastened to the brake shoes. It is pressed against the rotating brake drum to accomplish braking.

Brake shoe The part of the brake system upon which the brake lining is attached.

Breather tube A tube that connects the left and right bellows of a rack-and-pinion steering gear.

Brinelling A type of mechanical failure used to describe a dent in metal such as what occurs when a shock load is applied to a bearing. Named after Johann A. Brinell, a Swedish engineer.

Btu (British thermal unit) A unit of heat measurement.

Bump steer A term used to describe what occurs when the steering linkage is not level, causing the front tires to turn inward or outward as the wheels and suspension move up and down. Automotive chassis engineers call it *roll steer*.

Bump stop *See* Jounce bumper.

Caliper The U-shaped housing that contains the hydraulic cylinders and holds the pads on disc brake applications.

Camber The inward or outward tilt of the wheels from true vertical as viewed from the front or rear of the vehicle. Positive camber means the top of the wheel is out from center of the vehicle more than the bottom of the wheel.

Cardan joint A type of universal joint named for a sixteenth-century Italian mathematician.

Caster The forward or backward tilt of an imaginary line drawn through the steering axis as viewed from the side of the vehicle. Positive caster is where an imaginary line would contact the road surface in front of the contact patch of the tire.

Caster sweep A process used to measure caster during a wheel alignment procedure where the front wheels are rotated first inward, then outward, a specified amount.

Chatter Sudden grabbing and releasing of the drum when brakes are applied.

Center bolt A bolt used to hold the leaves of a leaf spring together in the center. Also called a *centering pin*.

Center support bearing A bearing used to support the center of a long drive shaft on a rear-wheel-drive vehicle. Also called a *steady bearing*.

Centering pin *See* Center bolt.

Centerline steering A term used to describe the position of the steering wheel while driving on a straight, level road. The steering wheel should be centered or within plus or minus 3° as specified by many vehicle manufacturers.

CFRC Carbon-fiber-reinforced carbon.

Chassis The frame, suspension, steering, and machinery of a motor vehicle.

Coefficient of friction A measure of the amount of friction usually measured from 0 to 1. A low number (0.3) indicates low friction and a high number (0.9) indicates high friction.

Coil spring A spring steel rod wound in a spiral (helix) shape. Used in both front and rear suspension systems.

Compensating port The port located in the master cylinder that allows excess fluid to return to the reservoir. *See also* Vent port.

Compensation A process used during a wheel alignment procedure where the sensors are calibrated to eliminate errors in the alignment readings that may be the result of a bent wheel or unequal installation of the sensor on the wheel of the vehicle.

Composite A term used to describe the combining of individual parts into a larger component. For example, a composite leaf spring is constructed of fiberglass and epoxy and a composite master brake cylinder contains both plastic parts (reservoir) and metal parts (cylinder housing).

Compression bumper   *See* Jounce bumper.

Compression rod   *See* Strut rod.

Concentric   Perfectly round—the relationship of two round parts on the same center.

Cone   The inner race or ring of a bearing.

Constant-velocity joint   Commonly called CV joints. CV joints are drive line joints that can transmit engine power through relatively large angles without a change in the velocity as is usually the case with conventional Cardan-type U-joints.

Controller   A term commonly used to describe a computer or an electronic control module.

Cotter key   A metal loop used to retain castle nuts by being installed through a hole. Size is measured by diameter and length (for example, ⅛ in. × 1½ in.). Also called a *cotter pin. Cotter* is an old English verb meaning "to close or fasten."

Coupling disc   *See* Flexible coupling.

Cow catcher   A large spring seat used on many General Motors MacPherson strut units. If the coil spring breaks, the cow catcher is designed to prevent one end of the spring from moving outward and cutting a tire.

Cross camber/caster   The difference of angle from one side of the vehicle to the other. Most manufacturers recommend a maximum difference side to side of ½° for camber and caster.

Cross steer   A type of steering linkage commonly used on light and medium trucks.

Cup   The outer race or ring of a bearing.

CV joints   Constant-velocity joints.

Cylinder hone   A tool that uses an abrasive to smooth out and bring to exact measurement such things as wheel cylinders.

Deflection   A bending or distorting motion. Usually applied to a brake drum when it is forced out of round during brake application.

Diaphragm   A flexible cloth–rubber sheet that is stretched across an area, separating two different compartments.

Diff   An abbreviation or slang for differential.

Differential   A mechanical unit containing gears that provides gear reduction and a change of direction of engine power and permits the drive wheels to rotate at different speeds as is required when turning a corner.

Directional stability   Ability of a vehicle to move forward in a straight line with a minimum of driver control. Cross winds and road irregularities will have little effect if directional stability is good.

Dog tracking   A term used to describe the condition where the rear wheels do not follow directly behind the front wheels. Named for the way that many species of dogs run, with their rear paws offset toward one side so that their rear paws will not hit their front paws while running.

DOT   Department of Transportation.

Double Cardan   A universal joint that uses two conventional Cardan joints together to allow the joint to operate at greater angles.

Double flare   A tubing end made such that the flare area has two wall thicknesses.

Drag link   A term used to describe a link in the center of the steering linkage usually called a *center link.*

Drag rod   *See* Strut rod.

Drift   A mild pull that does not cause a force on the steering wheel that the driver must counteract (also known as lead). Also refers to a tapered tool used to center a component in a bolt hole prior to installing the bolt.

Dropping point   The temperature at which a grease passes from a semisolid to a liquid state under conditions specified by ASTM.

Dry park test   A test of steering and/or suspension components. With the wheels in the straight-ahead position and the vehicle on flat level ground, have an assistant turn the steering wheel while looking and touching all steering and suspension components looking for any looseness.

Dual master cylinder   A two-compartment master cylinder.

Duo-Servo   Brand name of a Bendix dual-servo drum brake.

Durometer   The hardness rating of rubber products named for an instrument used to measure hardness that was developed about 1890.

Dust cap   A functional metal cap that keeps grease in and dirt out of wheel bearings. Also called a *grease cap.*

Dynamic balance   When the weight mass centerline of a tire is in the same plane as the centerline of the object.

EBCM   Electronic brake control module.

Eccentric   The relationship of two round parts having different centers. A part that contains two round surfaces, not on the same center.

EHCU   Electrohydraulic control unit.

Elastomer   Another term for rubber.

Emergency brake   *See* Parking brake.

Energized shoe   A brake shoe that receives greater applied force from wheel rotation.

Energy   Capacity for performing work.

EPA   Environmental Protection Agency.

EPR   Ethylene propylene rubber.

Fade   To grow weak; brakes becoming less effective.

Filler vent   A breather hole in the filler cap on the master cylinder.

Flexible coupling   A part of the steering mechanism between the steering column and the steering gear or rack-and-pinion assembly. Also called a *rag joint* or *steering coupling disc*. The purpose of the flexible coupling is to isolate noise vibration and harshness from being transmitted from the road and steering to the steering wheel.

Floating caliper   A type of caliper used with disc brakes that moves slightly to assure equal pad pressure on both sides of the rotor.

Flow control valve   A valve that regulates and controls the flow of power steering pump hydraulic fluids to the steering gear or rack-and-pinion assembly. The flow control valve is usually part of the power steering pump assembly.

FMSI   Friction Materials Standards Institute

Follower ball joint   A ball joint used in a suspension system to provide support and control without having the weight of the vehicle or the action of the springs transferred through the joint itself. Also called a *friction ball joint*.

Foot-pound   A measurement of torque. A 1-pound pull 1 foot from the center of an object.

Forward steer   *See* Front steer.

Foundation brake   *See* Service brake.

4 DR   Four-door model.

4 X 4   The term used to describe a four-wheel-drive vehicle. The first "4" indicates the number of wheels of the vehicle, and the second "4" indicates the number of wheels that are driven by the engine.

4 X 2   The term used to describe a two-wheel-drive vehicle. The "4" indicates the number of wheels of the vehicle and the "2" indicates the number of wheels that are driven by the engine.

4WAL   Four-wheel antilock.

Free play   The amount of movement that the steering wheel moves without moving the front wheels. The maximum allowable amount of free play is less than 2 in. for a parallelogram-type steering system and ⅜ in. for a rack-and-pinion steering system.

Frequency   The number of times a complete motion cycle takes place during a period of time (usually measured in seconds).

Friction   The resistance to sliding of two bodies in contact with each other.

Friction ball joint   Outer suspension pivot that does not support the weight of the vehicle. Also called a *follower ball joint*.

Front steer   A construction design of a vehicle that places the steering gear and steering linkage in front of the center line of the front wheels. Also called *forward steer*.

FWD   Front-wheel drive.

Galvanized steel   Steel with a zinc coating to help protect the steel from rust and corrosion.

Garter spring   A spring used in a seal to help keep the lip of the seal in contact with the moving part.

Gland nut   The name commonly used to describe the large nut at the top of a MacPherson strut housing. This gland nut must be removed to replace a strut cartridge.

Glazed drum   A drum surface hardened excessively by intense heat.

Grab   Seizure of the drum on linings when brakes are applied.

Gram   A metric unit of weight measurement equal to $\frac{1}{1000}$ kilogram (1 ounce = 28 grams). An American dollar bill or paper clip weighs about 1 gram.

Grease cap   A functional metal cap that keeps grease in and dirt out of wheel bearings. Also called a *dust cap*.

Grease retainer   *See* Grease seal.

Grease seal   A seal used to prevent grease from escaping and to prevent dirt and moisture from entering.

Green tire   An uncured assembled tire. After the green tire is placed in a mold under heat and pressure, the rubber changes chemically and comes out of the mold formed and cured.

Grommet   An eyelet usually made from rubber used to protect, strengthen, or insulate around a hole or passage.

GVW   Abbreviation for gross vehicle weight. GVW is the weight of the vehicle plus the weight of all passengers and cargo up to the limit specified by the manufacturer.

Half shaft   Drive axles on a front-wheel-drive vehicle or from a stationary differential to the drive wheels.

Halogenated compounds Chemicals containing chlorine, fluorine, bromine, or iodine. These chemicals are generally considered to be hazardous, and any product containing these chemicals should be disposed of using approved procedures.

Haltenberger linkage A type of steering linkage commonly used on light trucks.

Hand brake *See* Parking brake.

HD Heavy duty.

Heat checked Cracks in the braking surface of a drum caused by excessive heat.

Helper springs Auxiliary or extra springs used in addition to the vehicle's original springs to restore proper ride height or to increase the load-carrying capacity of the vehicle.

HEPA High-efficiency particulate air filter.

Hertz A unit of measurement of frequency. One hertz is one cycle per second, abbreviated Hz. Named for Heinrich R. Hertz, a nineteenth-century German physicist.

Hooke's law The force characteristics of a spring discovered by Robert Hooke (1635–1703), an English physicist. Hooke's law states: The deflection (movement or deformation) of a spring is directly proportional to the applied force.

HSS High-strength steel; a low-carbon-alloy steel that uses various amounts of silicon, phosphorus, and manganese.

Hub cap A functional and decorative cover over the lug nut portion of the wheel. *See also* wheel cover.

Hydrophilic A term used to describe a type of rubber used in many all-season tires where the rubber has an affinity for (rather than repelling) water.

Hydrophobic A term used to describe the repelling of water.

Hydroplaning Condition that occurs when driving too fast on wet roads where the water on the road gets trapped between the tire and the road, forcing the tire onto a layer of water and off the road surface. All traction between the tire and the road is lost.

Hypoid gear set A ring gear and pinion gear set that meshes together below the centerline of the ring gear. This type of gear set allows the drive shaft to be lower in the vehicle, yet requires special hypoid gear lubricant.

Included angle SAI angle added to the camber angle of the same wheel.

Independent suspension A suspension system that allows a wheel to move up and down without undue effect on the opposite side.

Inlet port *See* Replenishing port.

Iron Refined metal from iron ore (ferrous oxide) in a furnace. *See also* Steel.

IRS Independent rear suspension.

ISO International Standards Organization.

Isolator bushing Rubber bushing used between the frame and the stabilizer bar. Also known as a *stabilizer bar bushing*.

Jounce A term used to describe up-and-down movement or to cause up-and-down motion.

Jounce bumper A rubber or urethane stop to limit upward suspension travel. Also called a *bump stop*, a *strike-out bumper, suspension bumper*, or *compression bumper*.

Kerf Large water grooves in the tread of a tire.

Kevlar Dupont brand name for aramid fibers.

Kinetic energy The energy in any moving object. The amount of energy depends on the weight (mass) of the object and the speed of the object.

King pin A pivot pin commonly used on solid axles or early model twin-I-beam axles that rotate in bushings and allow the front wheels to rotate. The knuckle pivots about the king pin.

King pin inclination Inclining the tops of the king pins toward each other creates a stabilizing force to the vehicle.

KPI King pin inclination [also known as *steering axis inclination (SAI)*].

Ladder frame A steel frame for a vehicle that uses cross braces along the length similar to the rungs of a ladder.

Lateral runout A measure of the amount a tire or wheel is moving side to side while being rotated. Excessive lateral runout can cause a shimmy-type vibration if the wheels are on the front axle.

Lead A mild pull that does not cause a force on the steering wheel that the driver must counteract (also known as *drift*).

Leaf spring A spring made of several pieces of flat spring steel.

Live axle A solid axle used on the drive wheels and contains the drive axles that propel the vehicle.

LMA Low moisture absorption type of brake fluid (DOT 4).

Load index An abbreviated method that uses a number to indicate the load-carrying capabilities of a tire.

Load-carrying ball joint A ball joint used in a suspension system to provide support and control and through which the weight (load) of the vehicle is transferred to the frame.

Löbro joint   A brand name for a CV joint.

Lock nut   *See* Prevailing torque nut.

LSD   An abbreviation commonly used for limited slip differentials.

LT   Light truck.

M & S   Mud and snow.

MacPherson strut   A type of front suspension with the shock absorber and coil spring in one unit which rotates when the wheels are turned. Assembly mounts to the vehicle body at that top and to one ball joint and control arm at the lower end. Named for its inventor, Earle S. MacPherson.

Master cylinder   The part of the brake hydraulic system where the pressure is generated.

Match mount   The process of mounting a tire on a wheel and aligning the valve stem with a mark on the tire. The mark on the tire represents the high point of the tire and the valve stem location represents the smallest diameter of the wheel.

Mechanical brakes   Brakes that are operated by a mechanical linkage or cable connecting the brakes to the brake pedal.

Memory steer   Memory steer is a lead or pull of a vehicle caused by faults in the steering or suspension system. If after making a turn, the vehicle tends to pull in the same direction as the last turn, the vehicle has memory steer.

Mesothelioma   A fatal type of cancer of the lining of the chest or abdominal cavity.

Metering valve (Hold-off valve)   A valve installed between the master cylinder and front disc brakes which prevents operation of front disc brakes until 75 to 125 psi is applied to overcome rear drum brake return spring pressure.

Millisecond   One thousandth (1/1000) of 1 second.

Minutes   A unit of measure of an angle. Sixty minutes equal 1 degree.

Moly grease   Grease containing molybdenum disulfide.

Morning sickness   A slang term used to describe temporary loss of power steering assist when cold, caused by wear in the control valve area of a power rack-and-pinion unit.

MSDS   Material safety data sheets.

Mu (μ)   A Greek lowercase letter that represents the coefficient of friction.

NAO   Nonasbestos organic.

NAS   Nonasbestos synthetic.

NHTSA   National Highway Traffic Safety Administration.

Nitrile   A type of rubber that is acceptable for use with petroleum.

NLGI   National Lubricating Grease Institute. Usually associated with grease. The higher the NLGI number, the firmer the grease. Number 000 is very fluid, whereas #5 is very firm. The consistency most recommended is NLGI #2 (soft).

Noise   Noise is the vibration of air caused by a body in motion.

NVH   Abbreviation for *noise, vibration, and harshness*.

OE   Original equipment.

OEM   Original equipment manufacturer.

Offset   The distance the center section (mounting pad) is offset from the centerline of the wheel.

Orbital steer   *See* Bump steer.

OSHA   Occupational Safety and Health Administration.

Over-center adjustment   An adjustment made to a steering gear while the steering is turned through its center straight-ahead position. Also known as a *sector lash adjustment*.

Overinflation   A term used to describe a tire with too much tire pressure (greater than maximum allowable pressure).

Oversteer   A term used to describe the handling of a vehicle where the driver must move the steering wheel in the opposite direction from normal while turning a corner. Oversteer handling is very dangerous. Most vehicle manufacturers design their vehicles to understeer rather than oversteer.

Oz-in.   Measurement of imbalance. 3 oz-in. means that an object is out of balance and would require a 1-oz weight placed 3 in. from the center of the rotating object or 3 oz 1 in. from the center or any other combination that when multiplied equals 3 oz-in.

Panhard rod   A horizontal steel rod or bar attached to the rear axle housing at one end and the frame at the other to keep the center of the body directly above the center of the rear axle during cornering and suspension motions. Also called a *track rod*.

Parallelogram   A geometric box shape where opposite sides are parallel (equal distance apart). A type of steering linkage used with a conventional steering gear that uses a pitman arm, center link, idler arm, and tie-rods.

Parking brake Components used to hold a vehicle on a 30° incline. Called an *emergency brake* before 1967, when dual master cylinders and split braking systems became law. The parking brake is also called the *hand brake.*

Pawl A lever for engaging in a notch. Used to rotate the notched star wheel on self-adjusting brakes.

Penetration A test for grease where the depth of a standard cone is dropped into a grease sample and its depth is measured.

Perimeter frame A steel structure for a vehicle that supports the body of the vehicle under the sides as well as the front and rear.

Phenolic brake pistons Hard type of plastic disc brake caliper pistons which do not rust or corrode.

Pickle fork A tapered fork used to separate chassis parts that are held together by a nut and a taper. Hitting the end of the pickle fork forces the wedge portion of the tool between the parts to be separated and "breaks the taper." A pickle fork tool is generally *not* recommended because the tool can tear or rip the grease boot of the part being separated.

Pitch The pitch of a threaded fastener refers to the number of threads per inch.

Pitman arm A short lever arm that is splined to the steering gear cross shaft. It transmits the steering force from the cross shaft to the steering linkage.

Pitman shaft *See* Sector shaft.

Platform The platform of a vehicle includes the basic structure (frame and/or major body panels) as well as the basic steering and suspension components. One platform may be the basis for several different brand vehicles.

Ppm Parts per million.

Pressure bleeder A device that forces pressure into the master cylinder so that when the bleeder screws are opened at the wheel cylinder, air will be forced from the system.

Pressure-differential switch Switch installed between the two separate braking circuits of a dual master to light the dashboard brake light in the event of a brake system failure, causing a *difference* in brake pressure.

Prevailing torque nut A special design of nut fastener that is deformed slightly or has other properties that permit the nut to remain attached to the fastener without loosening.

Primary shoe A brake shoe installed facing the front of the vehicle.

Prop shaft An abbreviation for *propeller shaft.*

Propeller shaft A term used by many manufacturers for a drive shaft.

Proportioning valve Valve installed between the master cylinder and rear brakes which limits the amount of pressure to the rear wheels to prevent rear wheel lockup.

Psi Pounds per square inch.

Pull Vehicle tends to go left or right while traveling on a straight, level road.

Push rod The link rod connecting the brake pedal to the master cylinder piston.

RABS Rear antilock braking system.

Race Inner and outer machined surface of a ball or roller bearing.

Rack and pinion A type of lightweight steering unit that connects the front wheels through tie rods to the end of a long shaft called a *rack.* When the driver moves the steering wheel, the force is transferred to the rack-and-pinion assembly. Inside a small pinion gear that meshes with gear teeth cuts into the rack.

Radial runout A measure of the amount a tire or wheel is out of round. Excessive radial runout can cause a tramp-type vibration.

Radial tire A tire whose carcass plies run straight across (or almost straight across) from bead to bead.

Radius rod A suspension component to control longitudinal (front-to-back) support and is usually attached with rubber bushings to the frame at one end and the axle or control arm at the other end. *See also* Strut rod.

Rag joint *See* Flexible coupling.

Ratio The expression for proportion. For example, in a typical rear axle assembly, the drive shaft rotates three times faster than the rear axles. Expressed as a ratio of 3:1 and read as "three to one." Power train ratios are always expressed as driving divided by driven gears.

RBS Rubber-bonded socket.

RBWL Red brake warning lamp.

Reaction disc A feature built into a power brake unit to provide the driver with a "feel" of the pedal.

Rear spacing *See* Back spacing.

Rear steer A construction design of a vehicle that places the steering gear and steering linkage behind the centerline of the front wheels.

Rebuilt *See* Remanufactured.

Recirculating ball-steering gear A steering gear that uses a series of ball bearings that feed through and around the grooves in the worm and nut.

Remanufactured   A term used to describe a process where a component is disassembled, cleaned, inspected, and reassembled using new or reconditioned parts. According to the Automotive Parts Rebuilders Association (APRA), this same procedure is also call *rebuilt*.

Replenishing port   The Society of Automotive Engineers (SAE) term for the rearward low-pressure master cylinder port. Also called *inlet port, bypass port, filler port,* or *breather port.*

Residual check valve   A valve in the outlet end of the master cylinder to keep the hydraulic system under a light pressure on drum brakes only.

RIM   Reaction injection molded.

RMP   Reaction moldable polymer.

Road crown   A roadway where the center is higher than the outside edges. Road crown is designed into many roads to drain water off the road surface.

Roll bar   *See* Stabilizer bar.

Roll steer   *See* Bump steer.

Run-flat tires   Tires specially designed to operate for reasonable distances and speeds without air inside to support the weight of the vehicle. Run-flat tires usually require the use of special rims designed to prevent the flat tire from coming off the wheel.

RWAL   Rear wheel antilock.

RWD   Rear wheel drive.

SAE   Society of Automotive Engineers.

Saginaw   Brand name of steering components manufactured in Saginaw, Michigan, USA.

SAI   Steering axis inclination (same as KPI).

SBR   Styrene–butadiene rubber.

Schrader valve   A type of valve used in tires, air conditioning, and fuel injection systems. Invented in 1844 by August Schrader.

Scoring   Grooves worn into the drum or disc braking surface.

Scrub radius   Refers to an imaginary line drawn through the steering axis that intersects the ground compared to the centerline of the tire. Zero scrub radius means the line intersects at the centerline of the tire. Positive scrub radius means that the line intersects below the road surface and negative scrub radius means the line intersects above the road surface. Called *steering offset* by some vehicle manufacturers.

Secondary shoe   The brake shoe installed facing the rear of the vehicle.

Sector lash   Refers to clearance (lash) between a section of gear (sector) on the pitman shaft in a steering gear. *See also* Over-center adjustment.

Sector shaft   The name for the output shaft of a conventional steering gear. As part of the sector shaft in a section of a gear that meshes with the worm gear and is rotated by the driver when the steering wheel is turned. Also called a *pitman shaft.*

SED   Sedan.

Self-adjusting brakes   Brakes that maintain the proper lining-to-drum clearance by automatic adjusting mechanism.

Self-energizing   A brake shoe that when applied develops a wedging action that assists the braking force applied by the wheel cylinder.

SEMA   Specialty Equipment Manufacturers Association.

Semimets   Semimetallic brake linings.

Service brake   The main driver-operated vehicle brakes.

Servo action   Brake construction having the end of the primary shoe bear against the secondary shoe. When the brakes are applied, the primary shoe applies force to the secondary shoe.

Setback   The amount the front wheels are set back from true parallel with the rear wheels. Positive setback means that the right front wheel is set back farther than the left. Setback can be measured as an angle formed by a line perpendicular (90°) to the front axles.

Shackle   A mounting that allows the end of a leaf spring to move forward and backward as the spring moves up and down during normal operation of the suspension.

Shim   A thin metal spacer.

Shimmy   A vibration that results in a rapid back-and-forth motion of the steering wheel. A bent wheel or a wheel assembly that is not correctly dynamically balanced are common causes of shimmy.

Shock absorber   A device used to control spring movement in the suspension system.

Shoe pad   A raised support on the backing plate against which the shoe edge rests; also called a *shoe ledge.*

Short/long arm suspension   Abbreviated SLA; a suspension system with a short upper control arm and a long lower control arm. The wheel changes very little in camber with a vertical deflection.

Sintered metal   *See* Sintering.

Sintering A process where metal particles are fused together without melting.

Sipes Small traction improving slits in the tread of a tire.

SLA Abbreviation for *short/long arm suspension*. Also called *double wishbone suspension*.

Slip angle The angle between the true centerline of the tire and the actual path followed by the tire while turning.

SMC Sheet molding compound.

Solid axle A solid supporting axle for both front or both rear wheels. Also referred to as a *straight axle* or *nonindependent axle*.

Space-frame construction A type of vehicle construction that uses the structure of the body to support the engine and drive train as well as the steering and suspension. The outside body panels are nonstructural.

Spalling A term used to describe a type of mechanical failure caused by metal fatigue. Metal cracks then break out into small chips, slabs, or scales of metal. This process of breaking up is called *spalling*.

Spider Center part of a wheel. Also known as the *center section*.

Spindle nut Nut used to retain and adjust the bearing clearance of the hub to the spindle.

Split mu A term used to describe two different friction (mu) surfaces under the wheels of a vehicle. Mu (μ) is the Greek letter symbol for the coefficient of friction.

Split system A divided hydraulic brake system.

Spongy pedal When there is air in the brake lines, the pedal will have a springy or spongy feeling when applied.

Sprung weight The weight of a vehicle that is supported by the suspension.

Squeal A high-pitched noise caused by high-frequency vibrations when brakes are applied.

Stabilizer bar A hardened steel bar connected to the frame and both lower control arms to prevent excessive body roll. Also called an *antisway* or *antiroll bar*.

Stabilizer link Usually consists of a bolt, spacer, and nut to connect (link) the end of the stabilizer bar to the lower control arm.

Star wheel A notched wheel with a left- or right-hand threaded member for adjusting brake shoes.

Static balance When a tire has an even distribution of weight about its axis, it is in static balance.

Steady bearing *See* Center support bearing.

Steel Refined iron metal with most of the carbon removed.

Steering arms Arms bolted to or forged as a part of the steering knuckles. They transmit the steering force from the tie rods to the knuckles causing the wheels to pivot.

Steering coupling disc *See* Flexible coupling.

Steering gear Gears on the end of the steering column that multiply the driver's force to turn the front wheels.

Steering knuckle The inner portion of the spindle that pivots on the king pin or ball joints.

Steering offset *See* Scrub radius.

Step-bore cylinder A wheel cylinder having a different diameter at each end.

Straight axle *See* Solid axle.

Strike-out bumper *See* Jounce bumper.

Strut rod Suspension member used to control forward/backward support to the control arms. Also called *tension or compression rod* (TC rod) or *drag rod*.

Strut rod bushing A rubber component used to insulate the attachment of the strut rod to the frame on the body of the vehicle.

Stud A short rod with threads on both ends.

Suspension Parts or linkages by which the wheels are attached to the frame or body of a vehicle. These parts or linkages support the vehicle and keep the wheels in proper alignment.

Suspension bumper *See* Jounce bumper.

Sway bar Shortened name for *antisway bar*. *See* Stabilizer bar.

Table The portion of the shoe to which the lining is attached.

Tandem cylinder A master cylinder with two pistons arranged one ahead of the other. One cylinder operates rear brakes and the other front brakes.

TC rod *See* Strut rod.

Tension rod *See* Strut rod.

Thrust angle The angle between the geometric centerline of the vehicle and the thrust line.

Thrust line The direction the rear wheels are pointed as determined by the rear wheel toe.

Tie rod A rod connecting the steering arms together.

Toe-in It is the difference in measurement between the front of the wheels and the back of the wheels (the front are closer than the back).

Toe-out The back of the tires are closer than the front.

Torque A twisting force that may or may not result in motion. Measured in pound-feet or newton-meters.

Torque steer Torque steer occurs in front-wheel-drive vehicles when engine torque causes a front wheel to change its angle (toe) from straight ahead. The resulting pulling effect of the vehicle is most noticeable during rapid acceleration, especially whenever upshifting of the transmission creates a sudden change in torque.

Torque wrench A wrench that registers the amount of applied torque.

Torsion bar A type of spring in the shape of a straight bar. One end is attached to the frame of the vehicle and the opposite end is attached to a control arm of the suspension. When the wheels hit a bump, the bar twists and then untwists.

Torx A type of fastener that features a star-shaped indentation for a tool. A registered trademark of the Camcar Division of Textron.

Total toe The total (combined) toe of both wheels either front or rear.

Track The distance between the centerline of the wheels as viewed from the front or rear.

Track rod A horizontal steel rod or bar attached to the rear axle housing at one end and the frame at the other to keep the center of the rear axle centered on the body. Also known as a *panhard rod*.

Tracking A term used to describe the fact that the rear wheels should track directly behind the front wheels.

Tramp A vibration usually caused by up-and-down motion of an out-of-balance or out-of-round wheel assembly.

Turning radius Refers to the angle of the steering knuckles, which allow the inside wheel to turn at a sharper angle than the outside wheel whenever turning a corner. Also known as *toe-out on turns* (TOOT) or *Ackerman angle*.

2 DR Two-door model.

UNC Unified national coarse.

Underinflation A term used to describe a tire with too little tire pressure (less than minimum allowable pressure).

Understeer A term used to describe the handling of a vehicle where the driver must turn the steering wheel more and more while turning a corner.

UNF Unified national fine.

Unit-body A type of vehicle construction first used by the Budd Company of Troy, Michigan, that does not use a separate frame. The body is built strong enough to support the engine and the power train, as well as the suspension and steering system. The outside body panels are part of the structure. *See also* Space-frame construction.

Unsprung weight The parts of a vehicle not supported by the suspension system. Examples of items that are typical unsprung weight include wheels, tires, and brakes.

Vacuum Any pressure less than atmospheric pressure (14.7 psi).

Vacuum booster A vacuum power brake unit.

Vacuum power unit A device utilizing engine manifold vacuum to assist application of the brakes—reducing pedal effort.

Vent port The Society of Automotive Engineers (SAE) term for the front port of a master cylinder; also called the *compensating port* or *bypass*.

Vibration An oscillation, shake, or movement that alternates in opposite directions.

VOC Volatile organic compounds.

VSS Vehicle speed sensor.

Vulcanization A process where heat and pressure combine to change the chemistry of rubber.

Wander A type of handling that requires constant steering wheel correction to keep the vehicle going straight.

Warning light A light on the instrument panel to alert the driver when one-half of a split hydraulic system fails as determined by the pressure differential switch.

Watt's link A type of track rod that uses two horizontal rods pivoting at the center of the rear axle.

Wear bars *See* Wear indicators.

Wear indicators Bald area across the tread of a tire when only $\frac{2}{32}$ in. or less of tread depth remains.

Wear indicator ball joint A ball-joint design with a raised area around the grease fitting. If the raised area is flush or recessed with the surrounding area of the ball joint, the joint is worn and must be replaced.

Web The stiffening member of the shoe to which the shoe table is attached.

Weight-carrying ball joint *See* Load-carrying ball joint.

Wheel cover A functional and decorative cover over the entire wheel. *See also* Hub cap.

Wheel cylinder   The part of the hydraulic system that receives pressure from the master cylinder and applies the brake shoes to the drums.

Wheelbase   The distance between the centerline of the two wheels as viewed from the side.

Wishbone suspension   *See* SLA.

W/O   Without.

Worm and roller   A steering gear that uses a worm gear on the steering shaft. A roller on one end of the cross shaft engages the worm.

Worm and sector   A steering gear that uses a worm gear that engages a sector gear on the cross shaft.

WSS   Wheel speed sensor.

Zerk   A name commonly used for a grease fitting. Named for its developer, Oscar U. Zerk, in 1922, an employee of the Alamite Corporation. A grease fitting is also called an *Alamite fitting*.

◄ Appendix I ►

ASE-STYLE SAMPLE TEST FOR BRAKE SYSTEMS

1. Technician A says that the more lines on the head of a bolt, the higher the grade. Technician B says that non-graded hardware store bolts can be used in the place of a missing brake bolt. Which technician is correct?

   a. A only

   b. B only

   c. Both A and B

   d. Neither A nor B

2. When working on a vehicle, safety experts recommend that the technician

   a. wear safety glasses.

   b. wear a bump cap.

   c. wear gloves.

   d. all of the above.

3. A 6-mm-diameter bolt requires a _____ wrench.

   a. 6-mm

   b. 8-mm

   c. 10-mm

   d. 12-mm

4. A vehicle component interchange manual is usually called

   a. a vehicle interchange manual.

   b. the Hollander interchange manual.

   c. Lester's interchange manual.

   d. an SAE interchange manual.

5. Rear brakes tend to lock up during hard braking before front brakes because

   a. the rear brakes are larger.

   b. the vehicle weight transfers forward.

   c. the tires have less traction.

   d. Both B and C are correct.

6. Most vehicle manufacturers recommend using _____ brake fluid.

   a. DOT 2

   b. DOT 3

   c. DOT 4

   d. DOT 5

7. Used brake fluid should be disposed of

   a. with waste oil.

   b. as hazardous waste.

   c. by burning in an EPA-certified facility.

   d. by recycling.

8. The rubber used in most brake system components will swell if exposed to

   a. engine oil or ATF.

   b. moisture in the air.

   c. DOT 5 brake fluid.

   d. water.

9. The edge code lettering on the side of friction material tells the technician

   a. the coefficient of friction code.

   b. the quality of the friction material.

   c. the temperature resistance rating.

   d. All of the above

10. Technician A says that linings with asbestos can be identified by the dark gray color. Technician B says that all brake pads and linings should be treated as if they do contain asbestos.

    a. A only

    b. B only

    c. Both A and B

    d. Neither A nor B

11. Technician A says that brake fluid should be filled to the top of the reservoir to be assured of proper brake pressure when the brakes are applied. Technician B says that the brake fluid level should be filled only to the maximum level line to allow for expansion when the brake fluid gets hot during normal operation. Which technician is correct?

    a. A only

    b. B only

    c. Both A and B

    d. Neither A nor B

12. Self-application of the brakes can occur if

    a. the master cylinder is overfilled.

    b. the vent port is clogged or covered.

    c. the replenishing port is clogged or covered.

    d. Both A and B

13. Technician A says that the brake pedal height should be checked as part of a thorough visual inspection of the brake system. Technician B says that the pedal free play and pedal reserve should be checked. Which technician is correct?

    a. A only

    b. B only

    c. Both A and B

    d. Neither A nor B

14. Two technicians are discussing overhauling master brake cylinders. Technician A says that the bore of an aluminum master cylinder cannot be honed because of the special anodized surface. Technician B says that many overhaul (OH) kits include replacement piston assemblies. Which technician is correct?

    a. A only

    b. B only

    c. Both A and B

    d. Neither A nor B

15. Technician A says that the red brake warning lamp on the dash will light if there is a hydraulic failure or low brake fluid level. Technician B says that the red brake warning lamp on the dash will light if the parking brake is on. Which technician is correct?

    a. A only

    b. B only

    c. Both A and B

    d. Neither A nor B

16. A vehicle tends to lock up the *front* wheels when being driven on slippery road surfaces. Technician A says that the metering valve may be defective. Technician B says that the proportioning valve may be defective. Which technician is correct?

    a. A only

    b. B only

    c. Both A and B

    d. Neither A nor B

17. A vehicle pulls to the left during braking. Technician A says that the metering valve may be defective. Technician B says the proportioning valve may be defective. Which technician is correct?

    a. A only

    b. B only

    c. Both A and B

    d. Neither A nor B

18. A vehicle tends to lock up the rear wheels during hard braking. Technician A says that the metering valve may be defective. Technician B says that the proportioning valve may be defective. Which technician is correct?

    a. A only

    b. B only

    c. Both A and B

    d. Neither A nor B

19. Two technicians are discussing loosening a stuck bleeder valve. Technician A says to use a six-point wrench and simply pull on the wrench until it loosens. Technician B says that a shock is usually necessary to loosen a stuck bleeder valve. Which technician is correct?

    a. A only

    b. B only

    c. Both A and B

    d. Neither A nor B

20. The proper brake bleeding sequence for a front/rear split hydraulic system is

    a. right front, right rear, left front, left rear.

    b. right rear, left front, right front, left rear.

    c. left front, left rear, right front, right rear.

    d. right rear, left rear, right front, left front.

21. Two technicians are discussing bleeding air from the brake hydraulic system. Technician A says to depress the brake pedal slowly and not to the floor, to prevent possible seal damage inside the master cylinder. Technician B says to wait 15 seconds between strokes of the brake pedal. Which technician is correct?

    a. A only

    b. B only

    c. Both A and B

    d. Neither A nor B

22. Two technicians are discussing wheel bearings. Technician A says that conventional tapered roller bearings as used on the front of most rear-wheel-drive vehicles should be slightly loose when adjusted properly. Technician B says that the spindle nut should not be tightened more than finger-tight as the final step. Which technician is correct?

    a. A only

    b. B only

    c. Both A and B

    d. Neither A nor B

23. Wheel bearings are being packed with grease. Technician A says to use grease that is labeled "GC". Technician B says to use grease with a NLGI number of 2. Which technician is correct?

    a. A only

    b. B only

    c. Both A and B

    d. Neither A nor B

24. Technician A says that brake drums should be labeled left and right before being removed from the vehicle so that they can be reinstalled in the same location. Technician B says that the hold-down pins may have to be cut off to remove a worn brake drum. Which technician is correct?

    a. A only

    b. B only

    c. Both A and B

    d. Neither A nor B

25. Technician A says that the backing plate should be lubricated with chassis grease on the shoe pads. Technician B says that the brakes will squeak when applied if the shoe pads are not lubricated. Which technician is correct?

    a. A only

    b. B only

    c. Both A and B

    d. Neither A nor B

26. Technician A says that most experts recommend replacing all drum brake hardware, including the springs, every time the brake linings are replaced. Technician B says that the star wheel adjuster must be cleaned and lubricated to assure proper operation. Which technician is correct?

    a. A only

    b. B only

    c. Both A and B

    d. Neither A nor B

27. New brake shoes are being installed and they do not touch the anchor pin at the top. Technician A says that the brake shoes are not the correct size. Technician B says that the parking brake cable may need to be loosened. Which technician is correct?

    a. A only

    b. B only

    c. Both A and B

    d. Neither A nor B

28. A star wheel adjuster is installed on the wrong side of the vehicle. Technician A says that the adjuster cannot operate at all if installed on the wrong side. Technician B says that the adjuster would cause the clearance to increase rather than decrease when activated. Which technician is correct?

    a. A only

    b. B only

    c. Both A and B

    d. Neither A nor B

29. Technician A says to use synthetic grease to lubricate the backing plate. Technician B says to use special lithium-based brake grease. Which technician is correct?

    a. A only

    b. B only

    c. Both A and B

    d. Neither A nor B

30. One disc brake pad is worn more than the other. Technician A says that the caliper piston may be stuck in the caliper bore. Technician B says that the caliper slides may need to be cleaned and lubricated. Which technician is correct?

    a. A only

    b. B only

    c. Both A and B

    d. Neither A nor B

31. Technician A says that all metal-to-metal contact areas of the disc brake system should be lubricated with special brake grease for proper operation. Technician B says that the lubrication helps reduce brake noise (squeal). Which technician is correct?

    a. A only

    b. B only

    c. Both A and B

    d. Neither A nor B

32. After a disc brake pad replacement, the brake pedal went to the floor the first time the brake pedal was depressed. The most likely cause was

    a. air in the lines.

    b. improper disc brake pad installation.

    c. lack of proper lubrication of the caliper slides.

    d. normal operation.

33. Technician A says that the parking brake cable adjustment should be performed before adjusting the rear brakes. Technician B says that the parking brake cable should allow for about 15 "clicks" before the parking brake holds. Which technician is correct?

    a. A only

    b. B only

    c. Both A and B

    d. Neither A nor B

34. Technician A says that rotor thickness variation is a major cause of a pulsating brake pedal. Technician B says that at least 0.015 in. (0.4 mm) should be left on a rotor after machining to allow for wear. Which technician is correct?

    a. A only

    b. B only

    c. Both A and B

    d. Neither A nor B

35. The brake pedal of a vehicle equipped with ABS pulsates rapidly during hard braking. Technician A says that the rotor may require machining. Technician B says the lateral runout of the disc brake rotors may be excessive. Which technician is correct?

a. A only

b. B only

c. Both A and B

d. Neither A nor B

36. Two technicians are discussing hard spots in brake drums. Technician A says that the drum should be replaced. Technician B says that the hard spots are caused by using riveted rather than bonded brake shoes. Which technician is correct?

a. A only

b. B only

c. Both A and B

d. Neither A nor B

37. Disc brake rotors should be machined if rusted.

a. True

b. False

38. Before checking the brake fluid level in a typical integral ABS, the technician should pump the brake pedal

a. 2 or 3 times

b. 3 or 4 times

c. 5 to 10 times

d. 25 times or more

39. Technician A says that wheel speed sensors should be cleaned regularly as part of normal vehicle service. Technician B says that wheel speed sensors are magnetic and can attract metal particles. Which technician is correct?

a. A only

b. B only

c. Both A and B

d. Neither A nor B

40. Technician A says that a jumper wire or key can be used to retrieve diagnostic trouble codes. Technician B says that some ABS trouble codes can be erased simply by driving the vehicle over a certain speed. Which technician is correct?

a. A only

b. B only

c. Both A and B

d. Neither A nor B

◀ Appendix 2 ▶

ASE-STYLE SAMPLE TEST FOR STEERING, SUSPENSION, AND ALIGNMENT SYSTEMS

1. A customer complains that the steering lacks power assist only when cold. Technician A says that the power steering fluid may be low. Technician B says that a worn housing around the rotary (spool) valve is the most likely cause. Which technician is correct?

   a. A only

   b. B only

   c. Both A and B

   d. Neither A nor B

2. A front-wheel-drive vehicle pulls toward the right during acceleration. The most likely cause is

   a. worn or defective tires.

   b. leaking or defective shock absorbers.

   c. normal torque steer.

   d. a defective power steering rack-and-pinion steering assembly.

3. When replacing a rubber-bonded socket (RBS) tie rod end, the technician should be sure to

   a. remove the original using a special tool.

   b. install and tighten the replacement with the front wheels in the straight-ahead position.

   c. grease the joint before installing on the vehicle.

   d. install the replacement using a special clamp vise.

4. Whenever installing a tire on a rim, do not exceed

   a. 25 psi.

   b. 30 psi.

   c. 35 psi.

   d. 40 psi.

5. Two technicians are discussing mounting a tire on a wheel. Technician A says that for best balance, the tire should be match mounted. Technician B says that silicone spray should be used to lubricate the tire bend. Which technician is correct?

   a. A only

   b. B only

   c. Both A and B

   d. Neither A nor B

6. Technician A says that radial tires should *only* be rotated front to rear, never side to side. Technician B says that radial tires should be rotated using the modified "X" method. Which technician is correct?

   a. A only

   b. B only

   c. Both A and B

   d. Neither A nor B

7. For a tire that has excessive radial runout, Technician A says that it should be broken down on a tire changing machine and the tire rotated 180° on the wheel and retested. Technician B says that the tire should be replaced. Which technician is correct?
   a. A only
   b. B only
   c. Both A and B
   d. Neither A nor B

8. Technician A says that overloading a vehicle can cause damage to the wheel bearings. Technician B says that tapered roller bearings used on a non-drive wheel should be adjusted handtight only after seating. Which technician is correct?
   a. A only
   b. B only
   c. Both A and B
   d. Neither A nor B

9. Defective wheel bearings usually sound like
   a. a growl.
   b. a rumble.
   c. snow tires.
   d. all of the above.

10. Defective outer CV joints usually make a clicking noise
    a. only when backing.
    b. while turning and moving.
    c. while turning only.
    d. during braking.

11. The proper lubricant usually specified for use in a differential is
    a. SAE 15–40 engine oil.
    b. SAE 80W-90 GL-5.
    c. STF.
    d. SAE 80W-140 GL-1

12. A vehicle owner complained that a severe vibration was felt throughout the entire vehicle only during rapid acceleration from a stop and up to about 20 mph (32 km/h). The most likely cause is
    a. unequal drive shaft working angles.
    b. a bent drive shaft.
    c. defective universal joints.
    d. a bent rim or a defective tire.

13. To remove a C-clip axle, what step does *not* need to be done?
    a. Remove the differential cover.
    b. Remove the axle flange bolts/nuts.
    c. Remove the pinion shaft.
    d. Remove the pinion shaft lock bolt.

14. Drive shaft working angles can be changed by
    a. replacing the U-joints.
    b. using shims or wedges under the transmission or rear axle.
    c. rotating the position of the drive shaft on the yoke.
    d. tightening the differential pinion nut.

15. A driver complains that the vehicle darts or moves first toward one side and then to the other side of the road. Technician A says that bump steer caused by an unlevel steering linkage could be the cause. Technician B says that a worn housing in the spool valve area of the power rack and pinion is the most likely cause. Which technician is correct?
    a. A only
    b. B only
    c. Both A and B
    d. Neither A nor B

16. A vehicle equipped with power rack-and-pinion steering is hard to steer when cold only. After a couple of miles of driving, the steering power assist returns to normal. The most likely cause of this temporary loss of power assist when cold is
    a. a worn power steering pump.
    b. worn grooves in the spool valve area of the rack and pinion steering unit.
    c. a loose or defective power steering pump drive belt.
    d. a defective power steering computer sensor.

17. A dry park test is performed
    a. on a frame-type lift with the wheels hanging free.
    b. by pulling and pushing on the wheels with the vehicle supported by a frame-type lift.
    c. on the ground or on a drive-on type lift and moving the steering wheel while observing for looseness.
    d. driving in a figure 8 in a parking lot.

18. On a parallelogram-type steering linkage, the part that usually needs replacement first is the
    a. pitman arm.
    b. outer tie-rod end(s).
    c. center link.
    d. idler arm.

19. What parts need to be added to a "short rack" to make a "long" rack-and-pinion steering unit?
    a. Bellows and ball socket assemblies
    b. Bellows and outer tie-rod ends
    c. Ball socket assemblies and outer tie-rod ends
    d. Outer tie-rod ends

20. The adjustment procedure for a typical integral power steering gear is
    a. over-center adjustment, then worm bearing preload.

   b. worm bearing preload, then the over-center adjustment.

21. A vehicle is sagging at the rear. Technician A says that standard replacement shock absorbers should restore proper ride (trim) height. Technician B says that replacement springs are needed to restore ride height properly. Which technician is correct?

   a. A only

   b. B only

   c. Both A and B

   d. Neither A nor B

22. Technician A says that indicator ball joints should be loaded with the weight of the vehicle on the ground to observe the wear indicator. Technician B says that the nonindicator ball joints should be inspected *unloaded*. Which technician is correct?

   a. A only

   b. B only

   c. Both A and B

   d. Neither A nor B

23. The maximum allowable axial play in a ball joint is usually

   a. 0.001 in (0.025 mm).

   b. 0.003 in. (0.076 mm).

   c. 0.030 in. (0.76 mm).

   d. 0.050 in. (1.27 mm).

24. The ball joint used on MacPherson strut suspension is load carrying.

   a. True

   b. False

25. Technician A says that tapered parts, such as tie rod ends, should be tightened to specifications, then loosened ¼ turn before installing the cotter key. Technician B says that the nut used to retain tapered parts should never be loosened after torquing, but rather tightened further, if necessary, to line up the cotter key hole. Which technician is correct?

   a. A only

   b. B only

   c. Both A and B

   d. Neither A nor B

26. When should the strut rod (retainer) nut be removed?

   a. After compressing the coil spring

   b. Before removing the MacPherson strut from the vehicle

   c. After removing the cartridge gland nut

   d. Before removing the brake hose from the strut housing clip

27. "Dog tracking" is often caused by broken or damaged

   a. stabilizer bar links.

   b. strut rod bushings.

   c. rear leaf springs.

   d. track (panhard) rod.

28. A pull toward one side during braking is one symptom of defective or worn

   a. stabilizer bar links.

   b. strut rod bushings.

   c. rear leaf springs.

   d. track (panhard) rods.

29. Oil is added to the MacPherson strut housing before installing a replacement cartridge to

   a. lubricate the cartridge.

   b. transfer heat from the cartridge to the outside strut housing.

   c. act as a shock damper.

   d. prevent unwanted vibrations.

30. A vehicle will pull toward the side with the

   a. most camber.

   b. least camber.

31. Excessive toe out will wear the _____ edges of both front tires.

   a. inside

   b. outside

32. A vehicle will pull toward the side with the

   a. most caster.

   b. least caster.

33. If the turning radius (TOOT) is out of specification, what should be replaced?

   a. Outer tie-rod ends

   b. Inner tie-rod ends

   c. Idler arm

   d. Steering knuckle

34. SAI and camber together form the:

   a. included angle.

   b. turning radius angle.

   c. scrub radius angle.

   d. setback angle.

35. The thrust angle is being corrected. The alignment technician should adjust which angle to reduce thrust angle?

   a. Rear camber

   b. Front SAI or included angle and camber

   c. Rear toe

   d. Rear caster

36. Strut rods adjust _____ if there is a nut on both sides of the frame bushings.

   a. camber

   b. caster

   c. SAI or included angle depending on the exact vehicle

   d. toe

Questions 37 through 40 will use the following specifications:

Front camber	$0.5 \pm 0.3°$
Front caster	3.5 to 4.5°
Toe	$0 \pm 0.1°$
Rear camber	$0 \pm 0.5°$
Rear toe	−0.1 to 0.1°

Alignment angles:

Front camber left	0.5°
Front camber right	−0.1°
Front caster left	3.8°
Front caster right	4.5°
Front toe left	−0.2°
Front toe right	+0.2°
Total toe	0.0°
Rear camber left	0.15°
Rear camber right	−0.11°
Rear toe left	0.04°
Rear toe right	0.14°

37. The first angle corrected should be
    a. right front camber.
    b. right rear camber.
    c. right rear toe.
    d. left front camber.

38. The present alignment will cause excessive tire wear to the inside of both front tires.
    a. True
    b. False

39. The present alignment will cause excessive tire wear to the rear tires.
    a. True
    b. False

40. With the present alignment, the vehicle will
    a. pull toward the right.
    b. go straight.
    c. pull toward the left.

◀ Appendix 3 ▶

LUG NUT TIGHTENING TORQUE CHART

Name	Years	Ft-lb torque
Acura		
All models	86–93	80
American Motors		
All models	70–87	75
Audi		
All models	78–93	81
BMW		
All models	78–93	65–79
320i	77–83	59–65
528L	79–81	59–65
Buick		
All models except:	76–93	100
Century	76–81	80*
Regal	78–86	80*
LeSabre	76–85	80†
*With aluminum wheels		90
†With 1/2-in. studs		100
Cadillac		
All models except:	76–93	100
Seville	1976	80
Chevrolet		
Geo Prizm	92–93	100
Geo Prizm	89–91	76
Geo Metro	89–93	43
Geo Storm	90–93	86.5
Sprint	85–88	50
Spectrum	85–88	65†

Name	Years	Ft-lb torque
Chevette	82–87	80
Chevette	76–81	70
Nova	85–89	76
Vega and Monza	76–80	80*
Cavalier	82–93	100
Celebrity	82–90	100
Citation	80–86	100
Camaro	89–93	100
Camaro	78–88	80*
Lumina	90–93	100
Malibu and Monte Carlo	76–88	80*
Malibu Wagon	76–86	80*
Impala and Caprice Sedan	77–90	80
Impala and Caprice Wagon	77–90	100
Caprice	91–93	100
Corvette	84–93	100
Corvette	76–83	90
Corsica and Beretta	87–93	100
*With aluminum wheels		100
†With aluminum wheels		86.5
Chevrolet/GMC light trucks and vans		
Geo Tracker	92–93	60
Geo Tracker	89–91	37–58
Lumina APV	90–93	100
Astro/Safari Van	85–93	100
S/10 and S/15 Pickup	80–88	80
T/10 and T/15 Pickup	88–93	100
S/10 and S/15 Blazer/Jimmy	88–93	80
T/10 and T/15 Blazer/Jimmy	80–88	100

Name	Years	Ft-lb torque
C/K Pickup, all except:	88–93	120
C/K Pickup dual rear wheels	88–93	140
V10 (4WD full size) Suburban and Blazer (aluminum wheels)	88–89	100
V10 (4WD full size) Suburban and Blazer (steel wheels)	88–89	90
V10 (4WD full size) Suburban (all)	90–93	100
R10 (2WD full size) Suburban and Blazer	1989	100
R/V20 (2WD, 4WD full size) Suburban	1989	120
G10, 20 (full size) Van	88–93	100
G30 (full size) Van Except:	88–93	120
G30 (full size) Van (dual rear wheels)	88–93	140
El Camino/Caballero, Sprint	67–87	90
Luv Pickup	76–82	90
C/K10 Blazer and Jimmy	71–87	90
Chevy and GMC Pickups 10/15, 20/25, 30/35 (single rear wheel with 7/16-in. and 1/2-in. studs)	71–87	90
Chevy and GMC Pickups 10/15, 20/25, 30/35 (single rear wheel with 9/16-in. studs)	71/87	120

Chrysler

Name	Years	Ft-lb torque
Concorde	1993	95
Chrysler T/C by Maserati	89–91	95
Conquest	87–89	65–80
LeBaron (fwd)	84–93	95
LeBaron (fwd)	82–83	80
New Yorker (fwd)	83–93	95
Town & Country (fwd)	84–88	95
Town & Country (fwd)	82–83	80
Fifth Avenue (rwd)	83–90	85
New Yorker (rwd)	76–82	85
Laser	84–86	95
Limousine	85–86	95
Executive Sedan	84–85	95
E-Class	83–84	80
Imperial	90–93	95
Imperial (rwd)	81–83	85
Cordoba	76–83	85
LeBaron (rwd)	76–81	85
Town & Country (rwd)	78–81	85
Newport	76–81	85

Daihatsu

Name	Years	Ft-lb torque
Charade	88–91	65–87

Name	Years	Ft-lb torque
Daihatsu light trucks and vans		
Rocky (all)	90–91	65–87
Dodge		
Intrepid	1993	95
Stealth	91–93	87–101
Spirit	89–93	95
Shadow	87–93	95
Colt	76–93	65–80
Lancer	85–89	95
Aries	84–89	95
Aries	81–83	80
Charger	84–87	95
Charger	82–83	80
Daytona	84–93	95
Omni	84–90	95
Omni	78–83	80
Vista	84–93	50–57†
600	84–88	95
600	1983	80
Diplomat	78–89	80
Dynasty	88–93	95
Monaco	90–91	54–72‡
Conquest	1986	65–80
Conquest	84–85	50–57†
400	82–83	80
Challenger	78–83	51–58*
Mirada	80–83	85
St. Regis	79–81	85
Aspen	76–80	85
*With aluminum wheels		58–73
†With aluminum wheels		65–80
‡With aluminum wheels		80–100

Dodge light trucks and vans

Name	Years	Ft-lb torque
Caravan, Ram Van (FWD)	84–93	95
Rampage (FWD)	82–84	90
Ramcharge AD, AW100	79–93	105
Wagons B100/150	72–93	105
Wagons B200/250	72–93	105
Wagons B300/350 1/2-in. studs	69–93	105
Wagons B300/350 5/8-in. studs	79–93	200
D50 Pickup	78–86	55
D50 Pickup	87–93	95
D100/150 Pickup	72–93	105
D200/250 Pickup	81–93	105
D300/350 Pickup 1/2-in. studs	79–93	105
D300/350 Pickup 5/8-in. studs	79–93	200
W100/150 Pickup	79–93	105
W200/250 Pickup	79–93	105
W300/350 Pickup 1/2-in. studs	79–93	105
W300/350 Pickup 5/8-in. studs	79–93	200
Dakota	87–93	85

Name	Years	Ft-lb torque
Ford		
Probe	89–91	65–87
Festiva	89–93	65–87
All models	84–93	85–105
Escort	81–83	80–105
EXP	82–83	80–105
Fiesta	78–80	63–85
Mustang	79–83	80–105
Pinto	76–80	80–105
Fairmont	78–83	80–105
Granada	76–82	80–105
LTD	79–83	80–105
LTD	76–78	70–115
Torino	76–79	80–105
LTD Crown Victoria	1983	80–105
Country Sedan and Squire	79–83	80–105
Thunderbird	80–83	80–105
Thunderbird	76–79	70–115
Ford light trucks and vans		
E150/F150 and Bronco	88–93	100
E250/E350, F250, F350	88–93	140
Aerostar	86–93	100
Bronco	88–93	135
Bronco	72–87	100
Bronco II	84–87	100
Explorer	91–93	100
Club Wagon E100/150	75–87	100
Club Wagon E200/250	76–87	100
Club Wagon E300/350 (single rear wheels)	76–87	145
Club Wagon E300/350 (dual rear wheels)	76–87	220
Econoline Van E100/150	75–81	100
Econoline Van E200/250	76–87	100
Econoline Van E300/350 (single rear wheels)	76–87	145
Econoline Van E300/350 (dual rear wheels)	76–87	220
Ranger Pickup	84–87	100
Courier Pickup	77–83	65
F100/150 Pickup	75–87	100
F200/250 Pickup	76–87	100
F300/350 Pickup (single rear wheels)	76–87	145
F300/350 Pickup (dual rear wheels)	76–87	220
Honda		
All models	84–93	80
Civic, all models	73–83	58
Accord, all models	82–93	80
Accord, all models	76–81	58
Prelude, all models	79–93	80

Name	Years	Ft-lb torque
Hyundai		
All models	90–93	65–80
Excel	86–89	50–57*
*With alloy wheels		65–72
Infiniti		
Q45	90–93	72–87
M30	90–93	72–87
Isuzu		
Stylus	1991	87
Impulse	83–91	87
I Mark	87–89	65*
I Mark	1986	90
I Mark	1985	50†
I Mark	82–84	50‡
*With aluminum wheels		86.5
†With aluminum wheels		86
‡With aluminum wheels		90
Isuzu light trucks and vans		
Pickup	91–93	72.3*
Pickup	1990	58–87**
Amigo	1991	72.3*
Amigo	89–90	58–87†
Rodeo	1991	72‡
Trooper	84–91	58–87†
*With aluminum wheels		86.8
†With aluminum wheels		80–94
‡With aluminum wheels		87
Jaguar		
All models	89–91	65–75
XJ6 and XJS	1988	75
All models	81–87	40–80*
*With allow wheels		50
Jeep-Eagle		
Vision	1993	95
Premier	89–91	54–72*
Talon	90–93	87–101
Summit	89–91	65–80
Medallion	1988	67
*With aluminum wheels		80–100
Jeep light trucks and vans		
Grand Cherokee and Grand Wagoneer	92–93	88
Wrangler, YJ	90–91	80
Cherokee, Comanche	84–91	75
Wagoneer and Grand Wagoneer	84–91	75
Trucks (under 8400 GVW)	84–89	75
Trucks (over 8400 GVW)	84–89	130
CJ Series	84–86	80
CJ Series	81–83	65–80
Cherokee, Wagoneer	81–83	65–90

Name	Years	Ft-lb torque
Trucks (under 8400 GVW)	81–83	65–90
Trucks (over 8400 GVW)	81–83	110–150
Lexus		
All models	90–93	76
Lincoln		
All models	84–93	85–105
Mark IV	80–83	80–105
Continental	76–83	80–105
Town Car	81–83	80–105
Versailles	77–80	80–105
Mazda		
All models except:	88–93	65–87
Navajo	91–93	100
323	86–87	65–87
GLC	81–85	65–80
GLC	77–80	65–80
GLC wagon	84–85	65–87
626	84–87	65–87
626	1983	65–80
626	79–82	65–80*
COSMO	76–78	65–72*
808	76–77	65–72*
RX7	84–87	65–87
RX7	79–83	65–80*
RX7	76–78	65–72*
RX3	76–78	65–72*
*With aluminum wheels		69–80
Mazda light trucks and vans		
B2600	87–93	65–87*
B2200	86–93	65–87*
B2000/B2200	80–85	72–80
*With aluminum wheels		87–108
Mercedes		
All models	76–93	81
Mercury		
All models	84–93	85–105
All models	76–83	80–105
Mitsubishi		
Sigma V6	89–90	65–80
Mirage	85–91	65–80
Precis	87–89	51–58*
Cordia/Tredia	83–88	50–57*
Eclipse	90–93	87–101
Galant	85–86	50–57†
Galant	1987	65–80
Galant	88–93	65–80
Starion	1983	50–57*
Starion	84–89	50–57†
*With aluminum wheels		57–72
†With aluminum wheels		66–80

Name	Years	Ft-lb torque
Mitsubishi light trucks and vans		
Van/wagons	89–90	87–101
Montero	89–91	75–87
Pickups	89–91	72–87
Pickups	83–87	65
Montero	83–87	65
Nissan/Datsun		
All models	91–93	72–87
Maxima	89–90	72–87
Maxima	87–88	72–89
Maxima	85–86	58–72
Pulsar SE, XE	87–90	72–87
Pulsar	83–86	58–72
Sentra, all models	87–90	72–87
Sentra	83–86	58–72
Stanza, all models	87–90	72–87
Stanza	82–86	58–72
210	79–82	58–72
310	79–82	58–72
510	78–81	58–72
810	1981	58–72
810	77–80	58–65
200 SX, all models	87–88	87–108
200 SX	80–86	58–72
200 SX	77–79	58–65
280 ZX	79–83	58–72
300 ZX and ZX Turbo	1990	72–87
300 ZX and ZX Turbo	87–89	87–108
300 ZX	84–86	58–72
Axxess	1990	72–87
240 SX	1989	72–87
Nissan/Datsun light trucks and vans		
All pickups and Pathfinder	89–93	87–108
Van	87–88	72–87
Van	1990	72–87
Oldsmobile		
All models except:	76–93	100
Starfire	76–80	80
Cutlass (RWD)	76–88	80
Cutlass Supreme (FWD)	88–93	100
Delta 88	77–85	80*
*With 1/2-in. studs		100
Oldsmobile light trucks and vans		
Silhouette	90–93	100
Bravado	92–93	95
Plymouth		
Colt	83–93	65–80
Sundance	87–93	95
Acclaim	89–93	95
Laser	90–91	87–101

Name	Years	Ft-lb torque
Caravelle	85–88	95
Horizon	84–90	95
Horizon	78–83	80
Turismo	84–87	95
Turismo	82–83	80
Vista	84–91	50–57†
Reliant	84–89	95
Reliant	81–83	80
Gran Fury	80–89	85
Conquest	1986	50–57†
Conquest	84–85	50–57
Sapporo	78–83	51–58*
Champ	79–82	51–58*
*With aluminum wheels		58–73
†With aluminum wheels		65–80
Plymouth light trucks and vans		
Voyager	84–93	95
Pontiac		
All models except:	80–88	100
T-1000	81–87	80
Sunbird	76–80	80
Firebird	76–88	80*
Grand Prix	76–87	80*
LeMans	89–93	65
Catalina	76–86	80†
Bonneville	76–86	80†
Parisienne	83–86	80†
*With aluminum wheels		90
†With 1/2-in. studs		100
Pontiac light trucks and vans		
Trans Sport	90–93	100
Porsche		
All models	79–93	94
Range-rover		
All models	91–93	90–95
Saab		
All models	88–93	80–90
All models	76–87	65–80
Saturn		
All models	91–93	100
Sterling		
All models	87–91	53

Name	Years	Ft-lb torque
Subaru		
All models	76–93	58–72
Toyota		
All models except:	84–93	76
Celica	70–85	66–86
Supra	79–85	66–86
Corolla	80–83	66–86
Corona	75–82	66–86
Cressida	78–85	66–86
Corona MK II	1976	65–94
Tercel	80–85	66–86
Starlet	81–85	66–86
Toyota light trucks and vans		
All models except:	88–93	76
Land Cruiser	88–91	116
Pickups	75–87	75
Land Cruiser	75–84	75
Van Wagon	84–86	75
Volkswagen		
All models except van	88–93	81
Golf	85–87	81
Rabbit, all models	76–84	73–87
Jetta	81–87	81
Scirocco	76–84	73–87
Quantum	82–86	81
Dasher	76–81	65
Scirocco	85–87	81
Volkswagen light trucks and vans		
Vanagon	80–93	123
Pickups	79–84	81
Transporter	77–79	95
Volvo		
All models	89–93	63
740 Series	85–88	63
760 Series	83–88	63
GLE	76–82	72–94
260 Series	75–84	72–95
240 Series	81–88	63
240 Series	75–80	72–95
Yugo		
All models with steel wheels	86–90	63
All models with alloy wheels	88–90	81

◀ Appendix 4 ▶

ENGLISH–METRIC (SI) CONVERSION*

Inches to millimeters						Millimeters to inches		
inches			inches				inches	
fraction	decimal	mm	fraction	decimal	mm	mm	decimal	fraction
1/64	0.016	0.40	17/32	0.531	13.49	0.5	0.020	1/64
1/32	0.031	0.79	9/16	0.563	14.29	1	0.039	3/64
3/64	0.047	1.19	19/32	0.594	15.08	2	0.079	5/64
1/16	0.063	1.59	5/8	0.625	15.88	3	0.118	1/8
5/64	0.078	1.98	21/32	0.656	16.67	4	0.157	5/32
3/32	0.094	2.38	11/16	0.688	17.46	5	0.197	13/64
7/64	0.109	2.78	23/32	0.719	18.26	6	0.236	15/64
1/8	0.125	3.18	3/4	0.750	19.05	7	0.276	9/32
5/32	0.156	3.97	25/32	0.781	19.84	8	0.315	5/16
3/16	0.188	4.76	13/16	0.813	20.64	9	0.354	23/64
7/32	0.219	5.56	27/32	0.844	21.43	10	0.394	25/64
1/4	0.250	6.35	7/8	0.875	22.23	11	0.433	7/16
9/32	0.281	7.14	29/32	0.906	23.02	12	0.472	15/32
5/16	0.313	7.94	15/16	0.938	23.81	13	0.512	33/64
11/32	0.344	8.73	31/32	0.969	24.61	14	0.551	35/64
3/8	0.375	9.53	1	1.000	25.4	15	0.591	19/32
13/32	0.406	10.32				16	0.630	5/8
7/16	0.438	11.11				17	0.669	43/64
15/32	0.469	11.91				18	0.709	45/64
1/2	0.500	12.70				19	0.748	3/4
						20	0.787	25/32
						21	0.827	53/64
						22	0.866	55/64
						23	0.906	29/32
						24	0.945	15/16
						25	0.984	63/64

*Courtesy of FMC.

Fraction/Decimal/Millimeter Conversion Chart

1mm = .03937"   .001" = .0254mm

Fraction	Decimal	Millimeters
1/64	0.015625	0.397
1/32	.03125	0.794
3/64	.046875	1.191
1/16	.0625	1.588
5/64	.078125	1.984
3/32	.09375	2.381
7/64	.109375	2.778
1/8	.1250	3.175
9/64	.140625	3.572
5/32	.15625	3.969
11/64	.171875	4.366
3/16	.1875	4.763
13/64	.203125	5.159
7/32	.21875	5.556
15/64	.234375	5.953
1/4	.2500	6.350
17/64	.265625	6.747
9/32	.28125	7.144
19/64	.296875	7.541
5/16	.3125	7.938
21/64	.328125	8.334
11/32	.34375	8.731
23/64	.359375	9.128
3/8	.3750	9.525
25/64	.390625	9.922
13/32	.40625	10.319
27/64	.421875	10.716
7/16	.4375	11.113
29/64	.453125	11.509
15/32	.46875	11.906
31/64	.484375	12.303
1/2	.5000	12.700

Fraction	Decimals	Millimeters
33/64	0.515625	13.097
17/32	.53125	13.494
35/64	.546875	13.891
9/16	.5625	14.288
37/64	.578125	14.684
19/32	.59375	15.081
39/64	.609375	15.478
5/8	.6250	15.875
41/64	.640625	16.272
21/32	.65625	16.669
43/64	.671875	17.066
11/16	.6875	17.463
45/64	.703125	17.859
23/32	.71875	18.256
47/64	.734375	18.653
3/4	.7500	19.050
49/64	.765625	19.447
25/32	.78125	19.844
51/64	.796875	20.241
13/16	.8125	20.638
53/64	.828125	21.034
27/32	.84375	21.431
55/64	.859375	21.828
7/8	.8750	22.225
57/64	.890625	22.622
29/32	.90625	23.019
59/64	.921875	23.416
15/16	.9375	23.813
61/64	.953125	24.209
31/32	.96875	24.606
63/64	.984375	25.003
1	1.000	25.400

MM	INCHES	MM	INCHES
.1	.0039	46	1.8110
.2	.0079	47	1.8504
.3	.0118	48	1.8898
.4	.0157	49	1.9291
.5	.0197	50	1.9685
.6	.0236	51	2.0079
.7	.0276	52	2.0472
.8	.0315	53	2.0866
.9	.0354	54	2.1260
1	.0394	55	2.1654
2	.0787	56	2.2047
3	.1181	57	2.2441
4	.1575	58	2.2835
5	.1969	59	2.3228
6	.2362	60	2.3622
7	.2756	61	2.4016
8	.3150	62	2.4409
9	.3543	63	2.4803
10	.3937	64	2.5197
11	.4331	65	2.5591
12	.4724	66	2.5984
13	.5118	67	2.6378
14	.5512	68	2.6772
15	.5906	69	2.7165
16	.6299	70	2.7559
17	.6693	71	2.7953
18	.7087	72	2.8346
19	.7480	73	2.8740
20	.7874	74	2.9134
21	.8268	75	2.9528
22	.8661	76	2.9921
23	.9055	77	3.0315
24	.9449	78	3.0709
25	.9843	79	3.1102
26	1.0236	80	3.1496
27	1.0630	81	3.1890
28	1.1024	82	3.2283
29	1.1417	83	3.2677
30	1.1811	84	3.3071
31	1.2205	85	3.3465
32	1.2598	86	3.3858
33	1.2992	87	3.4252
34	1.3386	88	3.4646
35	1.3780	89	3.5039
36	1.4173	90	3.5433
37	1.4567	91	3.5827
38	1.4961	92	3.6220
39	1.5354	93	3.6614
40	1.5748	94	3.7008
41	1.6142	95	3.7402
42	1.6535	96	3.7795
43	1.6929	97	3.8189
44	1.7323	98	3.8583
45	1.7717	99	3.8976
		100	3.9370

DECIMAL EQUIVALENTS

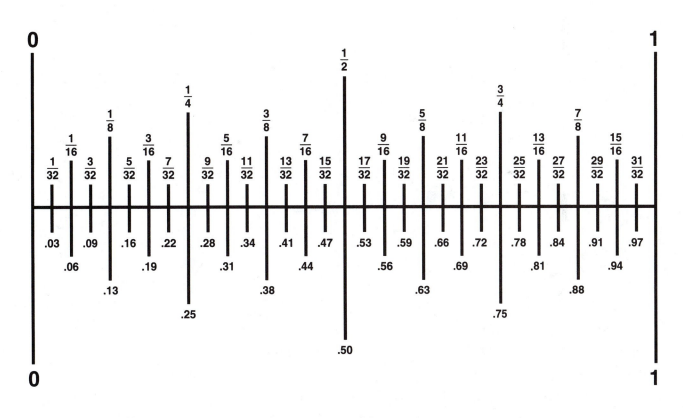

FRACTION/DECIMAL CONVERSION CHART

◀ Appendix 6 ▶

ALIGNMENT ANGLE CONVERSION

Angle Conversions from Degrees and Minutes

Degrees and minutes	Decimal degrees	Fractional degrees
0°05′	0.08°	$\frac{1}{12}$°
0°10′	0.16°	$\frac{1}{6}$°
0°15′	0.25°	$\frac{1}{4}$°
0°20′	0.33°	$\frac{1}{3}$°
0°25′	0.42°	$\frac{5}{12}$°
0°30′	0.50°	$\frac{1}{2}$°
0°35′	0.58°	$\frac{7}{12}$°
0°40′	0.67°	$\frac{2}{3}$°
0°45′	0.75°	$\frac{3}{4}$°
0°50′	0.83°	$\frac{5}{6}$°
0°55′	0.92°	$\frac{11}{12}$°
1°00′	1.00°	1°
(minutes of an hour)	(cents of a dollar)	

1 minute = 01′
60 minutes = 1°

Angle Conversions from Fractional Degrees

Fractional degrees	Degrees and minutes	Decimal degrees
$\frac{1}{8}$°	0°08′	0.125°
$\frac{1}{4}$°	0°15′	0.25°
$\frac{3}{8}$°	0°23′	0.375°
$\frac{1}{2}$°	0°30′	0.50°
$\frac{5}{8}$°	0°38′	0.625°
$\frac{3}{4}$°	0°45′	0.75°
$\frac{7}{8}$°	0°53′	0.875°
1°	1°00′	1.00°
(minutes of an hour)	(cents of a dollar)	

Angle Conversions from Decimal Degrees

Decimal degrees	Degrees and minutes	Fractional degrees
0.05°	0°03′	$\frac{1}{20}$°
0.10°	0°06′	$\frac{1}{10}$°
0.15°	0°09′	$\frac{3}{20}$°
0.20°	0°12′	$\frac{1}{5}$°
0.25°	0°15′	$\frac{1}{4}$°
0.30°	0°18′	$\frac{3}{10}$°
0.35°	0°21′	$\frac{7}{20}$°
0.40°	0°24′	$\frac{2}{5}$°
0.45°	0°27′	$\frac{9}{20}$°
0.50°	0°30′	$\frac{1}{2}$°
0.55°	0°33′	$\frac{11}{20}$°
0.60°	0°36′	$\frac{3}{5}$°
0.65°	0°39′	$\frac{13}{20}$°
0.70°	0°42′	$\frac{7}{10}$°
0.75°	0°45′	$\frac{3}{4}$°
0.80°	0°48′	$\frac{4}{5}$°
0.85°	0°51′	$\frac{17}{20}$°
0.90°	0°54′	$\frac{9}{10}$°
0.95°	0°57′	$\frac{19}{20}$°
1.00°	1°00′	1°
(cents of a dollar)	(minutes of an hour)	

Conversions from Fractional Inches

Fractional inches	Decimal inches	Millimeters
$\frac{1}{32}$	0.03	0.79
$\frac{1}{16}$	0.06	1.59
$\frac{3}{32}$	0.09	2.38
$\frac{1}{8}$	0.13	3.18
$\frac{5}{32}$	0.16	3.97
$\frac{3}{16}$	0.19	4.76
$\frac{7}{32}$	0.22	5.56
$\frac{1}{4}$	0.25	6.35
$\frac{9}{32}$	0.28	7.14
$\frac{5}{16}$	0.31	7.94
$\frac{11}{32}$	0.34	8.73
$\frac{3}{8}$	0.38	9.53
$\frac{13}{32}$	0.41	10.32
$\frac{7}{16}$	0.44	11.11
$\frac{15}{32}$	0.47	11.91
$\frac{1}{2}$	0.50	12.70

◀ Appendix 7 ▶

DOT TIRE CODES

The DOT manufacturer's code on this tire is FU (always the first two letters or numbers after the DOT lettering).

Code No.	Tire Manufacturer
AC	General Tire Company Charlotte, NC USA
AD	General Tire Company Mayfield, KY USA
AE	General Fabrica Espanola del Caucho S.A. (General Tire International) Torrelavego, Santander, Spain
AF	Manufactura Nacional de Borracha S.A.R.L. (General Tire International) Porto, Portugal

Code No.	Tire Manufacturer
AH	General Popo S.A. (General Tire International) Mexico City, Mexico
AJ	Uniroyal Tire Company Detroit, MI USA
AK	Uniroyal Tire Company Chicopee Falls, MA USA
AL	Uniroyal Goodrich Tire Co. Eau Claire, WI USA
AM	Uniroyal Tire Company Los Angeles, CA USA

Code No.	Tire Manufacturer	Code No.	Tire Manufacturer
AN	Uniroyal Goodrich Tire Co. Opelika, AL USA	BE	Uniroyal Goodrich Tire Co. Tuscaloosa, AL USA
AO	Wearwell Tire & Tube Co. Bhopal (MP), India	BF	Uniroyal Goodrich Tire Co. Woodburn, IN USA
AP	Uniroyal Goodrich Tire Co. Ardmore, OK USA	BH	Uniroyal Goodrich Tire Co. Kitchener, Ontario, Canada
AT	Avon Rubber Co., Ltd. Melksham, Wiltshire, England	BK	B. F. Goodrich do Brasil Campinas, Brazil
AU	Uniroyal Goodrich Tire Co. Kitchener, Ontario, Canada	BL	Industria Colobiana de Liantas S.A. (B. F. Goodrich International) Bogata, Colombia
AV	Seiberling Tire & Rubber Co. Barberton, OH USA	BM	B. F. Goodrich Australia Pty., Ltd. Campbellfield, Victoria, Australia
AW	Samson Tire & Rubber Co., Ltd. Tel Aviv, Israel	BN	B. F. Goodrich Philippines, Inc. Makti, Rizal, The Philippines
AX	Phoenix Gummiwerke, A.G. Hamburg, Germany	BO	General Tire Company Casablanca, Morocco
AY	Phoenix Gummiwerke, A.G. Reinsdorf, Germany	BT	Semperit, A. G. Traiskirehen, Austria
A1	Pneumatics Michelin Potiers, France	BU	Semperit Ireland Ltd. Dublin, Ireland
A2	Lee Tire & Rubber Company São Paulo, Brazil	BV	International Rubber Industries Jeffersontown, KY USA
A3	General Tire & Rubber Company Mount Vernon, IL USA	BW	The Gates Rubber Company Denver, CO USA
A4	Hung-A Industrial Co., Ltd. Pusan, Korea	BX	The Gates Rubber Company Nashville, TN USA
A5	Debickie Zaklady Opon Samochodowych Debica, Poland	BY	The Gates Rubber Company Littleton, CO USA
A6	Apollo Tyres, Ltd. Cochin, India	B1	Pneumatics Michelin La Rochesur-Yon, France
A7	Thai Bridgestone Tire Co. Ltd. Changwad Patoom-Thani, Thailand	B2	Dunlop Malaysian Industries Berhad Selongor, Malaysia
A8	P. T. Bridgestone Tire Indonesia Factory Jawa Barat, Indonesia	B3	Michelin Tire Mfg. Company of Canada, Ltd. Bridgewater, Nova Scotia, Canada
A9	General Tire Company Bryan, OH USA	B4	Taurus Hungarian Rubber Works Budapest, Hungary
BA	Uniroyal Goodrich Tire Company Akron, OH USA	B5	Olsztynskle Zaklady Opon Samochodowych Olsztyn, Poland
BB	B. F. Goodrich Tire Company Miami, OK USA	B6	Michelin Tire Corporation Spartanburg, SC USA
BC	B. F. Goodrich Tire Company Oaks, PA USA	B7	Michelin Tire Corporation Dothan, AL USA
BD	B. F. Goodrich Tire Company Miami, OK USA		

Code No.	Tire Manufacturer
B8	Cia Brasiliera de Pneumaticos Michelin Industria Rio De Janeiro, Brazil
B9	Michelin Tire Corporation Lexington, SC USA
CA	The Mohawk Rubber Company Akron, OH USA
CB	The Mohawk Rubber Company Helena, AR USA
CC	The Mohawk Rubber Company Salem, VA USA
CD	Alliance Tire & Rubber Co., Ltd. Hadera, Israel
CF	The Armstrong Rubber Company Des Moines, IA USA
CH	The Armstrong Rubber Company Hanford, CA USA
CJ	Inque Rubber Company, Ltd. Atsuta-ku, Nagoyo, Japan
CK	The Armstrong Rubber Company Madison, TN USA
CL	Continental A.C. Hannover, Germany
CM	Continental Gummi-Werke, A.G. Hannover, Germany
CN	Usine Francaise des Pneumatics Sarreguemines, France
CO	Goodyear Lastikleri Tas Adopozare, Turkey
CP	Continental Gummi-Werke, A.G. Korbach, Germany
CT	Continental Gummi-Werke, A.C. Dannenberg, Germany
CU	Continental Gummi-Werke, A.C. Hannover-Stocken, Germany
CV	The Armstrong Rubber Company Natchez, MS USA
CW	The Toyo Rubber Industry Co., Ltd. Itami, Hyogo, Japan
CX	The Toyo Rubber Industry Co., Ltd. Natori-Gun, Miyagi, Japan
CY	McCreary Tire & Rubber Company Indiana, PA USA
C1	Michelin (Nigeria) Ltd. Port-Harcourt, Nigeria

Code No.	Tire Manufacturer
C2	Kelly-Springfield Tire Company Americana, Sao Paulo, Brazil
C3	McCreary Tire & Rubber Company Baltimore, MD USA
C4	Dico Tire, Inc. Clinton, TN USA
C5	Poznanskie Zaklady Opon Samochodowych Poznon, Poland
C6	MITAS, n.p. Praha Prague, Czechoslovakia
C7	Ironsides Tire & Rubber Company Louisville, KY USA
C8	Hsin Chu Plant–Bridgestone Hsin Chu, Taiwan
C9	Seven Star Rubber Co., Ltd. Chang-Hua, Taiwan
DA	The Dunlop Tire & Rubber Co. Buffalo, NY USA
DB	The Dunlop Tire & Rubber Corp. Huntsville, AL USA
DC	Dunlop Tire Canada, Ltd. Whitby, Ontario, Canada
DD	The Dunlop Company, Ltd. Birmingham, England
DE	The Dunlop Company, Ltd. Washington, Durham, England
DK	Dunlop SA Montlucon, France
DL	Dunlop SA Amiens, France
DM	Dunlop A.G. Hanau Am Main, Germany
DN	Dunlop A.G. Wittlich, Germany
DO	Kelly-Springfield Lastikleri Tas Adapazari, Turkey
DV	N.V. Nederlandsch-Amerikaansche Enschnede, The Netherlands
DW	Rubberfabriek Vredestein Doetinchem, The Netherlands
DX	N.V. Bataafsche Rubber Ind. Radium Maastricht, The Netherlands
DY	Denman Tire Corporation Leavittsburg, OH USA

Code No.	Tire Manufacturer
D1	Viking Askim Askim, Norway
D2	Bridgestone/Firestone, Inc. LaVergne, TN USA
D3	United Tire & Rubber Mfg. Company Cobourg, Ontario, Canada
D4	Dunlop India, Ltd. West Bengal, India
D5	Dunlop India, Ltd. Madras, India
D6	Borovo Borovo, Yugoslavia
D7	Dunlop South Africa Ltd. Natal, Republic of South Africa
D8	Dunlop South Africa Ltd. Natal, Republic of South Africa
D9	United Tire & Rubber Co., Ltd. Rexdale, Ontario, Canada
EA	Metzler, A.G. Munchen, Germany
EB	Metzler, A.G. Zweigwerk Neustadt Neustadt Odenwald, Germany
EC	Metzler, A.G. Hochst, Germany
ED	Michelin–Okamoto Tire Corporation Tokyo, Japan
EE	Nitto Tire Company, Ltd. Kanagawa-ken, Japan
EF	Hung Ah Tire Company, Ltd. Seoul, Korea
EH	Bridgestone Tire Company, Ltd. Fukuoka-ken, Japan
EJ	Bridgestone Tire Company, Ltd. Saga-ken, Japan
EK	Bridgestone Tire Company, Ltd. Fukuoka-ken, Japan
EL	Bridgestone Tire Company, Ltd. Shiga-ken, Japan
EM	Bridgestone Tire Company, Ltd. Tokyo, Japan
EN	Bridgestone Tire Company, Ltd. Tochigi-ken, Japan
EO	Lee Lastikleri Tas Adapazari, Turkey
EP	Bridgestone Tire Company, Ltd. Tochigi-ken, Japan

Code No.	Tire Manufacturer
ET	Sumitomo Rubber Industries Fuklai, Kobe, Japan
EU	Sumitomo Rubber Industries Alchi Prefecture, Japan
EW	Pneumatiques Kleber, S.A. Toul (Meurthe-et-Moselle), France
EX	Pneumatiques Kleber, S.A. La Chapella St. Luc, France
EY	Pneumatiques Kleber, S.A. St. Ingbert (Saar), Germany
E1	Chung Hsing Industrial Co., Ltd. Taichung Hsien, Taiwan
E2	Industria de Pneumaticos Firestone Sao Paulo, Brazil
E3	Seiberling Tire & Rubber Company Lavergne, TN USA
E4	The Firestone Tire and Rubber Co. of New Zealand, Ltd. Papanui, Christ Church, New Zealand
E5	Firestone South Africa (Pty) Ltd. Port Elizabeth, South Africa
E6	Firestone-Tunisie S.A. Menzel-Bourguiba, Tunisia
E7	Firestone East Africa Ltd. Nairobi, Kenya
E8	Firestone Ghana Ltd. Accra, Ghana
E9	Firestone South Africa (Pty) Ltd. Brits, South Africa
FA	The Yokohama Rubber Company, Ltd. Hiratsuka, Kanagawa-Pref, Japan
FB	The Yokohama Rubber Company, Ltd. Watari-gun, Miye-Pref, Japan
FC	The Yokohama Rubber Company, Ltd. Mishima, Shizuoka-Pref, Japan
FD	The Yokohama Rubber Company, Ltd. Shinsharo, Aichi-Pref, Japan
FE	The Yokohama Rubber Company, Ltd. Ageo, Saitama-Pref, Japan
FF	Manufacture Francaise des Penumatiques Michelin Clermont-Ferrand, France
FH	Manufacture Francaise des Pneumatics Michelin Clermont-Ferrand, France

Code No.	Tire Manufacturer	Code No.	Tire Manufacturer
FJ	Manufacture Francaise des Pneumatics Michelin Bourges, France	F8	Vikrant Tyres Ltd. Mysore, Karnataka, India
FK	Manufacture Francaise des Pneumatics Michelin Cholet, France	F9	Dunlop New Zealand, Ltd. Upper Hutt, New Zealand
FL	Manufacture Francaise des Pneumatics Michelin Montceau-les-Mines, France	F0	Fidelity Tire Mfg. Company Natchez, MS USA
FM	Manufacture Francaise des Pneumatics Michelin Mesmin, Orleans, France	HA	S.A.F.E.N.M. (Michelin) Aranda, Spain
FN	Manufacture Francaise des Pneumatics Michelin Tours, France	HB	S.A.F.E.N.M. Lasarte (Michelin) Lasarte, Spain
FP	Ste. d'Applications Techniques Ind. Algers, Algeria	HC	S.A.F.E.N.M. (Michelin) Vitoria, Spain
FT	Michelin Reifenwerke, A.G. Bad Kreuznach, Germany	HD	Societa per Azoni, Michelin Cuneo, Italy
FU	Michelin Reifenwerke, A.G. Bamberg, Germany	HE	Societa per Azoni, S.P.A., Michelin Marengo, Italy
FV	Michelin Reifenwerke, A.G. Homburg, Germany	HF	Societa per Azoni, Michelin Turin (Dora), Italy
FW	Michelin Reifenwerke, A.G. Karlsruhe, Germany	HH	Societa per Azoni, Michelin Turin (Stura), Italy
FX	S.A. Belge du Pneumatique Michelin Zuen, Belgium	HJ	Michelin Tyre Company, Ltd. Ballymena, North Ireland
FY	N.V. Nederlandsche Banden, Industries Michelin S'Hertogenbosch, Bois-le-duc, Holland	HK	Michelin (Belfast) Ltd. Belfast, North Ireland
F1	Michelin Tyre Company, Ltd. Dundee, Scotland	HL	Michelin Tyre Company, Ltd. Burnley, England
F2	C.A. Firestone Venezolana Valencia, Venezuela	HM	Michelin Tyre Company, Ltd. Stoke-on-Trent, England
F3	Manufacture Francaise Pneumatiques Michelin Roanne, France	HN	Michelin Tire Mfg. Co. of Canada New Glasgow, Nova Scotia, Canada
F4	CNB-Companhia Nacional de Borrachas Oporto, Portugal	HP	Manufacture Saigonnaise des Pneumatiques Michelin Saigon, Vietnam
F5	FATE S.A.I.C.I. Buenos Aires, Argentina	HT	Ceat, S.P.A. Pneumatici via Leoncavallo Torino, Italy
F6	Torrelavega Torrelavega, Spain	HU	Ceat 10036 Settimo Torinese, Italy
F7	Puente San Miguel Firestone Torrelavega, Spain	HV	Gentyre S.P.A. Frosinone, Italy
		HW	Barum Tire Co. Otokovice, Czechoslovakia
		HX	The Dayton Tire & Rubber Company Dayton, OH USA
		HY	Bridgestone-Firestone Inc. Oklahoma City, OK USA

Code No.	Tire Manufacturer	Code No.	Tire Manufacturer
H1	DeLa S.A.F.E. Neumaticos Michelin Valladolid, Spain	JM	The Lee Tire & Rubber Company (Goodyear Tire & Rubber Co.) New Bedford, MA USA
H2	Kumho & Co., Inc. Tire Division Kwangsan-gun, Chonnam, Korea	JN	The Lee Tire & Rubber Company (Goodyear Tire & Rubber Co.) Topeka, KS USA
H3	Sava Industrija Gumijevih Skofjeloska, Kranj, Yugoslavia	JP	The Lee Tire & Rubber Company (Goodyear Tire & Rubber Co.) Tyler, TX USA
H4	Bridgestone-Houf Yamaguchi-Ken, Japan	JT	The Lee Tire & Rubber Company (Goodyear Tire & Rubber Co.) Union City, TN USA
H5	Hutchinson-MAPA Chalette Sur Loing, France	JU	The Lee Tire & Rubber Company (Goodyear Tire & Rubber Co.) Medicine Hat, Alberta, Canada
H6	Shin Hung Rubber Co. Ltd. Kyung Nam, Korea		
H7	Li Hsin Rubber Industrial Co. Chi-Hu, Chang-Hwa, Taiwan	JV	The Lee Tire & Rubber Company (Goodyear Tire & Rubber Co.) Toronto 14, Ontario, Canada
H9	Reifen-Berg Clevischer Ring, Germany	JW	The Lee Tire & Rubber Company (Seiberling Rubber Co. of Canada) Toronto 9, Ontario, Canada
H0	The General Tyre & Rubber Co. Karachi, Pakistan	JX	The Lee Tire & Rubber Company (Goodyear Tire & Rubber Co.) Valleyfield, Quebec, Canada
JA	The Lee Tire & Rubber Co. (Goodyear Co., Plant 1) Akron, OH USA	JY	The Lee Tire & Rubber Company (Neumaaticos Goodyear S.A.) Hurlingham, F.C.N.S.M., Argentina
JB	The Lee Tire & Rubber Co. (Goodyear Co., Plant 2) Akron, OH USA	J1	Phillips Petroleum Company Bartlesville, OK USA
JC	The Lee Tire & Rubber Company Conshohocken, PA USA	J2	Bridgestone Singapore Company Jurong Town, Singapore
JD	The Lee Tire & Rubber Company (Kelly-Springfield Tire Co.) Cumberland, MD USA	J3	Gumarne 1 Maja Puchov, Czechoslavakia
JE	The Lee Tire & Rubber Company (Goodyear Tire & Rubber Co.) Danville, VA USA	J4	Rubena, N.P. Nachod, Czechoslavakia
JF	The Lee Tire & Rubber Company (Goodyear Tire & Rubber Co.) Fayetteville, NC USA	J5	The Lee Tire & Rubber Company Logan, OH USA
JH	The Lee Tire & Rubber Company (Goodyear Tire & Rubber Co.) Freeport, IL USA	J6	Jaroslavi Tire Company Jaroslavi, C.I.S.
JJ	The Lee Tire & Rubber Company (Goodyear Tire & Rubber Co.) Gadsden, AL USA	J7	R & J Mfg. Corporation Plymouth, IN USA
JK	The Lee Tire & Rubber Company (Goodyear Tire & Rubber Co.) Jackson, MI USA	J8	Da Chung Hua Rubber Ind. Co. Shanghai, China
JL	The Lee Tire & Rubber Company (Goodyear Tire & Rubber Co.) Los Angeles, CA USA	J9	P.T. Intirub Besar, Jakarta, Indonesia
		J0	Korea Inocee Kasei Co., Ltd. Masan, Korea

Code No.	Tire Manufacturer	Code No.	Tire Manufacturer
KA	The Lee Tire & Rubber Company (Goodyear Tire & Rubber Co.) Granville, New South Wales, Australia	KW	The Lee Tire & Rubber Company (Compania Goodyear del Peru) Lima, Peru
KB	The Lee Tire & Rubber Company (Goodyear Tire & Rubber Co.) Thomastown, Victoria, Australia	KX	The Lee Tire & Rubber Company (The Goodyear Tire & Rubber Co.) Las Pinas, Rizal, The Philippines
KC	The Lee Tire & Rubber Company (Companhia Goodyear do Brasil) Sao Paulo, Brazil	KY	The Lee Tire & Rubber Company [Goodyear Tire & Rubber Co. (GB)] Glasgow, Scotland
KD	The Lee Tire & Rubber Company (Goodyear de Colombia S.A.) Cali, Colombia	K1	Phillips Petroleum Company Stow, OH USA
KE	The Lee Tire & Rubber Company (Goodyear Congo) Republic of the Congo	K2	The Lee Tire & Rubber Company Madisonville, KY USA
KF	The Lee Tire & Rubber Company (Compagnie Francaise Goodyear) 80 Amiens-Somme, France	K2	Kenda Rubber Ind. Co., Ltd. Yualin, Taiwan
KH	The Lee Tire & Rubber Company (Deut-che Goodyear G.M.B.H.) Phillipsburg Brucksal, Germany	K4	Uniroyal S.A. (Uniroyal Goodrich) Queretaro, Mexico
KJ	The Lee Tire & Rubber Company (Goodyear International) 64 Fulda, Germany	K5	VEB Reifenkombinat Furstenwalde Furstenwalde, Germany
KK	The Lee Tire & Rubber Company (Goodyear Hellas S.A.I.C.) Thessaloniki, Greece	K6	The Lee Tire & Rubber Company Lawton, OK USA
KL	The Lee Tire & Rubber Company (Goodyear International) Guatemala City, Guatemala	K7	The Lee Tire & Rubber Company Santiago, Chile
KM	The Lee Tire & Rubber Company Grand Duchy of Luxembourg	K8	The Kelly-Springfield Tire Co. Selangor, Malaysia
KN	The Lee Tire & Rubber Company (Goodyear India Ltd. Factory) District Gurgaon, India	K9	Natier Tire & Rubber Co., Ltd. Shetou, Changhua, Taiwan
KP	The Lee Tire & Rubber Company (The Goodyear Tire & Rubber Co.) Bogor, Republic of Indonesia	K0	Michelin Korea Tire Co., Ltd. San-Kun, Kyungsangnam, Korea
KT	The Lee Tire & Rubber Company (Goodyear Italiana) Latina, Italy	LA	The Lee Tire & Rubber Company (Goodyear Tyre & Rubber Co.) Uitenhage, South Africa
KU	The Lee Tire & Rubber Company (Goodyear Jamaica Ltd.) Jamaica, West Indies	LB	The Lee Tire & Rubber Company (Goodyear Gummi Fabics Aktirbolag) Norrkoping, Sweden
KV	The Lee Tire and Rubber Company (Compania Hulera Goodyear) Mexico City, Mexico	LC	The Lee Tire & Rubber Company (Goodyear International) Bangkok, Thailand
		LD	The Lee Tire & Rubber Company (Goodyear International) Kocaeli, Turkey
		LE	The Lee Tire & Rubber Company (CA Goodyear de Venezuela) Valencia, Edo Carabobo, Venezuela
		LF	The Lee Tire & Rubber Company [Goodyear Tire & Rubber Co. (GB)] Wolverhampton, England

Code No.	Tire Manufacturer	Code No.	Tire Manufacturer
LJ	Uniroyal Englebert Belgique S.A. Herstal-les-Liege, Belgium	MB	Goodyear Tire & Rubber Co. (Plant 2) Akron, OH USA
LK	Productors Niaconal de Llantas S.A. Cali, Colombia	MC	The Goodyear Tire & Rubber Co. Danville, VA USA
LL	Uniroyal Englebert France S.A. Compiegne, France	MD	The Goodyear Tire & Rubber Co. Gadsden, AL USA
LM	Uniroyal Englebert Deutschland D.A. Aachen, Germany	ME	The Goodyear Tire & Rubber Co. Jackson, MI USA
LN	Uniroyal SA (Uniroyal Goodrich) Mexico City, Mexico	MF	The Goodyear Tire & Rubber Co. Los Angeles, CA USA
LP	Uniroyal Ltd., Tire & Gen. Products Newbridge, Midlothian, Scotland	MH	The Goodyear Tire & Rubber Co. New Bedford, ME USA
LT	Uniroyal Endustri Turk Anonim Adapazari, Turkey	MJ	The Goodyear Tire & Rubber Co. Topeka, KS USA
LU	Uniroyal, C.A. Valencia, Venezuela	MK	The Goodyear Tire & Rubber Co. Union City, TN USA
LV	General Tire Canada Ltd. Barrie, Ontario, Canada	ML	The Goodyear Tire & Rubber Co. Cumberland, MD USA
LW	Trelleborg Gummifariks Akiebolag Trelleborg, Sweden	MM	The Goodyear Tire & Rubber Co. Fayetteville, NC USA
LX	Mitsuboshi Belting Ltd. Kobe, Japan	MN	The Goodyear Tire & Rubber Company Freeport, IL USA
LY	Mitsuboshi Belting Ltd. Shikoku, Japan	MO	Neumaticos De Chile S.A. Coquimbo, Chile
L1	Goodyear Taiwan Ltd. Taipei, Taiwan	MP	The Goodyear Tire & Rubber Company Tyler, TX USA
L2	WUON Poong Ind. Co., Ltd. Pusan, Korea	MT	The Goodyear Tire & Rubber Co. Conshohocken, PA USA
L3	Tong Shin Chemical Products Co. Seoul, Korea	MU	Neumaticos Goodyear, S.A. Hurlingham, F.C.N.G.S.M., Argentina
L4	Centrala Ind. de Prelucrare Cauciuc Oltentei, Romania	MV	The Goodyear Tyre & Rubber Co. Ltd. New South Wales, Australia
L5	BRISA Bridgestone Sabanci P.K. Izmit, Turkey	MW	The Goodyear Tyre & Rubber Co., Ltd. Thomastown, Victoria, Australia
L6	MODI Rubber Limited Meerut UP, India	MX	Companhia Goodyear do Brasil São Paulo, Brazil
L7	Intreprinderea De Anvelope Zalau Zalau, Judetul Saloj, Romania	MY	Goodyear de Colombia, S.A. Cali, Colombia
L8	Dunlop Zimbabwe Ltd. Domington, Bulawaye, Zimbabwe	M1	Goodyear Maroc, S.A. Casablanca, Morocco
L9	Panther Tyres, Ltd. Aintree, Liverpool, England	M2	The Goodyear Tire & Rubber Co. Madisonville, KY USA
L0	Mfg. Francaise Des Pneumatiques Clermont Ferrand Cedex, France	M3	Michelin Tire Corporation Greenville, SC USA
MA	Goodyear Tire & Rubber Co. (Plant 1) Akron, OH USA	M4	The Goodyear Tire & Rubber Co. Logan, OH USA

Code No.	Tire Manufacturer	Code No.	Tire Manufacturer
M5	Michelin Tire Mfg. Co. of Canada Kentville, Nova Scotia, Canada	NX	Goodyear Gummi Faabriks Aktirbolag Norrkoping, Sweden
M6	Goodyear Tire & Rubber Co. Lawton, OK USA	NY	Goodyear (Thailand) Ltd. Bangkok, Thailand
M7	Goodyear de Chile, S.A.I.C. Santiago, Chile	N1	Maloja AG. Pneu-Und Gummiwerke Gelterkinden, Switzerland
M8	Premier Tyres Limited Kerala State, India	N2	Hurtubise Nutread Inc. Tonawanda, NY USA
M9	Uniroyal Tire Corp. Middlebury, CT USA	N3	Nitto Tire Co., Ltd. Tohin-Cho, Inabe-Gun Mie-Ken, Japan
NA	Goodyear Congo Republic of the Congo	N4	Centrala Ind. de Prelucrare Cauciue Republic Socialista Romania
NB	The Goodyear Tyre & Rubber Co. Wolverhampton, England	N5	Pneumant, VEB Reifenwerk Riesa Riesa, Germany
NC	Compagnie Francaise Goodyear Amiens-Somme, France	N6	Pneumant Germany
ND	Deutsche Goodyear GMBH Phillipsburg Bruchsal, Germany	N7	Intreprinderea De Anvelope Caracal Caracal, Judetul Olt., Romania
NE	Gummiwerke Fuida GMBH Fulda, Germany	N8	Lee Tire & Rubber Co. (Goodyear) Selangor, Malaysia
NF	Goodyear Hellas S.A.I.C. Thessaloniki, Greece	N9	Cla Pneus Tropical Feira de Santana, Bahla, Brazil
NH	Gran Industria de Neumatticos Guatemala City, Guatemala	PA	Goodyear Lastikleri Tas Kocaeli, Turkey
NJ	Goodyear S.A. Grand Duchy of Luxembourge	PB	CA Goodyear de Venezuela Valencia, Edo Carabobo, Venezuela
NK	Goodyear India Ltd. Distric Gurgaon, India	PC	Goodyear Tire & Rubber Co. Medicine Hat, Alberta, Canada
NL	The Goodyear Tire & Rubber Co., Ltd. Bogor, Republic of Indonesia	PD	Goodyear Tire & Rubber Co. Valleyfield, Quebec, Canada
NM	Goodyear Italiana S.P.A. Latina, Italy	PE	Seiberling Rubber Co. Toronto, Ontario, Canada
NN	Goodyear Jamaica Ltd. Jamaica, West Indies	PF	The Goodyear Tire & Rubber Co. Toronto, Ontario, Canada
NO	South Pacific Tyres Victoria, Australia	PH	Kelly-Springfield Tire Co. Cumberland, MD USA
NP	Compania Hulera Goodyear Oxo Mexico City, Mexico	PI	Gislaved Daek AB Gislaved, Sweden
NT	Compania Goodyear del Peru Lima, Peru	PJ	Kelly-Springfield Tire Co. Fayetteville, NC USA
NU	Goodyear Tire & Rubber Co. Las Pinas, Rizal, The Philippines	PK	Kelly-Springfield Tire Co. Freeport, IL USA
NV	Goodyear Tyre & Rubber Co. Ltd. Glasgow, Scotland	PL	Kelly-Springfield Tire Co. Tyler, TX USA
NW	Goodyear Tyre & Rubber Co. Uitenhage, South Africa		

Code No.	Tire Manufacturer	Code No.	Tire Manufacturer
PM	Kelly-Springfield Tire Co. (Goodyear Tire & Rubber Co.) Conshohocken, PA USA	TA	Kelly-Springfield Tire Co. (Goodyear Tire & Rubber Co.) Union City, TN USA
PN	Kelly-Springfield Tire Co. (Goodyear Tire & Rubber Co.) Akron, OH USA	TB	Kelly-Springfield Tire Co. Hurlingham, F.C.N.G.S.M., Argentina
PO	South Pacific Tyres New South Wales, Australia	TC	Kelly-Springfield Tire Co. (Goodyear Tyre & Rubber Co.) New South Wales, Australia
PP	Kelly-Springfield Tire Co. (Goodyear Tire & Rubber Co.) Akron, OH USA	TD	Kelly-Springfield Tire Co. (Goodyear Tyre & Rubber Co.) Thomastown, Victoria, Australia
PT	Kelly-Springfield Tire Co. (Goodyear Tire & Rubber Co.) Danville, VA USA	TE	Kelly-Springfield Tire Co. (Companhia Goodyear do Brasil) Sao Paulo, Brazil
PU	Kelly-Springfield Tire Co. (Goodyear Tire & Rubber Co.) Gadsden, AL USA	TF	Kelly-Springfield Tire Co. (Goodyear de Colombia, S.A.) Yumbo, Call, Colombia
PV	Kelly-Springfield Tire Co. (Goodyear Tire & Rubber Co.) Jackson, MI USA	TH	Kelly-Springfield Tire Co. (Goodyear Congo) Republic of the Congo
PW	Kelly-Springfield Tire Co. (Goodyear Tire & Rubber Co.) Los Angeles, CA USA	TJ	Kelly-Springfield Tire Co. (Goodyear Tyre & Rubber Co.) Wolverhampton, England
PX	Kelly-Springfield Tire Co. (Goodyear Tire & Rubber Co.) New Bedford, ME USA	TK	Kelly-Springfield Tire Co. (Compagnie Francaise Goodyear S.A.) Amiens-Somme, France
PY	Kelly-Springfield Tire Co. (Goodyear Tire & Rubber Co.) Topeka, KS USA	TL	Kelly-Springfield Tire Co. (Deutsche Goodyear GMBH) Bruchsal, Germany
P1	Gummifabriken Gislaved Aktiebolag Gislaved, Sweden	TM	Kelly-Springfield Tire Co. (Goodyear International) Fulda, Germany
P2	Kelly-Springfield Tire Co. Madisonville, KY USA	TN	Kelly-Springfield Tire Co. (Goodyear Hellas S.A.I.C.) Thessaloniki, Greece
P3	Skepplanda Gummi, AB Alvangen, Sweden		
P4	Kelly-Springfield Tire Co. Logan, OH USA	TO	South Pacific Tyres Campbellfield, Victoria, Australia
P5	General Popo S.A. San Luis Potosi, Mexico	TP	The Kelly-Springfield Tire Co. (Goodyear International) Guatemala City, Guatemala
P6	Kelly-Springfield Tire Co. Lawton, OH USA	TT	The Kelly-Springfield Tire Co. (Goodyear S.A.) Grand Duchy of Luxembourg
P7	Kelly-Springfield Tire Co. Santiago, Chile		
P8	No. 2 Rubber Plant Oingdao Quingdao, Shandong, China	TU	Kelly-Springfield Tire Co. (Goodyear India Ltd.) District Gurgaon, India
P9	MRF, Ltd., PB No. 1 Ponda Goa, India		

Code No.	Tire Manufacturer	Code No.	Tire Manufacturer
TV	Kelly-Springfield Tire Co. (Goodyear Tire & Rubber Co.) Bogor, Republic of Indonesia	UF	Kelly-Springfield Tire Co. (Goodyear Thailand Ltd.) Bangkok, Thailand
TW	Kelly-Springfield Tire Co. (Goodyear Italiana) Latina, Italy	UH	Kelly-Springfield Tire Co. (Goodyear Lastikleri Tas Pk 2 Izmit) Kocaeli, Turkey
TX	Kelly-Springfield Tire Co. (Goodyear Jamaica Ltd.) Jamaica, West Indies	UJ	Kelly-Springfield Tire Co. (CA Goodyear de Venezuela) Edo Carabob, Venezuela
TY	Kelly-Springfield Tire Co. (Compania Hulera Goodyear) Mexico City, D.F. Mexico	UK	Kelly-Springfield Tire Co. (Goodyear Tire & Rubber Co.) Medicine Hat, Alberta, Canada
T1	Hankook Tire Mfg. Co. Ltd. Seoul, Korea	UL	Kelly-Springfield Tire Co. (Goodyear Tire & Rubber Co.) Valleyfield, Quebec, Canada
T2	Ozos (Uniroyal) Olsztyn, Poland	UM	Kelly-Springfield Tire Co. (Seiberling Rubber Co. of Canada) Toronto, Ontario, Canada
T3	Debickle Zattldy Opon Samochodowych (Uniroyal AG) Debica, Poland	UN	Kelly-Springfield Tire Co. (Goodyear Tire & Rubber Co.) Toronto, Ontario, Canada
T4	S.A. Carideng (Rubberfactory) Lanaken, Belgium	UO	South Pacific Tyres Thomastown, Vic., Australia
T5	Tigar-Pirot Pirot, Yugoslavia	UP	Cooper Tire & Rubber Co. Findlay, OH USA
T6	Hulera Tomel S.A. San Antonia, Mexico, D.F.	UT	Cooper Tire & Rubber Co. Texarkana, AR USA
T7	Hankook Tire Mfg. Co., Ltd. Daejun, Korea	UU	Carlisle Tire & Rubber Division Carlisle, PA USA
T8	Goodyear Malaysia Berhad Selangor, Malaysia	UV	Kzowa Rubber Ind. Co., Ltd. Nishinariku, Osaka, Japan
T9	MRF, Ltd. Arkonam, India	UW	Okada Tire Industry Ltd. Katsushika-ku, Tokyo, Japan
UA	Kelly-Springfield Tire Co. (Compania Goodyear del Peru) Lima, Peru	UX	Federal Corporation Taipei, Taiwan
UB	Kelly-Springfield Tire Co. (Goodyear Tire & Rubber Co.) Las Pinas, Rizal, The Philippines	UY	Cheng Shin Rubber Ind. Co. Ltd. Meci Kong, Teipi, Taiwan
UC	Kelly-Springfield Tire Co. (Goodyear Tyre & Rubber Co.) Glasgow, Scotland	U1	Lein Shin Tire Co. Ltd. Taipei, Taiwan
UD	Kelly-Springfield Tire Co. (Goodyear Tyre & Rubber Co.) Uitenhage, South Africa	U2	Sumitomo Rubber Industries, Ltd. Shirakawa City Fukushima Prefecture, Japan
UE	Kelly-Springfield Tire Co. (Goodyear Gummi Fabriks Aktiebolag) Norrkoping, Sweden	U3	Miloje Zakic, Krusevac, Yugoslavia
		U4	Geo. Byers Sons, Inc. Columbus, OH USA
		U5	Farbenfabriken Bayer GmbH Leverkusen, Germany

Code No.	Tire Manufacturer
U6	Pneumant Germany
U7	Pneumant Germany
U8	Hsin-Fung Fac. of Nankang Rubber Co. Taiwan Province, People's Republic of China
U9	Cooper Tire & Rubber Co. Tupelo, MS USA
VA	The Firestone Tire & Rubber Co. Akron, OH USA
VB	The Firestone Tire & Rubber Co. Akron, OH USA
VC	The Firestone Tire & Rubber Co. Albany, GA USA
VD	Bridgestone-Firestone Inc. Decatur, IL USA
VE	Bridgestone-Firestone Inc. Des Moines, IA USA
VF	The Firestone Tire & Rubber Co. South Gate, CA USA
VH	The Firestone Tire & Rubber Co. Memphis, TN USA
VJ	The Firestone Tire & Rubber Co. Pottstown, PA USA
VK	The Firestone Tire & Rubber Co. Salinas, CA USA
VL	The Firestone Tire & Rubber Co. Hamilton, Ontario, Canada
VM	The Firestone Tire & Rubber Co. Calgary, Alberta, Canada
VN	Bridgestone-Firestone Inc. Joliette, Quebec, Canada
VO	South Pacific Tyres Upper Hutt, Wellington, New Zealand
VP	The Firestone Tire & Rubber Co. Bari, Italy
VT	The Firestone Tire & Rubber Co. Bilbao, Spain
VU	Universal Tire Company Lancaster, PA USA
VV	The Firestone Tire & Rubber Co. Viskafors, Sweden
VW	Ohtsu Tire & Rubber Co. Ltd. Osaka, Japan

Code No.	Tire Manufacturer
VX	Firestone Tyre & Rubber Co., Ltd. Brentford, Middlesex, England
VY	Firestone Tyre & Rubber Co. Ltd. Denbighshire, Wrexhan, Wales
V1	Livingston's Tire Shop Hubbard, OH USA
V2	Vsesojuznoe Ojedinenic Avtoexport Volzhsk, C.I.S.
V3	TA Hsin Rubber Tire Co. Ltd. Taipei Hsieng, Taiwan
V4	Ohtsu Tire & Rubber Co. Miyazaki Prefecture, Japan
V5	Firestone El Centenario S.A. Mexico City, Mexico
V6	Firestone Cuernavaca Cuenavaca, Mexico
V7	Vsesojuznoe Ojedinenie Avtoexport Voronezh, C.I.S.
V8	Boras Gummifabrik AB Boras, Sweden
V9	M & H Tire Co. Gardner, MA USA
WA	Firestone–France S.A. Bethune, France
WB	Industria Akron De Costa Rico San Jose, Costa Rica
WC	Firestone Australia (Pty) Ltd. Sydney, New South Wales, Australia
WD	Fabrik fur Firestone Produkte A.G. Prattlen, Switzerland
WE	Nankang Rubber Tire Corp. West Taipei, Taiwan
WF	The Firestone Tire & Rubber Co. Burgos, Spain
WH	The Firestone Tire & Rubber Co. Boras, Sweden
WM	Dunlop Olympic Tyres W. Footscray, Victoria, Australia
WO	Inocce Rubber Co., Inc. Phatumtani, Thailand
WT	Madras Rubber Factory Ltd. Madras, India
WU	Ceat Typres of India Ltd. Bhandup, Bombay, India
WV	General Rubber Corporation Taipei, Taiwan

Code No.	Tire Manufacturer
WW	Euzkadi Co. Hulera Euzkadi, S.A. Mexico City, Mexico
WX	Euzkadi Co. Hulera Euzkadi, A.A. La Presa, Edo de, Mexico
WY	Euzkadi, Co. Hulera Euzkadi, S.A. Guadalijara, Jalisco, Mexico
W1	Bridgestone/Firestone, Inc. LaVergne, TN USA
W2	Bridgestone–Firestone Inc. Wilson, NC USA
W3	Vredestein Doetinchem B.V. Doetinehem, The Netherlands
W4	Dunlop Olympic Tyres Somerton, Victoria, Australia
W5	Firestone de la Argentina Province de Buenos Aires, Argentina
W6	Philtread Tire & Rubber Corp. Makati, Rizal, The Philippines
W7	Firestone Portuguesa, S.A.R.L. Alcochete, Portugal
W8	Siam Tyre Co., Ltd. Bangkok, Thailand
W9	Ind. de Pneumaticos Firestone S.A. Rio de Janeiro, Brazil
XA	Industrie Pirelli S.p.A., V. le Milan, Italy
XB	Industri Pirelli S.p.A. Settimo, Torinese, Torino, Italy
XC	Industrie Pirelli S.p.A. Tivoil-Roma, Italy
XD	Industrie Pirelli S.p.A. Messina, Italy
XE	Industrie Pirelli S.p.A. Ferrandina, Italy
XF	Productos Pirelli S.A. Barcelona, Spain
XH	Pirelli Hallas, S.A. Patrasso, Greece
XJ	Turk Pirelli Lastilkeri, S.A. Istanbul, Turkey
XK	Pirelli, S.A. São Paulo, Brazil
XL	Pirelli, S.A. Campinas, Brazil
XM	Pirelli Co. Platennse de Neumaticos, Merio, Buenos Aires, Argentina

Code No.	Tire Manufacturer
XN	Pirelli, Ltd. Carlisle, England
XP	Pirelli Ltd. Burton-on-Trent, England
XT	Veith-Pirelli, A.G. Sandbach, Germany
XV	Dayton Tire & Rubber Co. (Firestone Tire & Rubber Co.) Hamilton, Ontario, Canada
XW	Dayton Tire & Rubber Co. (Firestone Tire & Rubber Co.) Calgary, Alberta, Canada
XX	Bandag, Inc. Muscatine, IA USA
XY	Dayton Tire & Rubber Co. (Firestone Tire & Rubber Co.) Joliette, Quebec, Canada
X1	Tong Shin Chemical Prod. Co., Ltd. Seoul, Korea
X2	Hwa Fong Rubber Ind. Co. Ltd. Yualin, Taiwan
X3	Vaesojuznoe Ojedinenic Avtoexport Belaya Tserkov, C.I.S.
X4	Pars Tyre Co. Saveh, Iran
X5	J.K. Industries Ltds. Kankroli, Rajasthan, India
X6	Vsesojuznoe Ojedinenie Avtoexport Bobruysk, C.I.S.
X7	Vsesjuznoe Ojedinenie Avtoexport Chimkentsky, C.I.S.
X8	Vsesojuznoe Ojedinenie Avtoexport Dnepropetrovski, C.I.S.
X9	Vsesojuznoe Ojedinenie Avtoexport Moscow, Russia
X0	Vsesojuznoe Ojedinenie Avtoexport Mizhenkamsk, C.I.S.
YA	The Dayton Tire & Rubber Co. (Firestone Tire & Rubber Co.) Akron, OH USA
YB	The Dayton Tire & Rubber Co. (Firestone Tire & Rubber Co.) Akron, OH USA
YC	The Dayton Tire & Rubber Co. (Firestone Tire & Rubber Co.) Albany, GA USA

Code No.	Tire Manufacturer	Code No.	Tire Manufacturer
YD	The Dayton Tire & Rubber Co. (Firestone Tire & Rubber Co.) Decatur, IL USA	YY	Seiberling Tire & Rubber Co. (Firestone Tire & Rubber Co.) Salinas, CA USA
YE	The Dayton Tire & Rubber Co. (Firestone Tire & Rubber Co.) Des Moines, IA USA	Y1	Companhia Goodyear Do Brasil São Paulo, Brazil
YF	The Dayton Tire & Rubber Co. (Firestone Tire & Rubber Co.) South Gate, CA USA	Y2	Dayton Tire & Rubber Co. Wilson, NC USA
YH	The Dayton Tire & Rubber Co. (Firestone Tire & Rubber Co.) Memphis, TN USA	Y3	Seiberling Tire & Rubber Co. Wilson, NC USA
YJ	The Dayton Tire & Rubber Co. (Firestone Tire & Rubber Co.) Pottstown, PA USA	Y4	Dayton Tire & Rubber Co. Barberton, OH USA
YK	The Dayton Tire & Rubber Co. (Firestone Tire & Rubber Co.) Salinas, CA USA	Y5	Shanghai Tsen Tai Rubber Factory Shanghai, People's Republic of China
YL	Oy Nokia A.B. Nokia, Finland	Y6	Sime Tyres Inter. Sdn. Bhd. Kedah Darulaman, Malaysia
YM	Seiberling Tire & Rubber Co. (Firestone Tire & Rubber Co.) Akron, OH USA	Y7	Bridgestone/Firestone, Inc. LaVergne, TN USA
YN	Seiberling Tire & Rubber Co. (Firestone Tire & Rubber Co.) Akron, OH USA	Y8	Bombay Tyres International Ltd. Bombay, India
YO	Kumho and Co., Inc. Chunnam, Korea	Y9	P.T. Gadjah Tungual Jawa Barat, Republic of Indonesia
YP	Seiberling Tire & Rubber Co. (Firestone Tire & Rubber Co.) Albany, GA USA	1A	Union Rubber Ind. Co., Ltd. Taipei, Taiwan, People's Republic of China
YT	Seiberling Tire & Rubber Co. (Firestone Tire & Rubber Co.) Decatur, IL USA	2A	Jiuh Shuenn Enterprises Co., Ltd. Wufeng, Taichung, Taiwan, People's Republic of China
YU	Seiberling Tire & Rubber Co. (Firestone Tire & Rubber Co.) Des Moines, IA USA	3A	Hualin Rubber Plant People's Republic of China
YV	Seiberling Tire & Rubber Co. (Firestone Tire & Rubber Co.) South Gate, CA USA	4A	Vee Rubber Co., Ltd. Samutsakaom Province, Thailand
YW	Seiberling Tire & Rubber Co. (Firestone Tire & Rubber Co.) Memphis, TN USA	5A	Vee Rubber Inter. Co., Ltd. Bangkok, Thailand
YX	Seiberling Tire & Rubber Co. (Firestone Tire & Rubber Co.) Pottstown, PA USA	6A	Roadstone Tyre & Rubber Co., Ltd. Nontaburi Bangkok, Thailand
		7A	Kings Tire Ind. Co., Ltd. Taiwan, People's Republic of China
		8A	Zapater, Diaz I.C.S.A. Tire Co. Buenos Aires, Argentina
		9A	Siemese Rubber Co., Ltd. Bangkok, Thailand
		0A	Siam Rubber Ltd. Part. Samutsakhon, Thailand
		1B	Neumaticos De Venezuela C.A. Estadro Carabobo, Venezuela

Code No.	Tire Manufacturer	Code No.	Tire Manufacturer
2B	Deestone Limited Samutsakom, Thailand	5C	The Firestone Tire & Rubber Co. Bridgestone Brand Bilbao, Spain
3B	Tianjin/United Tire & Rubber Tianjin, P.R. China	6C	The Firestone Tire & Rubber Co. Bridgestone Brand Burgos, Spain
4B	Goodyear Canada, Inc. Napanee, Ontario, Canada	7C	The Firestone Tire & Rubber Co. Papanui, Christ Church, New Zealand
5B	The Kelly-Springfield Tire Co. Goodyear Canada, Inc. Napanee, Ontario, Canada	8C	Firestone–France S.A. Bethune, France
6B	Mt. Vernon Plant of General Tire Mt. Vernon, IL USA	9C	The Firestone Tire & Rubber Co. Bridgestone Brand Bari, Italy
7B	Bridgestone–Firestone Inc. Decatur, IL USA	0C	Michelin Siam Co., Ltd. Chonburi, Thailand
8B	Bridgestone–Firestone Inc. Des Moines, IA USA	1D	Bridgestone Shimonoseki Plant Yamaguchi-ken, Japan
9B	Bridgestone–Firestone, Inc. Joliette, Quebec, Canada	2D	Silverstone Tire & Rubber Co. Taiping, Perak, Malaysia
0B	Bridgestone–Firestone Inc. Wilson, NC USA	3D	Cooper Tire & Rubber Co. Albany, GA USA
1C	Bridgestone–Firestone Inc. Oklahoma City, OK USA	4D	Bridgestone/Firestone, Inc. Firestone Brand Morrison, TN USA
2C	Bridgestone/Firestone, Inc. Morrison, TN USA	5D	Bridgestone/Firestone, Inc. Dayton Brand Morrison, TN USA
3C	Mt. Vernon Plant of General Tire Mt. Vernon, IL USA		
4C	South Pacific Tyres Somerton, Victoria, Australia		

"MACHINE TO" DIMENSION FOR DRUMS AND ROTORS

Disc & Drum Brake
Service Specifications 1972–1994
For Passenger Cars & Light Trucks

Produced by **CHILTON DATALOG** *for*

HUNTER
Engineering Company

CAUTION: If state inspection regulations exceed the O.E. manufacturer's specifications for lining
or rotor/drum reserve, the state inspection regulation must be followed.
NOTE: Machining specifications in this catalog pertain to original equipment drums and rotors.
Some after market drum and rotor specifications may differ from the O.E.

ABBREVIATIONS

B - Large   **FT.- LBS.-** Foot-Pounds   **T.I.R.-** Total Indicatior Reading   **IN.- LBS.-** Inch-Pound
L- Lower   **S.** Small   **U-** Upper

Appendix 7 reproduced courtesy of Hunter Engineering Co.

PASSENGER CARS

VEHICLE	BRAKE SHOE	BRAKE DRUM		BRAKE PAD	BRAKE ROTOR	
		DIAMETER			MIN. THICKNESS	
YEAR, MAKE, MODEL	O.E. Minimum Lining Thickness	Standard Size	Machine To	O.E. Minimum Lining Thickness	Machine To	Discard At
ACURA						
94–90 Integra: Front	—	—	—	.060	.750	—
Rear	—	—	—	.060	.310	—
89–86 Integra: Front	—	—	—	.120	.670	—
Rear	—	—	—	.063	.310	—
94–93 Legend: Sedan Front	—	—	—	.060	.830	—
Coupe Front	—	—	—	.060	1.02	—
Rear	—	—	—	.060	.300	—
92–91 Legend: Coupe & Sedan Front	—	—	—	.060	.830	—
Rear	—	—	—	.060	.300	—
90–86 Legend: Coupe & Sedan Front	—	—	—	.120①	.750	—
Rear	—	—	—	.060	.310	—
94–91 NSX: Front	—	—	—	.060	1.020	—
Rear	—	—	—	.060	.750	—
94–92 Vigor: Front	—	—	—	.060	.830	—
Rear	—	—	—	.060	.310	—

① Sedan shown; Coupe .060".   ② 1990-89 16 ft/lbs.   ③ 1990-89 .006".

AMERICAN MOTORS						
83–82 Concord, Spirit, SX-4:	.030	9.000	9.060	.030③	.815	.810
83–82 Concord Wagon:	.030	10.000	10.060	.030③	.815	.810
81 Concord, Spirit, SX-4:	.030	9.000	9.060	.030③	.815	.810
81 Concord Wagon, Eagle Wagon:	.030	10.000	10.060	.030③	.815	.810
80–78 Concord, Eagle, AMX:	.030	10.000	10.060	.062	.815	.810
88–82 Eagle:	.030	10.000	10.060	.030③	.815	.810
80–78 Gremlin, Spirit L4 eng.	.030	9.000	9.060	.062	.815	.810
L6 eng.	.030	10.000	10.060	.062	.815	.810
77 Gremlin, Hornet	.030	10.000	10.060	.062	.815	.810
76–75 Gremlin, Hornet: L6 eng.	.030	9.000	9.060	.062	1.130	1.120
V8 eng.	.030	10.000	10.060	.062	1.130	1.120
74–72 Gremlin, Hornet, Javelin:						
L6 eng. w/drum brakes.........	.030	9.000	9.060	—	—	—
L6 eng. w/disc brakes	.030	9.000	9.060	.062	—	.940
L6 eng. w/disc brakes (72)	.030	10.000	10.060	.062	—	.940
V8 eng. w/drum brakes	.030	10.000	10.060	—	—	—
V8 eng. w/disc brakes	.030	10.000	10.060	.062	—	.940
78 Matador:	.030	10.000	10.060	.062	—	1.120
77 Matador:	.030	10.000	10.060	.062	—	1.120
76–75 Matador:	.030	10.000	10.060	.062	1.130	1.120
74–72 Matador, Ambassador:	.030	10.000	10.060	.062	—	.940
80–78 Pacer:	.030	10.000	10.060	.062	.815	.810
77 Pacer:	.030	10.000	10.060	.062	.815	.810
76–75 Pacer: Front w/drum brakes	.030	10.000	10.060	—	—	—
Rear w/drum brakes	.030	9.000	9.060	—	—	—
w/disc brakes	.030	9.000	9.060	.062	—	1.120

① Eagle .004".   ② 1972 12 in/lbs.   ③ .030" over rivet head; if bonded lining use .062".

AUDI

94–92 100, 100 Quattro: Front	—	—	—	.079	—	.906
Rear	—	—	—	.079	—	.315
91–89 100: Front	—	—	—	.078	.807	.787
Rear	—	—	—	.276⑤	.335	.315
91–89 100 Quattro: Front	—	—	—	.078	.945	.905
Rear	—	—	—	.276⑤	.335	.315
91–89 200: Front	—	—	—	.078	.945	.905
Rear	—	—	—	.276⑤	.335	.315
91–89 200 Quattro:						
Front w/single piston...........	—	—	—	.078	.945	.905
Front w/dual piston	—	—	—	.078	.945	.905
Rear	—	—	—	.276⑤	.335	.315
87–85 4000, Coupe GT:	.097	—	7.894	.079	.728	.709
84–80 4000, Coupe:	.097	—	7.894	.079	.413	.394
87–84 4000 Quattro: Front	—	—	—	.079	.728	.709
Rear	—	—	—	.079	.335	.315
88–81 5000:	.098	9.055	9.075	.079	.807	.787
80–78 5000:	.098	9.055	9.094	.078	.807	.787
88–80 5000 Turbo,						
5000 Turbo Quattro,						
5000 Quattro:						
Front w/single piston...........	—	—	—	.079	.807	.787
w/dual piston	—	—	—	.079	.984	.905
Rear w/Girling.................	—	—	—	.079	.335	.315
w/Teves....................	—	—	—	.276⑤	.335	.315
94–88 80, 90: Front	—	—	—	.078	.807	.787
Rear	—	—	—	.078	.334	.315
94–88 80 Quattro, 90 Quattro: Front	—	—	—	.078	.807	.787
Rear	—	—	—	.078	.334	.315
79–78 Fox:	.097	7.870	7.900	.078	.413	.393
77–73 Fox:	.097	7.870	7.900	.078	.413	.393
91–90 Quattro Coupe: Front	—	—	—	.078	.945	.905
Rear	—	—	—	.078	.335	.315
85–83 Quattro Turbo Coupe: Front	—	—	—	.079	.807	.787
Rear	—	—	—	.079	.335	.315
94–92 S4: Front	—	—	—	.079	—	.906
Rear	—	—	—	.079	—	.709
94–90 V8 Quattro: Front	—	—	—	.080	.945	.905
Rear	—	—	—	.080	.728	.710

⑤ Measurement of lining & metal.

BMW

76–72 2002:	.120	9.060	9.100	.080	.354	—
74–72 2002tii:	.120	9.060	9.100	.080	.459	—
94–93 318i: Front	—	—	—	.079	.409	.394
Rear	—	—	—	.079	.331	.315
92–91 318i: Front	—	—	—	.079	.437	.421
Rear	—	—	—	.079	.331	.315
91–88 318is, 325i, 325is, 325ix:						
Front	—	—	—	.079	.921	.906
Rear	—	—	—	.079	.331	.315
85–84 318i:	.060	—	9.035	.080	.437	.421
83–82 320i: w/ATE	.118	9.842	9.882	.118	.846	.827
w/Girling Caliper	.118	9.842	9.882	.118	.480	.461
81–77 320i:	.118	9.842	9.882	.118	.846	.827
94–92 325i, 325is: Front	—	—	—	.079	.803	.787
Rear	—	—	—	.079	.331	.315
88–84 325e, 325es: Front	—	—	—	.080	.803	.787
Rear	—	—	—	.080	.331	.315
86–85 524td: Front	—	—	—	.079	.409	.394
Rear	—	—	—	.079	.331	.315
94–83 525i, 528e, 533i, 535i,						
535is: Front	—	—	—	.080	.803	.787
Rear	—	—	—	.080	.331	.315
82 528e: Front	—	—	—	.138	.803	—
Rear	—	—	—	.138	.331	—
81–79 528i: Front	—	—	—	.080	.846	.827
Rear	—	—	—	.080	.354	.335
78–77 530i: Front	—	—	—	.080	.480	.461
Rear	—	—	—	.080	.354	.335
79–77 630CSi: Front	—	—	—	.080	.846	.827
Rear	—	—	—	.080	.728	.709
89–83 633CSi, 635CSi: Front	—	—	—	.079	.921	.906
Rear	—	—	—	.079	.331	.315
82–79 633CSi: Front	—	—	—	.080	.846	.827
Rear	—	—	—	.080	.728	.709
92–83 733i, 735i, 735iL, L7: Front	—	—	—	.080	1.039	1.024
Rear	—	—	—	.080	.409	.394
82–78 733i: Front	—	—	—	.080	.846	.827
Rear	—	—	—	.080	.374	.354

(continues)

BMW (Continued)

94–88 740i, 740iL, 750iL, 850i, 850Ci: Front	—	—	—	.079	1.118	1.102
Rear	—	—	—	.079	.724	.709
91–90 M3: Front	—	—	—	.079	.921	.906
Rear	—	—	—	.079	.331	.315
89–88 M3: Front	—	—	—	.079	.921	.906
Rear	—	—	—	.079	.409	.394
87 M3: Front	—	—	—	.079	.921	.906
Rear	—	—	—	.079	.409	.394
93–90 M5: Front	—	—	—	.079	①	1.039
Rear	—	—	—	.079	.724	.709
89–87 M5: Front	—	—	—	.079	1.118	1.102
Rear	—	—	—	.079	.331	.315
89–87 M6: Front	—	—	—	.079	1.118	1.102
Rear	—	—	—	.079	.331	.315

① Do not machine

BUICK

94–88 Century: w/H.D.	③	8.863	8.880	.030	.972	.957
exc. H.D.	③	8.863	8.880	.030	.830	.815
87–86 Century: w/H.D.	③	8.863	8.877	.030	.972	.957
exc. H.D.	③	8.863	8.877	.030	.830	.815
85 Century: w/H.D.	③	8.863	8.920	.030	.972	.957
exc. H.D.	③	8.863	8.920	.030	.830	.815
84 Century: w/H.D.	③	8.863	8.883	.030	.972	.957
exc. H.D.	③	8.863	8.883	.030	.830	.815
83–82 Century:	③	7.879	7.899	.030	.830	.815
81–79 Century, Regal, LeSabre:	③	9.500	9.560	.030	.980	.965
78–77 Century, Regal, LeSabre:	③	9.500	9.560	.030	.980	.965
75 Century, Regal:	③	9.500	9.560	.030	.980	.965
74–73 Century, Regal:	③	9.500	9.560	.030	.980	.965
90–83 Electra, Estate Wagon: (RWD)	③	11.000	11.060	.030	.980	.965
82–79 Electra, Estate Wagon:	③	11.000	11.060	.030	.980	.965
78–77 Electra, Estate Wagon, Riviera:	③	11.000	11.060	.030	.980	.965
76–75 Electra, Custom, LeSabre, Estate Wagon, Riviera:	③	12.000	12.060	.030	1.230	1.215
74–72 Electra, Custom, LeSabre, Centurion, Riviera, Wildcat:	③	11.000	11.060	.030	1.230	1.215
74–73 Estate Wagon:	③	12.000	12.060	.030	1.230	1.215
94–92 LeSabre: (FWD)	.030	8.860	8.880	.030	1.224	1.209
91 LeSabre: (FWD)	③	8.860	8.880	.030	.972	.957
90–83 LeSabre, Electra: (FWD)	③	8.858	8.880	.030	.972	.957
94–92 Park Avenue:	.030	8.860	8.880	.030	1.224	1.209
91 Park Avenue:	③	8.860	8.880	.030	—	1.200
94–88 Regal: (FWD) Front	—	—	—	.030	.984	.972
Rear	—	—	—	.030	.441	.429
87–83 Regal, LeSabre: (RWD)	③	9.500	9.560	.030	.980	.965
82 Regal, LeSabre:	③	9.500	9.560	.030	.980	.965
93–92 Riviera: Front	—	—	—	.030	1.250	1.209
Rear	—	—	—	.030	.423	.374
91–86 Riviera, Reatta: Front	—	—	—	.030	.971	.956
Rear	—	—	—	.030	.444	.429
85–83 Riviera: exc. rear disc brakes	③	9.500	9.560	.030	.980	.965
w/rear disc brakes	—	—	—	.030	.980	.965
82–79 Riviera: exc. rear disc brakes	③	9.500	9.560	.030	.980	.965
w/rear disc brakes	—	—	—	.030	.980	.965
94 Roadmaster:	.030	11.000	11.060	.030	.980	.965
w/rear disc brakes	—	—	—	.030	.735	.728
93–91 Roadmaster:	.030	11.000	11.060	.030	.980	.965
89–82 Skyhawk: vented rotor (89-82)	③	7.880	7.899	.030③	.830	.815
solid rotor (82)	③	7.880	7.899	.030③	.444	.429
80–76 Skyhawk:	③	9.500	9.560	.030	.830	.815
75 Skyhawk:	③	9.000	9.060	.030	.455	.440
94–93 Skylark:	③	7.874	7.899	.030	.751	.736
92–91 Skylark:	③	7.879	7.899	.030	.786	.736
90–80 Skylark, Regal: (FWD)	③	7.880	7.899	.030	.830	.815
79–77 Skylark:	③	9.500	9.560	.030	.980	.965
76–75 Skylark, Apollo:	③	9.500	9.560	.030	.980	.965
72 Skylark, Gran Sport, GS, Sports Wagon:	③	9.500	9.560	.030	.980	.965

③ .030″ over rivet head; if bonded lining use .062″.

CADILLAC

Year	Model						
93	Allante: Front	—	—	—	.030	1.250	1.209
	Rear	—	—	—	.030	.423	.374
92–87	Allante: Front	—	—	—	.030	.971	.956
	Rear	—	—	—	.030	.444	.429
94–93	Fleetwood Brougham: (RWD) ..	①	11.000	11.060	.030	.980	.965
92–82	Brougham, Fleetwood,						
	Deville: (RWD)	①	11.000	11.060	.030	.972	.957
81–79	Brougham, Fleetwood, Deville,						
	Seville: (RWD)	①	11.000	11.060	.062	.980	.965
	w/rear disc brakes (79)	—	—	—	.062	.980	.965
78–77	Brougham, Seville: Front	—	—	—	.062	.980	.965
	Rear	—	—	—	.062	.910	.905
78–77	DeVille:	①	11.000	11.060	.062	.980	.965
76	Seville:	①	11.000	11.060	.062	.980	.965
76–74	Calais, Deville, Brougham,						
	Fleetwood 75,						
	Commercial Ch.:	①	12.000	12.060	.062	1.220	1.215
73–72	Calais, Deville,						
	Fleetwood 60, Fleetwood 75,						
	Commercial Ch.:	①	12.000	12.060	.062	1.220	1.215
88–82	Cimarron:	①	7.880	7.899	.030	.830	.815
94–91	Commercial Chassis: (FWD)	①	8.860	8.880	.030	1.224	1.209
90–86	Commercial Chassis: (FWD)	①	7.880	7.899	.030⑤	.972	.965
94–92	Eldorado, Seville: Front	—	—	—	.030	1.250	1.209
	Rear	—	—	—	.030	.423	.374
91–86	Eldorado, Seville: Front	—	—	—	.030	.971	.956
	Rear	—	—	—	.030	.444	.429
85–82	Eldorado, Seville: Front	—	—	—	.030	.980	.965
	Rear	—	—	—	.030	.980	.965
81–79	Eldorado, Seville: Front	—	—	—	.062	.980	.965
	Rear	—	—	—	.062	.980	.965
78–77	Eldorado: Front	—	—	—	.062	—	1.170
	Rear	—	—	—	.062	—	1.170
76	Eldorado: Front	—	—	—	.062	1.205	1.190
	Rear	—	—	—	.062	1.205	1.190
75–72	Eldorado:	①	11.000	11.060	.062	1.205	1.190
94–91	Fleetwood, DeVille: (FWD)	①	8.860	8.880	.030	1.224	1.209
90–86	Fleetwood, Deville: (FWD)	①	8.863	8.917	.030	.972	.965
85	Fleetwood, Deville: (FWD)	①	8.858	8.880	.030	.972	.957
85–77	Fleetwood Limo, Comm. Ch.:						
	(RWD)	①	12.000	12.060	.062	1.230	1.215

① .030″ over rivet head; if bonded lining use .062″.
⑤ Inner Pad shown; outer .062″ (thickness over steel)

CHEVROLET

Year	Model						
94–93	Beretta, Corsica:	④	7.874	7.899	.030	.751	.736
92	Beretta, Corsica:	④	7.880	7.889	.030	.796	.736
91–87	Beretta, Corsica:	④	7.880	7.899	.030	.830	.815
94–93	Camaro:	.030	9.500	9.560	.030	1.250	1.209
	w/rear disc brakes	—	—	—	.030	.733	.724
92–89	Camaro: exc. H.D.	④	9.500	9.560	.030	.980	.965
	H.D. Front	—	—	—	.030	.980	.965
	H.D. Rear	—	—	—	.030	.744	.724
88–86	Camaro: w/rear drum brakes	④	9.500	9.560	.030	.980	.965
	w/rear disc brakes	—	—	—	.030	.986	.956
85–82	Camaro: w/rear drum brakes	④	9.500	9.560	.030	.980	.965
	w/rear disc brakes	—	—	—	.030	.980	.965
81–79	Camaro, Nova, Malibu,						
	Monte Carlo:	④	9.500	9.560	.030	.980	.965
78–77	Camaro, Nova:	④	9.500	9.560	.030④	.980	.965
76–73	Camaro, Chevelle,						
	Monte Carlo:	④	9.500	9.560	.030④	.980	.965
72	Camaro, Chevelle:	④	9.500	9.560	.030④	.980	.965
94	Caprice, Impala:	④	9.500	9.560	.030	.980	.965
	w/11″ rear drum brakes	④	11.000	11.060	.030	.980	.965
	w/rear disc brakes	—	—	—	.030	.735	.728
93–79	Caprice, Impala:	④	9.500	9.560	.030	.980	.965
	w/11″ rear drum brakes	④	11.000	11.060	.030	.980	.965
78	Caprice, Impala: exc. wagon	④	9.500	9.560	.030	.980	.965
	wagon	④	11.000	11.060	.030	.980	.965
77	Caprice, Impala: exc. wagon	④	9.500	9.560	.030	.980	.965
	wagon	④	11.000	11.060	.030	.980	.965
76–72	Caprice, Impala, Bel Air:	④	11.000	11.060	.030④	1.230	1.215
94–93	Cavalier:	.030	7.874	7.899	.030	.751	.736
92	Cavalier:	④	7.880	7.889	.030	.751	.736
91–82	Cavalier: vented rotor (91-82)	④	7.880	7.899	.030	.830	.815
	solid rotor (82)	④	7.880	7.899	.030	.444	.429

④ .030″ over rivet head; if bonded lining use .062″.

(continues)

CHEVROLET (Continued)

90–87 Celebrity: H.D. Front	—	—	—	.030	.972	.957
exc. H.D. Front	—	—	—	.030	.830	.815
Rear coupe/sedan	④	8.863	8.920	—	—	—
Rear wagon	④	8.863	8.877	—	—	—
w/rear disc brakes	—	—	—	.030	.702	.681
86–85 Celebrity: w/ H.D.	④	8.863	8.920	.030	.972	.957
exc. H.D.	④	8.863	8.920	.030	.830	.815
84–82 Celebrity:	④	7.879	7.899	.030	.830	.815
87–86 Chevette:	④	7.880	7.899	.030	.404	.374
85–84 Chevette:	④	7.874	7.899	.030	.430	.374
83 Chevette:	④	7.874	7.899	.030	.390	.374
82–78 Chevette:	④	7.874	7.899	.030	.390	.374
77 Chevette:	④	7.880	7.899	.030	.456	.441
76 Chevette:	④	7.870	7.899	.030	.448	.433
85–80 Citation:	④	7.880	7.899	.030	.830	.815
94–88 Corvette: Front exc. H.D.	—	—	—	.030	.744	.724
Front H.D.	—	—	—	.030	1.059	1.039
Rear exc. H.D.	—	—	—	.030	.744	.724
Rear H.D.	—	—	—	.030	1.059	1.039
87–84 Corvette: Front	—	—	—	.062	.724	—
Rear	—	—	—	.062	.724	—
82–77 Corvette: Front	—	—	—	.030④	1.230	1.215
Rear	—	—	—	.030④	1.230	1.215
76–72 Corvette: Front	—	—	—	.030④	1.230	1.215
Rear	—	—	—	.030④	1.230	1.215
94–90 Lumina: Front	—	—	—	.030	.984	.972
Rear	—	—	—	.030	.441	.429
88–82 Malibu, Monte Carlo:	④	9.500	9.560	.030	.980	.965
78–77 Malibu, Monte Carlo, Chevelle:	④	11.000	11.060	.030	.980	.965
80–76 Monza:	④	9.500	9.560	.030	.830	.815
75 Monza:	④	9.000	9.060	.030	.455	.440
88–85 Nova:	.039	7.874	7.913	.039	—	.492
w/rear disc brakes	—	—	—	.039	—	.315
76–72 Nova:	④	9.500	9.560	.030④	.980	.965
88–85 Spectrum: exc. Turbo	.031	7.090	7.140	.039	—	.378
w/Turbo	.039	7.090	7.140	.039	—	.650
88–85 Sprint:	.110⑩	7.090	7.160	.315	—	.315
77–76 Vega:	④	9.500	9.560	.030	.455	.440
75–72 Vega:	④	9.000	9.060	.030	.455	.440

④ .030" over rivet head; if bonded lining use .062"   ⑩ Measurement of lining & Metal

CHRYSLER CORP. — Chrysler, Dodge, Eagle, Plymouth

94–90 Acclaim, Spirit:	.030	8.661	8.691	.030	.912	.882
w/rear disc brakes				.281	.439	409
89 Acclaim, Spirit	.030	8.661	8.691	.030	.865	.803
89–83 Aries, Reliant, Lancer, LeBaron, 400, 600:						
w/7 7/8" rear brakes	.030	7.874	7.904	.030	.912	.882
w/8 21/32" rear brakes	.030	8.661	8.691	.030	.912	.882
82 Aries, Reliant, LeBaron, 400:	.030	7.870	7.900	.030	.912	.882
81 Aries, Reliant:	.030	7.870	7.900	.030	.912	.882
80–76 Arrow: w/rear drum brakes	.040	9.000	9.050	.040	—	.450
w/rear disc brakes	—	—	—	.040	—	.330
81–78 Aspen, Volare:	.030	10.000	10.060	.030	.970	.940
77–76 Aspen, Volare: exc. wagon	.030	10.000	10.060	.030	.970	.940
wagon	.030	11.000	11.060	.030	.970	.940
83–78 Challenger: w/rear drum brakes	.040	9.000	9.050	.040	—	.430
w/rear disc brakes	—	—	—	.040	—	.330
82–79 Champ: (FWD)	.040	7.100	7.150	.040	—	.450
74–73 Charger, Coronet, Challenger, Belvedere, Satellite, Barracuda: w/front disc brakes	.030	10.000	10.060	.030	.970	.940
72 Charger, Coronet, Crestwood, SE, Super Bee, Belvedere, Satelite, GTX, Regent, Road Runner, Sebring, Barracuda, Challenger:						
w/front disc brakes	.030	10.000	10.060	.030	.970	.940
w/10" drum brakes	.030	10.000	10.060	—	—	—
w/11" drum brakes	.030	11.000	11.060	—	—	—
94 Colt, Eagle Summit						
Coupe w/1.5L engine	.040	7.100	7.200	.080	—	.450
Coupe w/1.8L engine	.040	7.100	7.200	.080	—	.650
Sedan	.040	8.000	8.100	.080	—	.650
Sedan w/rear disc brakes				.080	—	.330

(continues)

CHRYSLER CORP. — Chrysler, Dodge, Eagle, Plymouth (Continued)

Year	Model						
93	Colt, Eagle Summit: Coupe	.040	7.100	7.200	.080	—	.449
	Sedan	.040	8.000	8.100	.080	—	.646
	w/rear disc brakes	—	—	—	.080	—	.331
92–91	Colt, Eagle Summit: Hatchback	.040	7.100	7.200	.080	—	.449
	Sedan	.040	7.100	7.200	.080	—	.646
90–89	Colt, Eagle Summit:						
	exc. rear disc brakes	.040	7.100	7.200	.080	—	.449
	w/rear disc brakes Front	—	—	—	.080	—	.882
	w/rear disc brakes Rear	—	—	—	.080	—	.331
88	Colt exc. Turbo	.040	7.100	7.200	.080	—	.450
	Turbo	.040	7.100	7.200	.080	—	.645
87–85	Colt exc. Turbo	.040	7.100	7.150	.040	—	.450
87–84	Colt Turbo	.040	7.100	7.150	.040	—	.645
84–79	Colt (FWD) exc. Turbo	.040	7.100	7.150	.040	—	.450
80–78	Colt (RWD)exc. Wagon	.040	9.000	9.050	.080	—	.450
77–73	Colt	.040	9.000	9.050	.080	—	.450
72	Colt	.040	9.000	9.050	.080	—	.374
94–92	Colt Vista, Eagle Summit SW:						
	w/8" drum brakes	.040	8.000	8.071	.080	—	.882
	w/9" drum brakes	.040	9.000	9.079	.080	—	.882
	w/rear disc brakes	—	—	—	.080	—	.331
91–89	Colt Vista, Eagle Vista: (4x2)	.040	8.000	8.100	.080	—	.650
	(4x4)	.040	9.000	9.100	.080	—	.880
88	Colt Vista: (4x2)	.040	8.000	8.100	.080	—	.650
	(4x4)	.040	9.000	9.100	.080	—	.880
87–84	Colt Vista: (4x2)	.040	8.000	8.150	.040	—	.650
	(4x4)	.040	9.000	9.078	.040	—	.880
90–89	Colt Wagon: (4x2)	.040	8.000	8.100	.080	—	.449
	(4x4)	.040	8.000	8.100	.080	—	.646
88	Colt Wagon:	.040	8.000	8.100	.080	—	.450
80–78	Colt Wagon: w/rear drum brakes	.040	9.000	9.050	.040	—	.330
	w/rear disc brakes	—	—	—	.040	—	.330
94–93	Concorde, Intrepid, Vision:	.030	8.661	8.691	.030	—	.882
	w/rear disc brakes	—	—	—	.030	—	.409
89–88	Conquest Front	—	—	—	.080	—	.880
	Rear	—	—	—	.080	—	.650
87–84	Conquest Front	—	—	—	.040	—	.880
	Rear	—	—	—	.040	—	.650
83–82	Cordoba, Mirada:	.030	10.000	10.060	.125	.970	.940
	w/11" rear brakes	.030	11.000	11.060	.125	.970	.940
81–78	Cordoba, Mirada:	.030	10.000	10.060	.030	.970	.940
	w/11" rear brakes	.030	11.000	11.060	.030	.970	.940
77–76	Cordoba, Charger SE:	.030	11.000	11.060	.030	.970	.940
75	Cordoba, Charger:	.030	10.000	10.060	.030	.970	.940
76–75	Dart, Valiant w/disc brakes	.030	10.000	10.060	.030	.970	.940
	w/front drum brakes Front	.030	10.000	10.060	—	—	—
	Rear	.030	9.000	9.060	—	—	—
74–73	Dart, Valiant w/disc brakes	.030	10.000	10.060	.030	.970	.940
	w/front drum brakes Front	.030	10.000	10.060	—	—	—
	Rear	.030	9.000	9.060	—	—	—
72	Dart, Valiant, GT, GTS, Swinger, Demon, Duster, Scamp, Signet: w/disc brakes	.030	10.000	10.060	.030	.790	.780
	w/10" drum brakes	.030	10.000	10.060	—	—	—
	w/9" drum brakes	.030	9.000	9.060	—	—	—
93–89	Daytona:	—	—	—	.030	.912	.882
	Rear w/solid rotor	—	—	—	.030	.439	.409
	Rear w/vented rotor...........	—	—	—	.030	.827	.797
88	Daytona:	.030	8.661	8.691	.030	.912	.882
	w/rear disc brakes	—	—	—	.030	.321	.291
87–84	Daytona, Laser: (FWD)						
	exc. rear disc brakes..........	.030	8.661	8.691	.030	.912	.882
	w/rear disc brakes	—	—	—	.030	.321	.291
93–92	Dynasty, New Yorker, Imperial, Fifth Ave.:	.030	8.661	8.691	.030	.912	.882
	w/rear disc brakes	—	—	—	.030	.439	.409
91–89	Dynasty, New Yorker, Imperial, Fifth Ave.:	.030	8.661	8.691	.030	.912	.882
	w/rear disc brakes	—	—	—	.030	.369	.339
88	Dynasty, New Yorker:	.030	8.661	8.691	.030	.912	.882
	w/rear disc brakes	—	—	—	.030	.321	.291
88	Eagle:	.030	10.000	10.060	.030②	.815	.810
78	Fury, Monaco:	.030	10.000	10.060	.030	.970	.940
	w/11" rear brakes	.030	11.000	11.060	.030	.970	.940
77–76	Fury, Monaco, Coronet	.030	11.000	11.060	.030	.970	.940
75	Fury, Monaco, Coronet						
	exc. wagon	.030	10.000	10.060	.030	.970	.940
	wagon	.030	11.000	11.060	.030	.970	.940

② .030" over rivet head; if bonded lining use .062".

(continues)

CHRYSLER CORP. — Chrysler, Dodge, Eagle, Plymouth (Continued)

Model						
74–72 Fury, Monaco, Polara, Newport, New Yorker, Town & Country, V.I.P.:	030	11.000	11.060	.030	1.195	.180
89–82 Gran Fury, Diplomat, Fifth Ave., Newport, New Yorker, Imperial, Caravelle: (RWD)	030	10.000	10.060	.125	.970	.940
w/11" rear brakes	030	11.000	11.060	.125	.970	.940
81–78 Gran Fury, Diplomat, LeBaron, St. Regis, Magnum, Newport, New Yorker, Imperial:	030	10.000	10.060	.030	.970	.940
w/11" rear brakes	030	11.000	11.060	.030	.970	.940
77–75 Gran Fury, Royal Monaco, New Yorker, Newport, Town & Country:	030	11.000	11.060	.030	1.195	.180
75–74 Imperial: Front	—	—	—	.030	1.195	.180
Rear	—	—	—	.030	.970	.940
73–72 Imperial:	030	11.000	11.060	.030	1.195	.180
94–90 Laser, Talon: Front	—	—	—	.080	.912	.882
Rear	—	—	—	.080	.360	.331
94–87 Lebaron:	030	8.661	8.691	.281	.912	.882
Rear w/solid rotor				.281	.439	.409
Rear w/vented rotor...........				.281	.827	.797
89–88 Medallion:	098	9.000	9.030	.256	⑨	.697
92–91 Monaco, Premier:	132	8.858	8.917	.160	—	.890
w/rear disc brakes	—	—	—	.062	—	.374
90–88 Monaco, Premier:	132	8.858	8.917	.236	⑨	.807
94 New Yorker, LHS: Front	—	—	—	.030	—	.882
Rear	—	—	—	.030	—	.882
88 New Yorker Turbo, Town & Country SW, Caravelle:	030	8.661	8.691	.030	.912	.882
87–83 New Yorker Turbo, Town & Country SW, Caravelle: (FWD) exc. rear disc brakes...........	030	8.661	8.691	.030	.912	.882
w/rear disc brakes	—	—	—	.030	.321	.291
90–83 Omni, Charger, Horizon, Turismo:	.030	7.874	7.904	.030	.461	.431
82–78 Omni, Horizon:	.030	7.870	7.900	.030	.461	.431
83–78 Sapporo: w/rear drum brakes	.040	9.000	9.050	.040	—	.430
w/rear disc brakes	—	—	—	.040	—	.330
94–89 Shadow, Sundance: w/7 ⅞" rear brakes	.030	7.874	7.904	.030	.912	.882
88–86 Shadow, Sundance: w/7 ⅞" rear brakes	.030	7.874	7.904	.030	.912	.882
w/8 ²¹⁄₃₂" rear brakes	.030	8.661	8.691	.030	.912	.882
94–91 Stealth: (FWD) Front	—	—	—	.080	—	.880
(FWD) Rear	—	—	—	.080	—	.650
(AWD) Front	—	—	—	.080	—	1.120
(AWD) Rear	—	—	—	.080	—	.720
91–89 TC Maserati:	—	—	—	.030	.912	.882
w/rear disc brakes	—	—	—	.030	.321	.291
94 Viper: Front	—	—	—	.100	1.197	1.167
Rear	—	—	—	.100	.803	.773

⑨ Machining not recommended

DAIHATSU

Model						
92–91 Charade: 1.0 eng.	.039	7.087	7.126	.120	—	.390
1.3 eng.	.039	7.874	7.913	.120	—	.670
89–88 Charade:	.040	7.090	7.126	.120	—	.390

FIAT

Model						
82–75 124, 2000: Front	—	—	—	.080	.368	.354
Rear	—	—	—	.080	.372	.354
74–72 124: Front	—	—	—	.080	.368	.354
Rear	—	—	—	.080	.372	.354
79–73 128:	.060	7.300	7.332	.080	.368	.354
78–75 131:	.181	9.000	9.030	.060	.368	.354
81–79 Brava:	.181	9.000	9.030	.060	—	.386
82–79 Strada:	.060	7.293	7.336	.060	.368	.350
82–74 X 1/9: Front	—	—	—	.080	.368	.354
Rear	—	—	—	.080	.368	.354

FORD MOTOR CO. — Ford, Lincoln, Mercury

Year	Model						
94	Aspire:						
	wautomatic transmission	.040	7.870	7.930	.080	.817	.780
	wmanual transmission	.040	7.870	7.930	.080	.660	.630
94–91	Capri: Front	—	—	—	.120	.660	.630
	Rear	—	—	—	.120	.380	.350
77–76	Capri:	.030	9.000	9.050	.100	.460	.450
74–72	Capri:	.030	9.000	9.050	.100	—	⑨
94–93	Continental: Front	—	—	—	.040	—	.974
	Rear	—	—	—	.123	—	.500
92	Continental: Front	—	—	—	.125	—	.974
	Rear	—	—	—	.123	—	.900
91	Continental: Front	—	—	—	.125	—	.970
	Rear	—	—	—	.123	—	.900
90–88	Continental: Front	—	—	—	.125	—	.970
	Rear	—	—	—	.125	—	.974
87–84	Continental: Front	—	—	—	.125	—	.972
	Rear	—	—	—	.125	—	.895
83–82	Continental: Front	—	—	—	.125	—	.972
	Rear	—	—	—	.125	—	.895
81–80	Continental:						
	w/10" rear brakes	.030	10.000	10.060	.125	—	.972
	w/11" rear brakes	.030	11.030	11.090	.125	—	.972
79–73	Continental:						
	w/10" rear brakes	.030	10.000	10.090	.125	—	1.120
	w/11" rear brakes	.030	11.030	11.090	.125	—	1.120
72	Continental:	.030	11.030	11.090	.030	1.135	1.120
94–93	Cougar, Thunderbird:	.030	9.800	9.860	.040	—	.974
	w/rear disc brakes	—	—	—	.123	—	.657
92	Cougar, Thunderbird:	.030	9.800	9.860	.125	—	.974
	w/rear disc brakes	—	—	—	.123	—	.900
91–90	Cougar, Thunderbird:	.030	9.800	9.860	.125	—	.935
	w/rear disc brakes	—	—	—	.123	—	.900
89	Cougar, Thunderbird:	.030	9.800	9.860	.125	—	.935
	w/rear disc brakes	—	—	—	.123	—	.895
88–87	Cougar, Thunderbird:						
	exc. Turbo w/9" rear brakes	.030	9.000	9.060	.125	—	.810
	w/10" rear brakes	.030	10.000	10.060	.125	—	.810
	w/Turbo	.030	9.000	9.060	.125	—	.972
	w/rear disc brakes	—	—	—	.125	—	.895
86–83	Cougar, Thunderbird, XR-7:						
	w/9" rear brakes	.030	9.000	9.060	.125	—	.810
	w/10" rear brakes	.030	10.000	10.060	.125	—	.810
82–81	Cougar, Thunderbird, XR-7:						
	w/9" rear brakes	.030	9.000	9.060	.125	—	.810
	w/10" rear brakes	.030	10.000	10.060	.125	—	.810
80	Cougar, Thunderbird:	.030	9.000	9.060	.125	—	.810
79	Cougar, Thunderbird, LTD II:	.030	11.030	11.090	.125	—	1.120
78–77	Cougar, Thunderbird, LTD II:	.030	11.030	11.090	.125	—	1.120
76–74	Cougar: w/10" rear brakes	.030	10.000	10.060	.030	—	1.120
	w/11" rear brakes	.030	11.030	11.090	.030	—	1.120
73–72	Cougar, Mustang:	.030	10.000	10.060	.030	.890	.875
94–92	Crown Victoria, Grand Marquis: Front	—	—	—	.125	—	.974
	Rear	—	—	—	.125	—	.440
91	Crown Victoria, Grand Marquis:						
	w/10" rear brakes	.030②	10.000	10.060	.125	—	.972
	w/11" rear brakes	.030②	11.030	11.090	.125	—	.972
90–83	Crown Victoria, Grand Marquis:						
	w/11" rear brakes	.030	11.030	11.090	.125	—	.972
	w/10" rear brakes	.030	10.000	10.060	.125	—	.972
78–74	Custom, Country Sedan, Country Squire, Galaxie, LTD, Colony Park, Marquis:						
	exc. rear disc brakes	.030	11.030	11.090	.125	—	1.120
	w/rear disc brakes	—	—	—	.030	—	.895
73	Custom, Country Sedan, Country Squire, Galaxie, LTD, Colony Park, Marquis:	.030	11.030	11.090	.125	—	1.120
72	Custom, Country Sedan, Country Squire, Galaxie, LTD, Colony Park, Marquis:	.030	11.030	11.090	.030	1.135	1.120
94–93	Escort, Tracer:	.040	7.870	7.910	.080	.820	.790
	w/rear disc brakes	—	—	—	.040	.310	.280
92–91	Escort, Tracer:	.040	9.000	9.040	.080	.820	.790
	w/rear disc brakes	—	—	—	.040	.310	.280
90–89	Escort: w/7" rear brakes	.030②	7.090	7.149	.125	—	.882
	w/8" rear brakes	.030②	8.000	8.060	.125	—	.882
88–81	Escort, EXP, Lynx, LN7:						
	w/7" rear brakes	.030②	7.090	7.149	.125	—	.882
	w/8" rear brakes	.030②	8.000	8.060	.125	—	.882

② .030" over rivet head; if bonded lining use .062".   ⑨ Machining not recommended.

(continues)

FORD MOTOR CO. — Ford, Lincoln, Mercury (Continued)

86–83 Fairmont, Zephyr, LTD,						
Marquis: w/9″ rear brakes	.030	9.000	9.060	.125	—	.810
w/10″ rear brakes	.030	10.000	10.060	.125	—	.810
82–79 Fairmont, Zephyr:						
w/9″ rear brakes	.030	9.000	9.060	.125	—	.810
w/10″ rear brakes	.030	10.000	10.060	.125	—	.810
78 Fairmont, Zephyr: exc. Wagon	.030	9.000	9.060	.125	—	.810
Wagon	.030	10.000	10.060	.125	—	.810
93–88 Festiva:	.040	6.690	6.750	.120	.463	.433
80–78 Fiesta:	.060	7.000	—	.060	—	.340
82–81 Granada: w/9″ rear brakes	.030	9.000	9.060	.125	—	.810
w/10″ rear brakes	.030	10.000	10.060	.125	—	.810
80–79 Granada, Monarch:	.030	10.000	10.060	.125	—	.810
78–77 Granada, Monarch:	.030	10.000	10.060	.125	—	.810
76–75 Granada, Monarch:						
exc. rear disc brakes	.030	10.000	10.060	.125	—	.810
w/rear disc brakes	—	—	—	.125	—	.895
94–93 Mark VIII: Front	—	—	—	.125	—	.974
Rear	—	—	—	.125	—	.657
92–91 Mark VII: Front	—	—	—	.125	—	.972
Rear	—	—	—	.123	—	.890
90–88 Mark VII: Front	—	—	—	.125	—	.972
Rear	—	—	—	.125	—	.895
87–84 Mark VII: Front	—	—	—	.125	—	.972
Rear	—	—	—	.125	—	.895
83–80 Mark VI: w/10″ rear brakes	.030	10.000	10.060	.125	—	.972
w/11″ rear brakes	.030	11.030	11.090	.125	—	.972
79–77 Mark V: exc. rear disc brakes	.030	11.030	11.090	.125	—	1.120
w/rear disc brakes	—	—	—	.125	—	.895
77–73 Mark IV: exc. rear disc brakes	.030	11.030	11.090	.125	—	1.120
w/rear disc brakes	—	—	—	.125	—	.895
72 Mark IV:	.030	11.030	11.090	.030	1.135	1.120
78–77 Maverick, Comet	.030	10.000	10.060	.125	—	.810
76–75 Maverick, Comet	.030	10.000	10.060	.125	—	.810
74 Maverick, Comet	.030	10.000	10.060	.125	—	.810
73–72 Maverick, Comet:						
w/9″ drum brakes	.030	9.000	9.060	—	—	—
w/10″ drum brakes	.030	10.000	10.060	—	—	—
94 Mustang: exc. Cobra Front	—	—	—	.040	—	.970
Cobra Front	—	—	—	.040	—	1.040
Rear	—	—	—	.123	—	.500
93–89 Mustang: exc. 5.0L eng.	.030②	9.000	9.060	.125	—	.810
w/5.0L eng.	.030②	9.000	9.060	.125	—	.972
88–87 Mustang: exc. 5.0L eng.						
w/9″ rear brakes	.030	9.000	9.060	.125	—	.810
w/10″ rear brakes	.030	10.000	10.060	.125	—	.810
w/5.0L eng.						
exc. rear disc brakes...........	.030	10.000	10.060	.125	—	.972
w/rear disc brakes	—	—	—	.125	—	.895
86–83 Mustang, Capri:						
exc. 5.0L eng. or SVO						
w/9″ rear brakes	.030	9.000	9.060	.125	—	.810
w/10″ rear brakes	.030	10.000	10.060	.125	—	.810
86–83 Mustang: w/5.0L eng. or SVO						
exc. rear disc brakes...........	.030	10.000	10.060	.125	—	.972
w/rear disc brakes	—	—	—	.125	—	.895
82–79 Mustang, Capri:						
w/9″ rear brakes	.030	9.000	9.060	.125	—	.810
w/10″ rear brakes	.030	10.000	10.060	.125	—	.810
80–77 Pinto, Bobcat, Mustang II:	.030	9.000	9.060	.030	—	.810
76–75 Pinto, Bobcat, Mustang II:	.030	9.000	9.060	.030	—	.810
74 Pinto, Mustang II:	.030	9.000	9.060	.030	—	.875
73–72 Pinto:	.030	9.000	9.060	.030	.700	.685

② .030″ over rivet head; if bonded lining use .062″.

FORD MOTOR CO. — Ford, Lincoln, Mercury (Continued)

94–93 Probe:	.040	9.000	9.060	.040	.890	.863
w/rear disc brakes	—	—	—	.040	.345	.315
92–89 Probe:	.040	9.000	9.060	.120	.890	.860
w/rear disc brakes	—	—	—	.040	.345	.315
94–90 Taurus, Sable: Sedan	.030	8.858	8.918	.040	—	.974
Wagon	.030	9.842	9.902	.040	—	.974
w/rear disc brakes	—	—	—	.123	—	.900
89–88 Taurus, Sable: Sedan	.030	8.858	8.918	.125	—	.974
Wagon	.030	9.842	9.902	.125	—	.974
87–86 Taurus, Sable: Sedan	.030	8.858	8.918	.125	—	.896
Wagon	.030	9.842	9.902	.125	—	.896
94–91 Tempo, Topaz:	.060	8.060	8.120	.125	—	.882
90–89 Tempo, Topaz:	.030②	8.000	8.060	.125	—	.882
88–84 Tempo, Topaz:	.030②	8.000	8.060	.125	—	.882
76–73 Thunderbird:	.030	11.030	11.090	.125	—	.120
72 Thunderbird:	.030	11.030	11.090	.030	1.135	.120
76–74 Torino, Montego:						
w/10″ rear brakes	.030	10.000	10.060	.030	—	.120
w/11″ rear brakes	.030	11.030	11.090	.030	—	.120
73–72 Torino, Montego:	.030	10.000	10.060	.030	—	.120
94–91 Town Car: Front	—	—	—	.125	—	.974
Rear	—	—	—	.125	—	.440
90–83 Town Car: w/10″ rear brakes	.030	10.000	10.060	.125	—	.972
w/11″ rear brakes	.030	11.030	11.090	.125	—	.972
82–79 Town Car, LTD, Marquis:						
w/10″ rear brakes	.030	10.000	10.060	.125	—	.972
w/11″ rear brakes	.030	11.030	11.090	.125	—	.972
89–88 Tracer: exc. rear disc brakes	.040	7.870	7.910	.120	.660	.630
w/rear disc brakes	—	—	—	.120	.380	.350
80–79 Versailles:						
exc. rear disc brakes	.030	10.000	10.060	.125	—	.810
w/rear disc brakes	—	—	—	.125	—	.895
78–77 Versailles:	.030	10.000	10.060	.125	—	.810

② .030″ over rivet head; if bonded lining use .062″.

GEO

94–92 Metro:	.039	7.090	7.160	.120	—	.315
91 Metro: Hardtop	.110②	7.090	7.160	.315	—	.315
Convertible	.110②	7.090	7.160	.320	—	.630
90–89 Metro:	.110②	7.090	7.160	.315	—	.315
94–93 Prizm:	.039	7.874	7.913	.039	—	.787
92–89 Prizm:	.039	7.874	7.913	.030	—	.669
89 Spectrum: exc. Turbo	.039	7.090	7.140	.039	—	.378
Turbo	.039	7.090	7.140	.039	—	.650
93–90 Storm:	.039	7.870	7.930	.039	—	.811

② Measurement of lining & metal.

HONDA

94–90 Accord: exc. Wagon	.080	8.661	8.701	.063	.827	—
Wagon	.080	8.661	8.701	.063	.910	—
w/rear disc brakes	—	—	—	.063	.315	—
89 Accord: exc. Fuel Inj.	.080	7.870	7.910	.120	.670	—
w/Fuel Inj.	.080	7.870	7.910	.120	.750	—
w/rear disc brakes	—	—	—	.060	.310	—
88 Accord: exc. Fuel Inj.	.080	7.870	7.910	.120	.670	—
w/Fuel Inj.	.080	7.870	7.910	.060	.750	—
87–86 Accord:	.080	7.870	7.910	.120	.670	—
85–84 Accord:	.080	7.870	7.910	.059	.670	—
83–82 Accord:	.079	7.870	7.910	.063	.600	—
81–79 Accord:	.079	7.080	7.130	.063	.413	—
78–76 Accord:	.079	7.080	7.130	.063	.449	.437
91–90 Civic CRX: DX	.080	7.090	7.130	.120	.750	—
HF	.080	7.090	7.130	.120	.590	—
Si	.080	7.090	7.130	.120	.670	—
w/rear disc brakes	—	—	—	.120	.310	—
89 Civic CRX: w/DX, Si	.080	7.090	7.130	.120	.670	—
w/HF	.080	7.090	7.130	.120	.590	—
88 Civic CRX: w/DX, Si	.080	7.090	7.130	.120	.670	—
w/HF	.080	7.090	7.130	.120	.590	—
87–84 Civic CRX: w/1300 HF	.080	7.090	7.130	.120	.350	—
w/1500 exc. HF	.080	7.090	7.130	.120	.590	—
94–93 Civic del Sol:	.080	7.090	7.130	.060	.750	—
94–92 Civic Hatchback, Civic Sedan,						
Civic Coupe:	.080	7.090	7.130	.060	.750	—
w/H.D. rear drum brakes	.080	7.870	7.910	.060	.750	—
w/rear disc brakes	—	—	—	.060	.310	—
91–90 Civic Hatchback,						
Civic Sedan:	.080	7.087	7.126	.120	.750	—

(continues)

HONDA (Continued)

89 Civic Hatchback, Civic Sedan:	.080	7.090	7.130	.120	.670	—
88 Civic Hatchback, Civic Sedan:	.080	7.090	7.130	.120	.669	—
87–84 Civic Hatchback: w/1300	.080	7.090	7.130	.120	.390	—
w/1500	.080	7.090	7.130	.120	.590	—
83 Civic Hatchback: w/1300 4 SPD	.079	7.090	7.130	.063	.350	—
w/1300 5 SPD	.079	7.090	7.130	.063	.390	—
w/1500	.079	7.090	7.130	.063	.590	—
82 Civic Hatchback: w/1300 4 SPD	.079	7.090	7.130	.063	.350	—
exc. 1300 4 SPD	.079	7.090	7.130	.063	.390	—
81–80 Civic Hatchback, Civic CVCC:	.079	7.090	7.130	.063	.350	—
79–75 Civic Hatchback, Civic CVCC:	.079	7.080	7.130	.063	.354	.343
74–73 Civic Hatchback:	.079	7.080	7.130	.063	.354	.343
87–84 Civic Sedan:	.080	7.090	7.130	.120	.590	—
83 Civic Sedan:	.079	7.090	7.130	.063	.590	—
82–81 Civic Sedan:	.079	7.090	7.130	.063	.390	—
91–90 Civic Wagon, Wagovan: (4x2)	.080	7.874	7.913	.120	.750	—
(4x4)	.080	7.874	7.913	.120	.670	—
89 Civic Wagon, Wagovan: (4x2)	.080	7.870	7.910	.060	.669	—
(4x4)	.080	7.870	7.910	.120	.669	—
88 Civic Wagon, Wagovan: (4x2)	.080	7.870	7.910	.080	.669	—
(4x4)	.080	7.870	7.910	.120	.669	—
87–84 Civic Wagon:	.080	7.870	7.910	.120	.590	—
83–80 Civic Wagon:	.079	7.870	7.910	.063	.390	—
79–76 Civic Wagon:	.079	7.870	7.910	.063	.449	.437
94–92 Prelude: Front	—	—	—	.060	.830	—
Rear	—	—	—	.060	.310	—
91 Prelude: Front	—	—	—	.060	.750	—
Rear	—	—	—	.060	.310	—
90–88 Prelude: Front exc. Fuel Inj.	—	—	—	.120	.670	—
w/Fuel Inj.	—	—	—	.120	.750	—
Rear	—	—	—	.080	.310	—
87–86 Prelude: Front exc. Fuel Inj.	—	—	—	.118	.670	—
Rear	—	—	—	.063	.310	—
Front w/Fuel Inj.	—	—	—	.120	.670	—
Rear	—	—	—	.063	.310	—
85–84 Prelude: Front	—	—	—	.120	.670	—
Rear	—	—	—	.060	.310	—
83 Prelude:	.080	7.870	7.910	.120	.670	—
82–79 Prelude:	.079	7.080	7.130	.063	.413	—

HYUNDAI

94–92 Elantra:	.059	8.000	8.079	.079	—	.787
94–91 Excel:	.039	7.087	7.165	.039	—	.669
90 Excel:	.039	7.100	7.165	.039	—	.670
89 Excel:	.040	7.100	7.200	.040	—	.450
88 Excel: w/solid rotor	.040	7.100	7.200	.040	—	.450
w/vented rotor	.040	7.100	7.200	.040	—	.670
87–85 Excel:	.040	7.086	7.165	.040	—	.450
87–86 Pony Sedan:	.039	7.992	④	.079	—	.449
85–83 Pony Sedan:	.039	7.992	④	.079	—	.449
94–91 Scoupe:	.039	7.100	7.165	.039	—	.669
93–91 Sonata:	.031	8.858	8.936	.079	—	.797
w/rear disc brakes	—	—	—	.031	—	.413
90 Sonata:	.039	9.000	9.079	.079	—	.787
w/rear disc brakes	—	—	—	.031	—	.413
89 Sonata:	.059	9.000	9.079	.079	—	.787
87 Stellar:	.040	9.000	9.079	.040	—	.669
86–85 Stellar:	.060	9.000	—	.060	—	.450

④ Machining specification is not available; discard at 8.071".

INFINITI

94–91 G20: Front	—	—	—	.079	—	.787
Rear	—	—	—	.079	—	.310
94–93 J30: Front	—	—	—	.079	—	1.024
Rear	—	—	—	.079	—	.551
92–90 M30: Front	—	—	—	.079	—	.787
Rear	—	—	—	.079	—	.354
94–90 Q45: Front	—	—	—	.079	—	1.024
Rear	—	—	—	.079	—	.315

ISUZU

89–88 I-Mark:						
exc. Turbo or DOHC eng.	.040	7.090	7.140	.040	—	.378
w/Turbo or DOHC eng.	.040	7.090	7.140	.040	—	.650
87–85 I-Mark:	.039	7.090	7.140	.039	—	.378
84–81 I-Mark:	.039	9.000	9.040	.067	.354	.338
92–90 Impulse: Front	—	—	—	.039	—	.810
Rear	—	—	—	.039	—	.299
89–85 Impulse: Front	—	—	—	.040	.668	.654
Rear	—	—	—	.040	.668	.654
84–83 Impulse: Front	—	—	—	.125	.706	.654
Rear	—	—	—	.125	.706	.654
93–91 Stylus: w/rear drum brakes	.039	7.870	7.929	.039	—	.653
w/rear disc brakes	—	—	—	.039	—	.810

JAGUAR

94–88 XJ6, XJ12: Front	—	—	—	.125	—	②
Rear	—	—	—	.125	—	②
87–75 XJ6, Vanden Plas: Front	—	—	—	.125	—	②
Rear	—	—	—	.125	—	.450
74–72 XJ6: Front	—	—	—	.125	—	.450
Rear	—	—	—	.125	—	.450
94–76 XJS, Saloons: Front	—	—	—	.125	—	②
Rear w/overdrive	—	—	—	.125	—	②
Rear exc. Overdrive	—	—	—	.125	—	.450
79–78 XJ12: Front	—	—	—	.125	—	②
Rear	—	—	—	.125	—	.450
77–72 XJ12, V12: Front	—	—	—	.125	—	②
Rear	—	—	—	.125	—	.450

② Minimum thickness is stamped on rotor.

LEXUS

91–90 ES250: Front	—	—	—	.039	—	.945
Rear	—	—	—	.039	—	.354
94–92 ES300: Front	—	—	—	.039	—	1.063
Rear	—	—	—	.039	—	.354
94 GS300: Front	—	—	—	.039	—	1.181
Rear	—	—	—	.039	—	.591
94–93 LS400: Front	—	—	—	.039	—	1.181
Rear	—	—	—	.039	—	.591
92–91 LS400: Front	—	—	—	.039	—	1.024
Rear	—	—	—	.039	—	.591
90 LS400: Front	—	—	—	.039	—	.906
Rear	—	—	—	.039	—	.591
94–92 SC300: Front	—	—	—	.039	—	1.024
Rear	—	—	—	.039	—	.591
94–92 SC400: Front	—	—	—	.039	—	1.181
Rear	—	—	—	.039	—	.591

MAZDA

94–92 323, Protege:	.040	7.870	7.910	.080	—	.790
w/rear disc brakes	—	—	—	.040	—	.310
91–90 323, Protege:	.040	9.000	9.040	.080	—	.790
w/rear disc brakes	—	—	—	.040	—	.280
89–88 323 Hatchback, 323 Sedan:	.040	7.870	7.910	.080	—	.630
w/rear disc brakes	—	—	—	.040	—	.310
87–86 323 Hatchback, 323 Sedan:	.040	7.870	7.910	.118	—	.630
w/rear disc brakes	—	—	—	.040	—	.350
88–87 323 Wagon:	.040	9.000	9.040	.118②	—	.630
94–93 626, MX-6:	.040	9.000	9.059	.080	—	.870
w/rear disc brakes	—	—	—	.040	—	.310
92–88 626, MX-6:	.040	9.000	9.060	.080	—	.870
w/rear disc brakes	—	—	—	.040	—	.310
87–86 626:	.040	7.870	7.910	.118	—	.710
w/rear disc brakes	—	—	—	.040	—	.350
85 626: w/gas eng.	.040	7.870	7.910	.118	—	.470
w/diesel eng.	.040	7.870	7.910	.118	—	.710
84–83 626:	.040	7.874	7.913	.040	—	.490
82–79 626:	.040	9.000	9.040	.040	—	.472
77–74 808:	.040	7.874	7.914	.256④	—	.394
73–72 808:	.040	7.874	7.914	.256④	—	.394
94–92 929: Front	—	—	—	.040	—	.870
Rear	—	—	—	.040	—	.630
91–90 929: Front	—	—	—	.080	—	.790
Rear	—	—	—	.080	—	.630
89–88 929: Front	—	—	—	.080	—	.790
Rear	—	—	—	.080	—	.310
78–76 Cosmo: Front	—	—	—	.276④	—	.669
Rear	—	—	—	.276④	—	.354
85 GLC Hatchback, GLC Sedan: . . .	.040	7.090	7.130	.118	—	.390

(continues)

MAZDA (Continued)

Model						
84–81 GLC Hatchback, GLC Sedan: ...	.040	7.090	7.130	.040	—	.393
80–77 GLC Hatchback, GLC SEdan: ...	.040	7.874	7.914	.040	—	.472
83–81 GLC Wagon:	.040	7.874	7.913	.040	—	.472
94 Miata MX-5: Front	—	—	—	.040	—	.710
Rear	—	—	—	.040	—	.310
93–91 Miata MX-5: Front	—	—	—	.040	—	.630
Rear	—	—	—	.040	—	.280
90 Miata MX-5: Front	—	—	—	.040	—	.630
Rear	—	—	—	.080	—	.280
93–92 MX-3:	.040	7.870	7.910	.080	—	.790
w/rear disc brakes	—	—	—	.040	—	.310
73–72 RX-2:	.040	7.874	7.914	.276④	—	.433
77–75 RX-3:	.040	7.874	7.914	.276④	—	.394
73–72 RX-3:	.040	7.874	7.914	.256④	—	.394
78–74 RX-4:	.040	9.000	9.040	.276④	—	.433
94–93 RX-7: Front	—	—	—	.040	—	.790
Rear	—	—	—	.040	—	.710
91–89 RX-7: Front Std.	—	—	—	.080	—	.790
Front H.D.	—	—	—	.080	—	.790
Rear Std.	—	—	—	.040	—	.310
Rear H.D.	—	—	—	.040	—	.710
88–86 RX-7: Front w/14″ Wheels	—	—	—	.040	—	.790
Front w/15″ Wheels	—	—	—	.118	—	.790
Rear w/14″ Wheels	—	—	—	.040	—	.310
Rear w/15″ Wheels	—	—	—	.040	—	.710
85–80 RX-7:	.040	7.874	7.914	.040	—	.670
w/rear solid rotor	—	—	—	.040	—	.354
w/rear vented rotor	—	—	—	.040	—	.787
79 RX-7:	.040	7.874	7.914	.040	—	.670

④ Measurement of lining & metal.

MERCEDES BENZ

Model						
93–86 190E: 2.3, 2.3-16, 2.6 Front	—	—	—	.138	.787	.764
2.3, 2.3-16, 2.6 Rear	—	—	—	.138	.300	.287
90–84 190D,E: Front	—	—	—	.138	.374	.354
Rear	—	—	—	.079	.300	.287
94 220C, 280C, 300E Diesel:						
Front	—	—	—	.078	.787	.764
Rear	—	—	—	.078	.300	.287
93–86 260E, 300CE,TE,DT,E,TDT,D:						
Front	—	—	—	.079	.787	.764
Rear	—	—	—	.079	.300	.287
91–81 300 SD,SE,SEL,						
300 SDL Turbo Diesel,						
350 SD,SDL, 380 SE,SEC,SEL,						
420 SEL, 500 SEC,SEL,						
560 SEC,SEL:						
Front Fixed Caliper 60mm	—	—	—	.079	.787	.764
Fixed Caliper 57mm	—	—	—	.079	1.024	1.000
Floating Caliper	—	—	—	.138	.787	.764
Rear	—	—	—	.079	.339	.327
94 320E: Front	—	—	—	.078	.787	.764
Rear	—	—	—	.078	.866	.843
94 320E Cabriolet: Front	—	—	—	.078	.906	.882
Rear	—	—	—	.078	.300	.287
94 320E Wagon: Rear	—	—	—	.078	.697	.685
Front	—	—	—	.078	.906	.882
94 320S, 350S Turbo Diesel:						
Front 2 Piston Caliper	—	—	—	.078	1.023	1.000
Front 4 Piston Caliper	—	—	—	.078	1.102	1.079
Rear	—	—	—	.078	.413	.385
94 320SL, 500SL: Front	—	—	—	.078	1.026	1.000
Rear	—	—	—	.078	.300	.287
93–90 300SL, 500SL: Front	—	—	—	.079	1.024	1.000
Rear	—	—	—	.079	.300	.287
94 420S, 500S, 600S, 600SL:						
Front 2 Piston Caliper	—	—	—	.078	1.023	1.000
Front 4 Piston Caliper	—	—	—	.078	1.102	1.079
Rear	—	—	—	.078	.787	.764
94 500E: Front	—	—	—	.078	1.023	1.000
Rear	—	—	—	.078	.866	.843
93–92 400E, 500E: Front	—	—	—	.079	1.024	1.000
Rear	—	—	—	.079	.866	.842

(continues)

MERCEDES BENZ (Continued)

	Col1	Col2	Col3	Col4	Col5	Col6
94 600SL: Front	—	—	—	.078	1.122	1.102
Rear	—	—	—	.078	.787	.764
89–86 560SL: Front	—	—	—	.079	.787	.764
Rear	—	—	—	.079	.338	.327
85–73 380SL,SLC, 450SL,SLC: Front	—	—	—	.079	—	.811①
Rear	—	—	—	.079	—	.327
85–77 230, 240D, 300D,CD,TD, 280E,CE: Front	—	—	—	.079	—	.417
Rear	—	—	—	.079	—	.327
80 280SE, 300SD, 450SEL: Front	—	—	—	.079	—	.764
Rear	—	—	—	.079	—	.327
79–73 280S,SE, 300SD, 450SE,SEL, 6.9: Front	—	—	—	.079	—	.787
Rear	—	—	—	.079	—	.327
76–72 220,D, 220/8, 230, 240D, 250,C, 280,C, 300D: Front	—	—	—	.079	—	.432②
Rear	—	—	—	.079	—	.327
73–72 280,SE,SEL, 300SEL: Front	—	—	—	.079	—	.431③
Rear	—	—	—	.079	—	.327
72 600: Front	—	—	—	.354	—	.725
Rear	—	—	—	.079	—	.570

① w/57mm caliper piston shown; w/60mm caliper piston up to 3/80 .787'' from 3/80 .763''.
② w/57mm caliper piston shown; w/60mm caliper piston .417
③ 1st version shown; 2nd version .700''.

MERKUR

	Col1	Col2	Col3	Col4	Col5	Col6
90–88 Scorpio: Front	—	—	—	.140	—	.900
Rear	—	—	—	.150	—	.350
89–85 XR4Ti:	.040	10.000	10.060	.060	.927	.897

① 1987 .0004''.   ② 1990 .002''.

MITSUBISHI

	Col1	Col2	Col3	Col4	Col5	Col6
94–91 3000GT: (FWD) Front	—	—	—	.080	—	.880
(FWD) Rear	—	—	—	.080	—	.650
(AWD) Front	—	—	—	.080	—	1.120
(AWD) Rear	—	—	—	.080	—	.720
88 Cordia, Tredia: exc. Turbo	.040	8.000	8.100	.080	—	.650
w/Turbo	.040	8.000	8.100	.080	—	.880
87–84 Cordia, Tredia:	.040	8.000	8.050	.040	—	.650
83 Cordia, Tredia:	.040	8.000	8.050	.040	—	.450
94–92 Diamante: Front	—	—	—	.080	—	.880
exc. Wagon Rear	—	—	—	.080	—	.650
Wagon Rear	—	—	—	.080	—	.720
94–90 Eclipse: Front	—	—	—	.080	—	.882
Rear	—	—	—	.080	—	.331
94 Galant	.040	8.976	9.078	.080	—	.882
w/rear disc brakes Rear				.080	—	.331
93–89 Galant: w/rear drum brakes	.040	8.000	8.100	.080	—	.882
w/rear disc brakes	—	—	—	.080	—	.331
90–88 Galant Sigma: Front	—	—	—	.079	—	.882
Rear	—	—	—	.079	—	.646
87 Galant: w/rear drum brakes	.040	8.000	8.050	.040	—	.880
w/rear disc brakes	—	—	—	.040	—	.330
86–85 Galant: w/rear drum brakes	.040	8.000	8.050	.040	—	.650
w/rear disc brakes	—	—	—	.040	—	.330
94–93 Mirage: w/solid rotor	.040	7.100	7.200	.080	—	.449
w/vented rotor	.040	8.000	8.100	.080	—	.646
w/rear disc brakes	—	—	—	.080	—	.331
92–91 Mirage: w/solid rotor	.040	7.100	7.200	.080	—	.449
w/vented rotor	.040	7.100	7.200	.080	—	.646
w/rear disc brakes Front	—	—	—	.080	—	.882
Rear	—	—	—	.080	—	.331
90 Mirage: exc. disc brakes	.040	7.100	7.200	.080	—	.449
Front w/rear disc brakes	—	—	—	.080	—	.882
Rear	—	—	—	.080	—	.331
89 Mirage: exc. rear disc brakes	.040	7.100	7.200	.080	—	.449
Front w/rear disc brakes	—	—	—	.080	—	.882
Rear	—	—	—	.080	—	.331
88–85 Mirage: exc. Turbo	.040	7.100	7.150	.040④	—	.450
w/Turbo	.040	7.100	7.150	.040	—	.650
94–90 Precis:	.040	7.087	7.165	.040	—	.670
89–88 Precis: w/solid rotor	.040	7.100	7.200	.040	—	.450
w/vented rotor	.040	7.100	7.200	.040	—	.670
89–83 Starion: Front	—	—	—	.040④	—	.880
Rear	—	—	—	.040④	—	.650

④ 1989–88 .080''.

NISSAN (DATSUN)

Year	Model						
88–87	200SX: Front w/4 cyl. eng.	—	—	—	.079	—	.630
	Front w/6 cyl. eng.	—	—	—	.079	—	.787
	Rear	—	—	—	.079	—	.354
86–84	200SX: w/rear drum brakes	.059	9.000	9.055	.079	—	.630
	w/rear disc brakes	—	—	—	.079	—	.354
83	200SX: Front	—	—	—	.079	—	.413
	Rear	—	—	—	.079	—	.354
82	200SX: Front	—	—	—	.079	—	.413
	Rear	—	—	—	.079	—	.339
81	200SX: Front	—	—	—	.079	—	.413
	Rear	—	—	—	.063	—	.339
80	200SX: Front	—	—	—	.079	—	.413
	Rear	—	—	—	.079	—	.339
79	200SX:	.059	9.000	9.055	.060	—	.331
78–77	200SX:	.059	9.000	9.055	.063	—	.331
82–79	210:	.059	8.000	8.050	.063	—	.331
94–89	240SX: exc. ABS Front	—	—	—	.079	—	.709
	w/ABS Front	—	—	—	.079	—	.787
	Rear	—	—	—	.079	—	.315
83–82	280ZX: Front	—	—	—	.080	—	.709
	Rear	—	—	—	.080	—	.339
81–79	280ZX: Front	—	—	—	.080	—	.709
	Rear	—	—	—	.080	—	.339
78–72	280Z, 260Z, 240Z:	.059	9.000	9.055	.080	.423	.413
94–91	300ZX: Front	—	—	—	.079	—	1.102
	Rear	—	—	—	.079	—	.630
90	300ZX: Front	—	—	—	.079	—	.945
	Rear	—	—	—	.079	—	.630
89–87	300ZX: Front w/Turbo	—	—	—	.079	—	.945
	Front exc. Turbo	—	—	—	.079	—	.787
	Rear	—	—	—	.079	—	.709
86–84	300ZX: Front	—	—	—	.080	—	.787
	Rear	—	—	—	.080	—	.354
82–79	310:	.059	8.000	8.050	.079	—	.339
81–78	510:	.059	9.000	9.055	.080	—	.331
73–72	510:	.059	9.000	9.055	.040	—	.331
76–75	610:	.059	9.000	9.055	.063	—	.331
74–73	610:	.059	9.000	9.055	.040	—	.331
77–76	710:	.059	9.000	9.055	.063	—	.331
75–73	710:	.059	9.000	9.055	.063	.341	.331
88–85	810 Maxima: Front	—	—	—	.079	—	.787
	Rear	—	—	—	.079	—	.354
84–83	810 Maxima:	.059	9.000	9.055	.079	—	.630
	w/rear disc brakes	—	—	—	.079	—	.354
82	810 Maxima:	.059	9.000	9.055	.079	—	.630
	w/rear disc brakes	—	—	—	.079	—	.339
81	810:	.059	9.000	9.055	.079	—	.630
	w/rear disc brakes	—	—	—	.079	—	.339
80–77	810:	.059	9.000	9.055	.080	—	.413
73–72	1200:	.059	8.000	8.051	.063	—	.331
94–93	Altima:	.059	9.000	9.060	.079	—	.787
	w/rear disc brakes	—	—	—	.059	—	.315
78–74	B210:	.059	8.000	8.050	.063	—	.331
78–76	F10:	.039	8.000	8.051	.063	—	.339
94–92	Maxima:	.059	9.000	9.060	.079	—	.787
	w/rear disc brakes	—	—	—	.079	—	.315
91–90	Maxima:	.059	9.000	9.060	.079	—	.787
	w/rear disc brakes	—	—	—	.079	—	.354
89	Maxima:	.059	9.000	9.060	.079	—	.787
	w/rear disc brakes	—	—	—	.079	—	.315
90–87	Pulsar NX: exc. SE	.059	8.000	8.050	.079	—	.394
	SE	.059	8.000	8.050	.079	—	.630
86	Pulsar NX:	.059	8.000	8.050	.079	—	.433
85	Pulsar NX:	.059	8.000	8.050	.079	—	.394
84–83	Pulsar NX:	.059	7.090	7.130	.080	—	.394
94–91	Sentra, NX Coupe:						
	Front exc. SE	—	—	—	.079	—	.630
	Front SE	—	—	—	.079	—	.945
	w/rear drum brakes	.059	7.090	7.130	—	—	—
	w/rear disc brakes	—	—	—	.079	—	.236
90–88	Sentra: (4x2) exc. Wagon	.059	8.000	8.050	.079	—	.394
	Wagon	.059	8.000	8.050	.079	—	.630
	(4x4)	.059	9.000	9.050	.079	—	.630
87	Sentra: w/gas eng.	.059	8.000	8.050	.079	—	.394
	w/diesel eng.	.059	8.000	8.050	.079	—	.630
86–85	Sentra: w/gas eng. (1986)	.059	8.000	8.050	.079	—	.433
	w/gas eng. (1985)	.059	8.000	8.050	.079	—	.394
	w/diesel eng.	.059	8.000	8.050	.079	—	.630
84–83	Sentra: w/gas eng.	.059	7.090	7.130	.080	—	.394
	w/diesel eng.	.059	8.000	8.050	.080	—	.630

(continues)

NISSAN (DATSUN) (Continued)

92–91 Stanza:	.059	9.000	9.060	.079	—	.787
w/rear disc brakes	—	—	—	.079	—	.354
90 Stanza:	.059	9.000	9.060	.079	—	.787
w/rear disc brakes	—	—	—	.079	—	.354
89–88 Stanza:	.059	9.000	9.060	.079	—	.787
87 Stanza: exc. Wagon	.059	10.240	10.300	.079	—	.787
86–84 Stanza: exc. Wagon	.059	8.000	8.050	.080	—	.630
83–82 Stanza:	.059	8.000	8.050	.080	—	.630
88–86 Stanza Wagon: (4x2)	.059	9.000	9.060	.079	—	.787
(4x4)	.059	10.240	10.300	.079	—	.787

OLDSMOBILE

94–93 98:	.030	8.863	8.880	.030	1.224	1.209
92–91 98:	②	8.860	8.880	.030	1.224	1.209
94–93 Achieva:	.030	7.874	7.899	.030	.751	.736
92 Achieva:	.030	7.879	7.899	.030	.796	.736
91 Calais:	②	7.879	7.899	.030	.786	.736
90–80 Calais, Omega:	②	7.880	7.899	.030	.830	.815
94–88 Ciera, Cruiser: Front H.D.	—	—	—	.030	.972	.957
Front exc. H.D.	—	—	—	.030	.830	.815
Rear	②	8.863	8.877	—	—	—
87–86 Ciera: H.D.	②	8.863	8.877	.030	.972	.957
exc. H.D.	②	8.863	8.877	.030	.830	.815
85 Ciera: H.D.	②	8.863	8.920	.030	.972	.957
exc. H.D.	②	8.863	8.920	.030	.830	.815
84 Ciera: H.D.	②	8.863	8.883	.030	.972	.957
exc. H.D.	②	8.863	8.883	.030	.830	.815
83–82 Ciera:	②	7.879	7.899	.030	.830	.815
91–78 Custom Cruiser, Delta 88, 98: (Delta 88 w/403) (RWD)	②	11.000	11.060	.030	.980	.965
77 Custom Cruiser, Delta 88, 98: (Delta 88 w/403) (RWD)	②	11.000	11.060	.030	.980	.965
94–88 Cutlass: (FWD) Front	—	—	—	.030	.987	.972
Rear	—	—	—	.030	.444	.429
77–76 Cutlass:	②	11.000	11.060	.030	.980	.965
75–72 Cutlass, Omega: exc. Vista Cruiser	②	9.500	9.560	.062	.980	.965
94–93 Delta 88:	.030	8.863	8.880	.030	1.224	1.209
92 Delta 88:	②	8.860	8.880	.030	1.224	1.209
91 Delta 88:	②	8.860	8.880	.030	.972	.957
90–83 Delta 88, 98: (FWD)	②	8.858	8.880	.030	.972	.957
88–82 Delta 88, Cutlass: (RWD)	②	9.500	9.560	.030	.980	.965
81–78 Delta 88, Cutlass: (RWD) (Delta 88 w/o 403 eng.)	②	9.500	9.560	.030	.980	.965
77 Delta 88: (RWD) (w/o 403 eng.)	②	9.500	9.560	.030	.980	.965
76–72 Delta 88, 98: exc. Wagon & H.D. Pkg.	②	11.000	11.060	.062	1.230	1.215
76–72 Delta 88: Wagon & H.D. Pkg.	②	12.000	12.060	.062	1.230	1.215
88–82 Firenza: vented rotor (88-82)	②	7.880	7.899	.030	.830	.815
solid rotor (82)	②	7.880	7.899	.030	.444	.429
79–78 Omega: w/5 Speed	②	11.000	11.060	.030	.980	.965
w/o 5 Speed	②	9.500	9.560	.030	.980	.965
77–76 Omega: w/5 Speed	②	11.000	11.060	.030	.980	.965
w/o 5 Speed	②	9.500	9.560	.030	.980	.965
80–76 Starfire:	②	9.500	9.560	.062	.830	.815
75 Starfire:	②	9.000	9.060	.062	.455	.440
92 Toronado, Trofeo: Front	—	—	—	.030	1.250	1.209
Rear	—	—	—	.030	.423	.374
91–86 Toronado, Trofeo: Front	—	—	—	.030	.971	.956
Rear	—	—	—	.030	.444	.429
85–79 Toronado:	②	9.500	9.560	.030	.980	.965
w/rear disc brakes	—	—	—	.030	.980	.965
78–72 Toronado:	②	11.000	11.060	.062	1.185	1.170
75–72 Vista Cruiser:	②	11.000	11.060	.062	.980	.965

② .030″ over rivet head; if bonded lining use .062″.

OPEL

79–76 Coupe, Sedan:	.040	9.000	9.040	.067	—	.339
75 Manta, 1900 Wagon:	.040	9.000	9.040	.067	—	.465
74–72 1900 Coupe, 1900 Sedan, 1900 Wagon, Manta, GT:	.030	9.060	9.090	.067	.404	.394

PEUGEOT

92–89 405: Front	—	—	—	—	—	.728
Rear	—	—	—	—	—	.315
83–80 504: Front	—	—	—	—	.443	.423
w/10 in. drum Rear	—	10.039	10.079	—	—	—
w/11 in. drum Rear	—	11.023	11.063	—	—	—
w/10 mm caliper Rear	—	—	—	—	.354	.335
w/12 mm caliper Rear	—	—	—	—	.433	.413
91–80 505: w/solid rotors	—	—	10.098	—	—	.443
91–85 505: w/vented rotors	—	—	—	—	—	.709
91–80 505: w/rear disc brakes	—	—	—	—	—	.315
84–82 604: Front	—	—	—	—	—	.709
Rear	—	—	—	—	—	.394

PONTIAC

91–89 6000: Front w/H.D.	—	—	—	.030	.972	.957
exc. H.D.				.030	.830	.815
Rear coupe/sedan	④	8.863	8.920	—	—	—
wagon	④	8.863	8.877	—	—	—
w/rear disc brakes	—	—	—	.030	.702	.681
88–87 6000: Front w/H.D.	—	—	—	.030	.972	.957
exc. H.D.				.030	.830	.815
Rear coupe/sedan	④	8.863	8.920	—	—	—
wagon	④	8.863	8.877	—	—	—
w/rear disc brakes	—	—	—	.030	.444	.429
86–85 6000: exc. H.D.	④	8.863	8.920	.030	.830	.815
86 6000: w/H.D.	④	8.863	8.920	.030	.978	.931
85 6000: w/H.D.	④	8.863	8.920	.030	.972	.957
84 6000: w/H.D.	④	8.863	8.883	.030	.972	.957
exc. H.D.	④	8.863	8.883	.030	.830	.815
83–82 6000:	④	7.879	7.899	.030	.830	.815
94–92 Bonneville:	.030	8.860	8.880	.030	1.224	1.209
91–87 Bonneville:	④	8.860	8.880	.030	.972	.957
89–77 Bonneville, Catalina, LeMans, Grand Prix, Grand Am, Parisienne, Safari:						
(RWD) w/9.5" rear brakes	④	9.500	9.560	.030	.980	.965
(RWD) w/11" rear brakes	④	11.000	11.060	.030	.980	.965
76–72 Bonneville, Catalina, Gran Ville: exc. Wagon	.125	11.000	11.060	.125	1.230	1.215
Wagon, Grand Safari	.125	12.000	12.060	.125	1.230	1.215
88 Fiero: Front	—	—	—	.030	.702	.681
Rear	—	—	—	.030	.702	.681
87–84 Fiero: Front	—	—	—	.062	.386	.374
Rear	—	—	—	.062	.440	.430
94–93 Firebird:	.030	9.500	9.560	.030	1.250	1.209
w/rear disc brakes	—	—	—	.030	.733	.724
92–89 Firebird: exc. H.D.	④	9.500	9.560	.030	.980	.965
H.D. Front	—	—	—	.030	.980	.965
H.D. Rear	—	—	—	.030	.744	.724
88–86 Firebird: w/rear drum brakes	④	9.500	9.560	.030	.980	.965
w/rear disc brakes	—	—	—	.030	.986	.956
85–82 Firebird: w/rear drum brakes	④	9.500	9.560	.030	.980	.965
w/rear disc brakes	—	—	—	.030	.980	.965
81–77 Firebird, Ventura, Phoenix:	④	9.500	9.560	.030	.980	.965
w/rear disc (81-79)	—	—	—	.030	.921	.905
88–85 Firefly:	.110	7.090	7.160	.315	—	.315
94–93 Grand Am: (FWD)	④	7.874	7.899	.030	.751	.736
92–91 Grand Am: (FWD)	④	7.879	7.899	.030	.751	.736
90–80 Grand Am, Phoenix: (FWD)	④	7.880	7.899	.030	.830	.815
94–89 Grand Prix: (FWD) Front	—	—	—	.030	.984	.972
Rear	—	—	—	.030	.441	.429
88 Grand Prix: (FWD) Front	—	—	—	.030	1.019	.972
Rear	—	—	—	.030	.441	.429
93–88 LeMans: w/9" solid rotor	.030	7.870	7.900	.030	.420	.380
w/9" vented rotor	.030	7.870	7.900	.030	.669	.646
w/10" vented rotor	.030	7.870	7.900	.030	.870	.830
76–74 LeMans, Firebird, Grand Prix, Ventura: exc. Wagon	.125	9.500	9.560	.125	.980	.965
Wagon	.125	11.000	11.060	.125	.980	.965
73–72 LeMans, Firebird, Grand Prix:	.125	9.500	9.560	.125	.980	.965
94–93 Sunbird	.030	7.874	7.899	.030	.751	.736
92 Sunbird, J2000:	.030	7.880	7.900	.030	.751	.736

④ .030" over rivet head; if bonded lining use .062".

(continues)

PONTIAC (Continued)

91–82	Sunbird, J2000:					
	vented rotor (91-82)	④	7.880	7.899	.030	.830	.815
	solid rotor (82)	④	7.880	7.899	.030	.444	.429
80–76	Sunbird, Astre:	④	9.500	9.560	.030	.830	.815
89–85	Sunburst, Storm: exc. Turbo	.039	7.090	7.140	.039	—	.378
	Turbo	.039	7.090	7.140	.039	—	.650
87–86	T1000:	④	7.880	7.899	.030	.404	.374
85–83	T1000:	④	7.874	7.899	.030	.390	.374
82–81	T1000:	④	7.874	7.899	.030	.390	.374
73–72	Ventura:	.125	9.500	9.560	.125	—	.980

④ .030″ over rivet head; if bonded lining use .062″.

PORSCHE

94–90	911 Carrera 2,4: Front	—	—	—	.079	1.047	1.024
	Rear	—	—	—	.079	.890	.866
94	911 Turbo 3.6: Front	—	—	—	.079	1.200	1.180
	Rear	—	—	—	.079	1.050	1.020
89–84	911: exc. Turbo & Turbo Look						
	Front	—	—	—	.079	.890	.866
	Rear	—	—	—	.079	.890	.866
	Turbo & Turbo Look Front	—	—	—	.079	1.200	1.180
	Rear	—	—	—	.079	1.050	1.020
83–78	911: Front	—	—	—	.079	.750	.728
	Rear	—	—	—	.079	.732	.709
77–72	911: Front	—	—	—	.079	.732	.709
	Rear	—	—	—	.079	.732	.709
76–72	914: Front	—	—	—	.080	.394	.375
	Rear	—	—	—	.080	.351	.335
76–72	914, 916: Front	—	—	—	.080	.732	.709
	Rear	—	—	—	.080	.374	.354
88–86	924S: Front	—	—	—	.079	.751	.728
	Rear	—	—	—	.079	.751	.728
82–80	924: Turbo Front	—	—	—	.079	.751	.728
	Turbo Rear	—	—	—	.079	.755	.732
82–77	924: exc. Turbo	.098	9.055	9.094	.079	.472	.453
94–86	928S, 928S-4, 928GT: Front	—	—	—	.079	1.205	1.181
	Rear	—	—	—	.079	.890	.866
85–82	928, 928S: Front	—	—	—	.079	1.228	1.205
	Rear	—	—	—	.079	.756	.732
81–78	928: Front	—	—	—	.079	.756	.732
	Rear	—	—	—	.079	.756	.732
90	944: Turbo S Front	—	—	—	.079	1.205	1.180
	Turbo S Rear	—	—	—	.079	.890	.866
91–90	944 S2: Front	—	—	—	.079	1.047	1.024
	Rear	—	—	—	.079	.890	.866
90–83	944: exc. Turbo Front	—	—	—	.079	.752	.728
	exc. Turbo Rear	—	—	—	.079	.756	.732
90–86	944: Turbo Front	—	—	—	.079	1.047	1.024
	Turbo Rear	—	—	—	.079	.890	.866
94	968: Front	—	—	—	.079	1.050	1.020
	wsport suspension Front	—	—	—	.079	1.200	1.180
	Rear	—	—	—	.079	.890	.866

RENAULT

86–84	18i Sportwagon:	.188②	9.000	9.040	.359②	—	.709
83–80	18i:	.020	8.996	9.035	.276②	—	.433
87–83	Alliance, Encore:	.020	8.000	8.060	.276②	—	.433
85–80	Fuego: exc. Turbo	.020	8.996	9.035	.276②	—	.433
84–82	Fuego: Turbo	.020	8.996	9.035	.276②	—	.709
80–74	Gordini:	.203②	9.000	9.035	.276②	—	.354
87	GTA:	.020	8.000	8.060	.276②	—	.750
84–76	LeCar: (5 Series)	.203②	7.096	7.136	.275②	—	.354
89–88	Medallion:	.098	9.000	9.030	.256	①	.697

① Machining not recommended.   ② Measurement of lining & metal.   ③ .001″-.002″ end play.

SAAB

94 900, 900 S, 900 Turbo: Front	—	—	—	.200	.890	.870
Rear	—	—	—	.200	.330	.310
93–88 900, 900 S, 900 Turbo: Front	—	—	—	.040	.870	.850
Rear	—	—	—	.040	.320	.300
87–86 900: Turbo 16 Valve Front	—	—	—	.040	.744	.709
Rear	—	—	—	.040	.374	—
87–79 900: exc. Turbo 16 Valve						
Front	—	—	—	.040	.461	—
Rear	—	—	—	.040	.374	—
94–88 9000 Turbo: Front	—	—	—	.040	.950	.910
Rear	—	—	—	.040	.325	.300
87 9000 Turbo: Front	—	—	—	.040	.890	.850
Rear	—	—	—	.040	.325	.300
94–90 9000 S, 9000 CS, 9000 CD:						
Front	—	—	—	.040	.950	.910
Rear	—	—	—	.040	.325	.300
89–87 9000 S: Front	—	—	—	.040	.890	.850
Rear	—	—	—	.040	.325	.300
87–86 9000: Front	—	—	—	.039	.787	.768
Rear	—	—	—	.039	.295	.276
78–75 99: Front	—	—	—	.080	.461	—
Rear	—	—	—	.080	.374	—
74–70 99: Front	—	—	—	.080	.374	—
Rear	—	—	—	.080	.374	—

SATURN

94–91 SC, SC1, SC2, SL, SL1, SL2, SW1, SW2: w/rear drum brakes	—	7.870	7.900	—	.633	.625
w/rear disc brakes	—	—	—	—	.370	.350

STERLING

91–87 825: Front	—	—	—	.322①	—	.748
Rear	—	—	—	.283①	—	.314

① Measurement of lining & metal.

SUBARU

86–85 DL, GL:	.060	7.090	7.170	.295②	—	.630
w/rear disc brakes	—	—	—	.256②	—	.335
84–83 DL, GL: w/solid rotor	.060	7.090	7.120	.295②	—	.394
w/vented rotor	.060	7.090	7.120	.295②	—	.610
82–80 DL, GL:	.060	7.090	7.120	.295②	—	.394
79–75 DL, GL:	.060	7.090	7.120	.060	—	.330
74–72 DL, GL Coupe:	.060	7.090	7.120	.060	—	.330
74–72 DL, GL Sedan, Wagon: Front	.060	9.000	9.040	—	—	—
Rear	.060	7.090	7.120	—	—	—
94–93 Impreza: w/13" rotor Front	—	—	—	.295	—	.630
w/14" rotor Front	—	—	—	.259	—	.870
wrear drum brakes Rear	.059	9.000	9.079	—	—	—
wrear disc brakes Rear	—	—	—	.256	—	.335
94–89 Justy: w/12" Wheels	.067	7.090	7.170	.295②	—	.610
w/13" Wheels	.067	7.090	7.170	.315②	—	.610
88–87 Justy:	.067	7.090	7.170	.295②	—	.610
94–90 Legacy: Front	—	—	—	.295②	—	.870
Rear	—	—	—	.256②	—	.335
94–87 Loyale:	.060	7.090	7.170	.295②	—	.630
w/rear disc brakes	—	—	—	.256②	—	.335
94–92 SVX: Front	—	—	—	.295②	—	1.020
Rear	—	—	—	.256②	—	.335
91–85 XT:	.060	7.090	7.170	.295②	—	.630
w/rear disc brakes	—	—	—	.256②	—	.335
91–88 XT6: Front	—	—	—	.295②	—	.787
Rear	—	—	—	.315②	—	.335

② Measurement of lining & metal.

SUZUKI

94–93 Swift GA, Swift GS: Hatchback	.110①	7.090	7.160	.315①	—	.590
Sedan	.110①	7.870	7.950	.315①	—	.590
92–89 Swift GLX:	.110①	7.090	7.160	.315①	—	.590
94–89 Swift GT, Swift GTI: Front	—	—	—	.315①	—	.650
Rear	—	—	—	.236①	—	.315

① Measurement of lining & metal.

TOYOTA

94–92 Camry: w/rear drum brakes	.039	9.000	9.079	.039	—	1.024
w/rear disc brakes	—	—	—	.039	—	.354
91–88 Camry: w/rear drum brakes	.040	9.000	9.079	.040	—	.945
w/rear disc brakes	—	—	—	.040	—	.354
87 Camry:	.040	9.000	9.079	.040	—	.827
86–83 Camry:	.040	7.874	7.913	.040	—	.827
93–92 Celica: w/rear drum brakes	.039	7.874	7.913	.039	—	.906
w/rear disc brakes	—	—	—	.039	—	.354
91–90 Celica: w/rear drum brakes	.040	7.874	7.913	.040	—	.787
w/rear disc brakes	—	—	—	.040	—	.354
89–88 Celica: (4x2) exc. ABS	.040	7.874	7.913	.040	—	.827
(4x2)w/ABS Front	—	—	—	.040	—	.945
(4x2)w/ABS Rear	—	—	—	.040	—	.354
(4x4) Front	—	—	—	.040	—	.945
(4x4) Rear	—	—	—	.040	—	.354
87–86 Celica: w/rear drum brakes	.040	7.874	7.913	.040	—	.827
w/rear disc brakes	—	—	—	.040	—	.354
85–82 Celica: w/rear drum brakes	.040	9.000	9.040	.118	—	.750
w/rear disc brakes	—	—	—	.118	—	.670
81 Celica:	.040	9.000	9.040	.118	—	.450
80–77 Celica:	.040	9.000	9.040	.040	—	.450
76 Celica:	.040	9.000	9.040	.040	—	.350
75–72 Celica:	.040	9.000	9.040	.040	—	.350
94–93 Corolla:	.039	7.874	7.913	.039	—	.787
92 Corolla:	.039	7.874	7.913	.039	—	.669
91–90 Corolla: Front SOHC eng.	.040	7.874	7.913	.040	—	.669
Front DOHC eng.	.040	7.874	7.913	.040	—	.827
Rear	—	—	—	.040	—	.315
89–88 Corolla FX, Corolla FX16:						
w/rear drum brakes	.040	7.874	7.913	.040	—	.665⑤
w/rear disc brakes	—	—	—	.040	—	.315
87 Corolla FX16: Front	—	—	—	.040	—	.669
Rear	—	—	—	.040	—	.315
87–84 Corolla: exc. Coupe, FX16	.040	7.874	7.913	.040	—	.492
Coupe w/front disc brakes	.040	9.000	9.079	.040	—	.669
Coupe w/rear disc brakes	—	—	—	.040	—	.354
83–80 Corolla:	.040	9.000	9.040	.040	—	.453
79–75 Corolla 1200:	.040	7.874	7.914	.040	—	.350
74–72 Corolla 1200:	.040	7.874	7.914	.040	.360	.350
79–75 Corolla 1600:	.040	9.000	9.040	.040	—	.350
74–72 Corolla 1600:	.040	9.000	9.040	.040	—	.350
82–75 Corona:	.040	9.000	9.040	.040	—	.450
74 Corona: Deluxe	.040	9.000	9.040	.040	—	.450
Standard	.040	9.000	9.040	.040	—	.350
73–72 Corona:	.040	9.010	9.050	.040	—	.350
76–73 Corona Mark II:	.040	9.000	9.040	.040	—	.450
73–72 Corona Mark II:	.040	9.000	9.040	.040	—	.450
92–89 Cressida: Front	—	—	—	.040	—	.827
Rear	—	—	—	.040	—	.669
88–85 Cressida: w/rear drum brakes	.040	9.000	9.040	.040	—	.830
w/rear disc brakes	—	—	—	.040	—	.670
84–83 Cressida: w/rear drum brakes	.040	9.000	9.040	.040	—	.830
w/rear disc brakes	—	—	—	.040	—	.670
82–81 Cressida:	.040	9.000	9.040	.040	—	.669
80–78 Cressida:	.040	9.000	9.040	.040	—	.450
93–91 MR2: Front	—	—	—	.040	—	.945
Rear	—	—	—	.040	—	.591
89–87 MR2: Front	—	—	—	.118	—	.827
Rear	—	—	—	.040	—	.354
86–85 MR2: Front	—	—	—	.040	—	.670
Rear	—	—	—	.040	—	.354
94–92 Paseo:	.039	7.087	7.126	.039	—	.669
84–81 Starlet	.040	7.870	7.910	.040	—	.354
94–93 Supra: Front Non-turbo	—	—	—	.040	—	1.102
Front Turbo	—	—	—	.040	—	1.181
Rear	—	—	—	.040	—	.591
92–86 Supra: Front	—	—	—	.040	—	.827
Rear	—	—	—	.040	—	.669
86–82 Supra: Front	—	—	—	.118	—	.750
Rear	—	—	—	.118	—	.669
81–79 Supra: Front	—	—	—	.118	—	.450
Rear	—	—	—	.040	—	.354
94–91 Tercel:	.040	7.087	7.126	.040	—	.669
90–85 Tercel: Sedan	.040	7.087	7.126	.040	—	.394
Wagon	.040	7.874	7.913	.040	—	.394
84–83 Tercel: (4x2)	.040	7.087	7.126	.040	—	.394
(4x4)	.040	7.874	7.913	.040	—	.394
82–80 Tercel:	.040	7.087	—	.040	—	.354

⑤ FX, FX16 DOHC eng. shown; SOHC eng. .439″.

VOLKSWAGEN

Year/Model						
74–72 411, 412: Sedan	.100	9.768	9.803	.080	—	.393
78–72 Beetle: Front	.100	9.059	9.068	—	—	—
Rear	.100	9.055	9.094	—	—	—
92–87 Cabriolet	.098	7.087	7.106	.276④	—	.709
92–90 Corrado: Front	—	—	—	.276④	—	.787
Rear	—	—	—	.276④	—	.315
81–78 Dasher:	②	7.850	7.900	.080	.413	.393
77 Dasher:	②	7.850	7.900	.080	.413	.393
76–74 Dasher:	②	7.850	7.900	.080	.433	.413
91–87 Fox: exc. Wagon	.098	7.087	7.106	.276④	—	.393
Wagon	.098	7.874	7.894	.276④	—	.393
92–85 Golf, GTI, Jetta: exc. ABS						
w/solid front rotor	.098	7.087	7.106	.276④	—	.393
exc. ABS						
w/vented front rotor	.098	7.087	7.106	.276④	—	.709
w/rear disc brakes	—	—	—	.276④	—	.315
92–89 Jetta: w/ABS Front	—	—	—	.276④	—	.709
w/ABS Rear	—	—	—	.276④	—	.315
74–73 Karman Ghia:	.100	9.055	9.094	.080	—	.335
72 Karman Ghia:	.100	9.055	9.094	.080	—	.335
94–90 Passat: Front	—	—	—	.276④	—	.709
Rear	—	—	—	.276④	—	.315
88–82 Quantum: w/4 cyl. eng.	.098	7.874	7.894	.276④	.413	.393
w/5 cyl. eng.	.098	7.874	7.894	.276④	—	.709
88–85 Quantum Syncro: Front	—	—	—	.276④	—	.709
Rear	—	—	—	.276④	—	.315
86–85 Rabbit: Conv.	.098	7.086	7.150	.250④	.413	.393
84–83 Rabbit GTI:	.098	7.086	7.150	.375④	.728	.709
84–82 Rabbit, Jetta, Scirocco:	.098	7.086	7.105	.250④	.413	.393
81–80 Rabbit, Jetta, Scirocco:						
w/Girling Caliper	②	7.086	7.105	.080	.413	.393
w/Kelsey-Hayes Caliper	②	7.086	7.105	.080	.413	.393
79–77 Rabbit: w/drum brakes Front	.040	9.059	9.079	—	—	—
w/drum brakes Rear	②	7.086	7.105	—	—	—
79–75 Rabbit, Scirocco:						
w/Girling Caliper	②	7.086	7.105	.080	.452	.413
w/ATE Caliper	②	7.086	7.105	.080	.452	.413
89–86 Scirocco: Front exc. 16 V eng.	.098	7.087	7.106	.276④	—	.709
Front w/16 V eng..............	—	—	—	.276④	—	.709
Rear	—	—	—	.276④	-.	.315
85 Scirocco:	.098	7.086	7.150	.250④	.413	.393
73–72 Squareback, Fastback:	.100	9.768	9.803	.080	—	.393
79–72 Super Beetle: Front	.100	9.768	9.803	—	—	—
Rear	.100	9.055	9.094	—	—	—

② .098″ riveted; .059″ bonded.   ④ Measurement of lining & metal.

VOLVO

Year/Model						
74–72 140: Front	—	—	—	.062	.457	—
Rear w/Girling..................	—	—	—	.062	.331	—
Rear w/ATE	—	—	—	.062	.331	—
75–72 160: Front	—	—	—	.062	.900	—
Rear w/Girling..................	—	—	—	.062	.331	—
Rear w/ATE	—	—	—	.062	.331	—
73–72 1800: Sport Coupe Front	—	—	—	.062	—	.520
Rear	—	—	—	.062	—	.331
93–88 240: Front Solid Disc	—	—	—	.120	.500	—
Front Vented: ATE.............	—	—	—	.120	.898	—
Front Girling	—	—	—	.120	.803	—
Rear	—	—	—	.120	.330	—
87–75 240: Front w/ATE: vented disc	—	—	—	.062	.897	—
Front w/Girling: vented disc	—	—	—	.062	.818	—
Front Solid disc	—	—	—	.062	.519	—
Rear w/ATE	—	—	—	.062	.331	—
Rear w/Girling.................	—	—	—	.062	.331	—
82–76 260:						
Front w/vented disc brakes	—	—	—	.062	.818	—
Front w/solid disc brakes........	—	—	—	.062	.519	—
Rear w/ATE	—	—	—	.062	.331	—
Rear w/Girling..................	—	—	—	.062	.331	—
92–88 740: Front Solid disc............	—	—	—	.120	—	.433
Front Vented disc	—	—	—	.120	—	.788
Front Vented H.D.	—	—	—	.120	—	.910
Rear Standard	—	—	—	.078	—	.393
Rear w/Multi-Link	—	—	—	.078	—	.314

(continues)

VOLVO (Continued)

87–82 740, 760, 780:						
Front w/solid disc brakes	—	—	—	.118	—	.433
Front w/vented disc brakes	—	—	—	.118	—	.788
Rear	—	—	—	.078	—	.330
91–88 760, 780: Front	—	—	—	.120	—	.788
Rear	—	—	—	.078	—	.314
94–93 850: Front	—	—	—	.120	.937	.905
Rear	—	—	—	.120	.350	.330
94–91 940, 960:						
Front w/solid disc brakes	—	—	—	.120	—	.433
Front w/vented disc brakes	—	—	—	.120	—	.788
Front w/H.D. vented disc	—	—	—	.120	—	.910
Rear w/Multi-Link	—	—	—	.078	—	.314
Rear Standard	—	—	—	.078	—	.330

YUGO

92–86 Cabrio, GV, GVL, GVX:	.059	7.293	7.336	.059	.368	.354

LIGHT TRUCKS

CHRYSLER CORP. TRUCK — Chrysler, Dodge, Plymouth

82–79 Arrow Pickup:	.040	9.500	9.550	.040	—	.720
85–76 B100, B150, PB100, PB150:						
w/10" rear brakes	.030	10.000	10.060	.030	1.210	1.180
w/11" rear brakes	.030	11.000	11.060	.030	1.210	1.180
75–73 B100, PB100:						
w/10" rear brakes	.030	10.000	10.060	.030	1.210	1.180
w/11" rear brakes	.030	11.000	11.060	.030	1.210	1.180
72 B100: w/10" rear brakes	.030	10.000	10.060	.030	1.210	1.180
w/11" rear brakes	.030	11.000	11.060	.030	1.210	1.180
94–86 B150, B250:	.030	11.000	11.060	.125[9]	1.210	1.180
83–78 B200, B250, PB200, PB250:	.030	10.000	10.060	.030	1.210	1.180
77–76 B200, PB200:	.030	11.000	11.060	.030	1.210	1.180
75–73 B200, PB200:	.030	11.000	11.060	.030	1.210	1.180
72 B200: w/10" rear brakes	.030	10.000	10.060	.030	1.210	1.180
w/11" rear brakes	.030	11.000	11.060	.030	1.210	1.180
85–84 B250, PB250:	.030	11.000	11.060	.030	1.210	1.180
85–79 B300, B350, CB300, CB350, PB300, PB350: w/3,600lb. F.A.	.030	12.000	12.060	.030	1.210	1.180
w/4,000lb. F.A.	.030	12.000	12.060	.030	1.160	1.130
78–76 B300, CB300, PB300:	.030	12.000	12.060	.030	1.160	1.130
75–73 B300, CB300, PB300:	.030	12.000	12.060	.030	1.160	1.130
72 B300, CB300:	.030	12.000	12.060	.030	1.160	1.130
94–87 B350: w/3,600lb. F.A.	.030	12.000	12.060	.125[9]	1.210	1.180
w/4,000lb. F.A.	.030	12.000	12.060	.125[9]	1.155	1.125
86 B350: w/3,600lb. F.A.	.030	12.000	12.060	.125[9]	1.210	1.180
w/4,000lb. F.A.	.030	12.000	12.060	.125[9]	1.155	1.125
94–93 Caravan, Voyager: (FWD)	.030	9.000	9.060	.030	.912	.882
: (AWD)	.030	11.000	11.060	.030	.912	.882
92–91 Caravan, Voyager: (FWD)	.090	9.000	9.060	.090	.833	.803
(AWD)	.090	11.000	11.060	.090	.833	.803
90–84 Caravan, Mini Ram Van, Voyager:	.030	9.000	9.060	.030[7]	.833	.803
83–75 D100, D150:						
exc. w/9¼" rear axle	.030	10.000	10.060	.030	1.220	1.190
76–75 D100: w/9¼" rear axle	.030	11.000	11.060	.030	1.220	1.190
74–73 D100: exc. w/11" front brakes	.030	10.000	10.060	.030	1.220	1.190
73–72 D100: w/11" front brakes	.030	11.000	11.060	—	—	—
72 D100:	.030	10.000	10.060	.030	—	1.180
94–90 D150:	.030	11.000	11.060	.030	1.220	1.190
89–84 D150:	.030	11.000	11.060	.030	1.220	1.190
94–90 D200, D250: w/3,300lb. F.A.	.030	12.000	12.060	.030	1.220	1.190
94–90 D200, D250, D300, D350: W/4,000lb. F.A.	.030	12.000	12.060	.030	1.220	1.190
89–81 D200, D250: w/3,300lb. F.A.	.030	12.000	12.060	.030	1.220	1.190
89–79 D200, D250, D300, D350: w/4,000lb. F.A.	.030	12.000	12.060	.030	1.160	1.130
80–79 D200:						
w/3,300lb.F.A.over 6,200lbs.GVW	.030	12.000	12.060	.030	1.220	1.190
w/3,300lb. F.A. 6,200lbs. GVW	.030	12.120	12.180	.030	1.220	1.190
78–75 D200: w/6600lbs. GVW	.030	12.120	12.180	.030	1.160	1.130
78–75 D200, D300: over 6,000lbs. GVW	.030	12.000	12.060	.030	1.160	1.130
74 D200: w/6,000lbs. GVW	.030	12.120	12.180	.030	1.160	1.130
74 D200, D300: over 6,000lbs. GVW	.030	12.000	12.060	.030	1.160	1.130
73 D200: w/6,000lbs. GVW	.030	12.120	12.180	.030	—	1.125
w/12.57" rotor diameter	.030	12.000	12.060	.030	—	1.180
w/12.82" rotor diameter	.030	12.000	12.060	.030	—	1.125

[7] .030" over rivet head; if bonded lining use .062". [9] Combined shoe & lining thickness .3125".

CHRYSLER CORP. TRUCK — Chrysler, Dodge, Plymouth (Continued)

Year	Application						
72	D200: w/6,000lbs. GVW	.030	12.120	12.180	.030	—	1.125
	w/12.57" rotor diameter	.030	12.000	12.060	.030	—	1.180
	w/12.82" rotor diameter	.030	12.000	12.060	.030	—	1.125
73–72	D300:	.030	12.000	12.060	—	—	
94–87	Dakota:						
	(4x2) w/9" rear brakes	.030	9.000	9.060	.250	.841	.811
	(4x2) w/10" rear brakes	.030	10.000	10.060	.250	.841	.811
	(4x4) w/9" rear brakes	.030	9.000	9.060	.250	.841	.811
	(4x4) w/10" rear brakes	.030	10.000	10.060	.250	.841	.811
89–88	Raider:	.040	10.000	10.079	.079	—	.803
87	Raider:	.040	10.000	10.079	.040	—	.724
93–87	Ram 50: (4x2)	.040	10.000	10.079	.080	—	.803
	(4X4)	.040	10.000	10.079	.080	—	.803
86–83	Ram 50: w/9½" rear brakes	.040	9.500	9.570	.040	—	.720
	W/10" rear brakes	.040	10.000	10.070	.040	—	.720
82–79	D 50, Ram 50:	.040	9.500	9.550	.040	—	.720
93–90	Ramcharger: (4x2)	.030	11.000	11.060	.030	1.220	1.190
	(4x4)	.030	11.000	11.060	.030	1.220	1.190
89–84	Ramcharger: (4x2)	.030	11.000	11.060	.030	1.220	1.190
	(4x4)	.030	11.000	11.060	.030	1.220	1.190
83–79	Ramcharger: (4x2)	.030	10.000	10.060	.030	1.220	1.190
	(4x4)	.030	10.000	10.060	.030	1.220	1.190
78–74	Ramcharger: (4x2)	.030	11.000	11.060	.030	1.220	1.190
	(4x4)	.030	11.000	11.060	.030	1.220	1.190
84–83	Rampage, Scamp:	.030	7.874	7.904	.030	.461	.431
82	Rampage:	.030	7.780	7.900	.030	.461	.431
94–93	Town & Country Van: (FWD)	.030	9.000	9.060	.030	—	.882
	(AWD)	.030	11.000	11.060	.030	—	.882
92–90	Town & Country Van:	.030	9.000	9.060	.030⑦	.833	.803
83–79	W100, W150:	.030	10.000	10.060	.030	1.220	1.190
78–77	W100:	.030	11.000	11.060	.030	1.220	1.190
76–75	W100:	.030	11.000	11.060	.030	1.220	1.190
74–72	W100:	.030	11.000	11.060	—	—	—
94–90	W150:	.030	11.000	11.060	.030	1.220	1.190
89–84	W150:	.030	11.000	11.060	.030	1.220	1.190
94–90	W200, W250: exc. w/Spicer 60	.030	12.000	12.060	.030	1.220	1.190
94–79	W200, W250, W300, W350:						
	w/Spicer 60	.030	12.000	12.060	.030	1.160	1.130
89–79	W200, W250: exc. w/Spicer 60	.030	12.000	12.060	.030	1.160	1.130
78–75	W200: exc. w/Spicer 60	.030	12.000	12.060	.030	1.160	1.130
78–75	W200, W300: w/Spicer 60	.030	12.000	12.060	.030	1.160	1.130
74–72	W200: Front	.030	12.120	12.180	—	—	—
	Rear	.030	12.000	12.060	—	—	—
74–72	W200, W300: w/4,500lb. F.A.	.030	12.000	12.060	—	—	—

⑦ .030" over rivet head; if bonded lining use .062".

DAIHATSU TRUCK

Year	Application						
92–90	Rocky:	.039	10.000	10.060	.060	—	.450

FORD MOTOR CO. TRUCK

Year	Application						
94–86	Aerostar: w/9" rear brakes	.030	9.000	9.060	.030	—	.810
	w/10" rear brakes	.030	10.000	10.060	.030	—	.810
94	Bronco: exc. integral rotor	.030	11.031	11.091	.030	—	.960
	w/integral rotor	.030	11.031	11.091	.030	—	.960
93–90	Bronco: w/integral rotor	.030	11.031	11.091	.030	—	1.120
	exc. integral rotor	.030	11.031	11.091	.030	—	1.120
89–86	Bronco: w/integral rotor	.030	11.031	11.091	.030	—	1.120
	exc. integral rotor	.030	11.031	11.091	.030	—	1.120
85–81	Bronco:	.030	11.031	11.091	.030	—	1.120
80–76	Bronco:	.030	11.031	11.091	.030	—	1.120
75–72	Bronco: w/10" brakes	.030	10.000	10.060	—	—	—
	w/11" brakes	.030	11.000	11.060	—	—	—
82–79	Courier:	.039	10.236	10.244	.276⑭	.433	—
78–77	Courier:	.039	10.236	10.244	.314⑭	.433	—
76–74	Courier:	.039	10.236	10.244	—	—	—
73–72	Courier:	.039	10.236	10.244	—	—	—
94–90	E150: (4x2)	.030	11.031	11.091	.030	—	1.120
89–86	E150, F150: (4x2)	.030	11.031	11.091	.030	—	1.120
85–77	E100, E150, F100, F150:						
	w/4,600-4,900lbs. GVW	.030	10.000	10.060	.030	—	.810
		.030	11.031	11.091	.030	—	1.120
76–75	E100, E150, F100, F150:	.030	11.031	11.091	.030	—	1.120
74–72	E100, F100: w/10" brakes	.030	10.000	10.060	—	—	—
	w/11" brakes	.030	11.031	11.091	.030	—	1.120
94–90	E250, F250, F350:						
	(4x2) w/single rear wheels	.030	12.000	12.060	.030	—	1.180
	(4x2) w/dual rear wheels	.030	12.000	12.060	.030	—	1.180
89–86	E250, F250, F350:						
	(4x2) w/single rear wheels	.030	12.000	12.060	.030	—	1.180
	(4x2) w/dual rear wheels	.030	12.000	12.060	.030	—	1.180

⑭ Measurement of lining & metal.

FORD MOTOR CO. TRUCK (Continued)

85–78 E250, E350, F250, F350: 6,900lbs. GVW H.D.	.030	12.000	12.060	.030	—	1.180
77 E250, E350, F250, F350: 6,900lbs. GVW H.D.	.030	12.000	12.060	.030	—	1.214
76 E250, E350, F250, F350:						
w/12" rear brakes	.030	12.000	12.060	.030	—	1.180
w/12 1/8" rear brakes	.030	12.125	12.185	.030	—	1.180
75 E250, E350:	.030	12.000	12.060	.030	—	.940
74–72 E250:	.030	11.031	11.091	—	—	—
74–72 E350:	.030	12.000	12.060	—	—	—
94–90 F-Super Duty: Front	—	—	—	.030	—	1.430
Rear	—	—	—	.030	—	1.430
89–88 F-Super Duty: Front	—	—	—	.030	—	1.430
Rear	—	—	—	.030	—	1.430
94 F150: (4X2)	.030	11.031	11.091	.030	—	.960
(4x4) exc. integral rotor	.030	11.031	11.091	.030	—	.960
(4x4) w/integral rotor	.030	11.031	11.091	.030	—	.960
93–90 F150: (4x4) w/integral rotor	.030	11.031	11.091	.030	—	1.120
(4x4) exc. integral rotor	.030	11.031	11.091	.030	—	1.120
89–86 F150: (4x4) w/integral rotor	.030	11.031	11.091	.030	—	1.120
(4x4) exc. integral rotor	.030	11.031	11.091	.030	—	1.120
85–81 F100, F150, F200:						
(4x4) w/Dana 44IFS F.A.	.030	11.031	11.091	.030	—	1.120
80–76 F100, F150: (4x4)	.030	11.031	11.091	.030	—	1.120
75–72 F100: (4x4) Front	.030	11.000	11.060	—	—	—
(4x4) Rear	.030	11.031	11.091	—	—	—
94–86 F250, F350:						
(4X4) w/Dana 44IFS axle	.030	12.000	12.060	.030	—	1.180
(4x4) exc. Dana 44IFS axle	.030	12.000	12.060	.030	—	1.180
85–83 F250, F350:						
(4X4) exc.Dana 44IFS F.A.	.030	12.000	12.060	.030	—	1.180
85–77 F250: 6,900lbs. GVW std.	.030	12.000	12.060	.030	—	1.120
82–81 F250, F350: (4X4)	.030	12.000	12.060	.030	—	1.180
(4x4)	.030	12.000	12.060	.030	—	1.180
80–76 F250, F350: (4x4)	.030	12.000	12.060	.030	—	1.180
76–75 F250: 6,900lbs. GVW						
Std. w/12" rear brakes	.030	12.000	12.060	.030	—	1.120
w/12 1/8" rear brakes	.030	12.125	12.185	.030	—	1.120
75–72 F250, F350: w/12" rear brakes	.030	12.000	12.060	.030	—	.940
w/12 1/8" rear brakes	.030	12.125	12.185	.030	—	.940
75–72 F250: (4x4) Front	.030	12.125	12.185	—	—	—
(4x4) Rear	.030	12.000	12.060	—	—	—
74–73 F250: 6,900lbs. GVW						
Std. w/12" rear brakes	.030	12.000	12.060	.030	—	1.120
w/12 1/8" rear brakes	.030	12.125	12.185	.030	—	1.120
72 F250: std. w/12" brakes	.030	12.000	12.060	—	—	—
w/12 1/8" brakes	.030	12.125	12.185	—	—	—
75–72 F350: (4X4)	.030	12.000	12.060	.030	—	.940
94–91 Ranger, Explorer:						
(4x2) w/9" rear brakes	.030	9.000	9.060	.030	—	.810
(4x2) w/10" rear brakes	.030	10.000	10.060	.030	—	.810
(4x4) w/9" rear brakes	.030	9.000	9.060	.030	—	.810
(4x4) w/10" rear brakes	.030	10.000	10.060	.030	—	.810
90–83 Ranger, Bronco II:						
(4x2) w/9" rear brakes	.030	9.000	9.060	.030	—	.810
(4x2) w/10" rear brakes	.030	10.000	10.060	.030	—	.810
(4x4) w/9" rear brakes	.030	9.000	9.060	.030	—	.810
(4x4) w/10" rear brakes	.030	10.000	10.060	.030	—	.810

GENERAL MOTORS TRUCK — Chevrolet, GMC

94–85 Astro, Safari Van:						
W/1.04" rotor	.030	9.500	9.560	.030	.980	.965
w/1.25" rotor	.030	9.500	9.560	.030	1.230	1.215
94–92 Blazer, Jimmy, Pickup, Sonoma: S/T Series	.030	9.500	9.560	.030	.980	.965
91–88 Blazer, Jimmy, Pickup: (S/T Series)	.062	9.500	9.560	.030	.980	.965
87–83 Blazer, Jimmy, Pickup: (S/T Series)	.062	9.500	9.560	.030	.980	.965
94–92 Blazer, Yukon: C/K Series	.030	10.000	10.051	.030	1.230	1.215
91–88 Blazer, Jimmy: (Full Size)	.030	11.150	11.210	.030	1.230	1.215
87–83 Blazer, Jimmy: (Full Size)	.062	11.150	11.210	.030	1.230	1.215
82–76 Blazer, Jimmy: (4x2)	.062	11.150	11.210	.030	1.230	1.215
(4x4)	.062	11.150	11.210	.030	1.230	1.215
75–74 Blazer, Jimmy: (4x2)	.030	11.000	11.060	.030	1.230	1.215
(4x4) w/11" rear brakes	.030	11.000	11.060	.030	1.230	1.215
(4x4) w/11 1/8" rear brakes	.030	11.150	11.210	.030	1.230	1.215
73–72 Blazer, Jimmy: (4x2)	.030	11.000	11.060	.030	1.230	1.215
(4x4)	.030	11.000	11.060	.030	1.230	1.215

(continues)

GENERAL MOTORS TRUCK — Chevrolet, GMC (Continued)

91–88 C10, C15, K10, K15:						
w/1.0" rotor	.142	10.000	10.051	.030	.980	.965
w/1.25" rotor	.142	10.000	10.051	.030	1.230	1.215
87–76 C10, C15, R10, R15, Suburban R10, Suburban R15:						
w/1.0" rotor						
w/11" rear brakes	.062	11.000	11.060	.030	.980	.965
w/11 ⅛" rear brakes	.062	11.150	11.210	.030	.980	.965
w/1.25" rotor						
w/11" rear brakes	.062	11.000	11.060	.030	1.230	1.215
w/11 ⅛" rear brakes	.062	11.150	11.210	.030	1.230	1.215
75–74 C10, C15, Suburban C10, Suburban C15:						
w/11" rear brakes	.030	11.000	11.060	.030	1.230	1.215
w/11 ⅛" rear brakes	.030	11.050	11.210	.030	1.230	1.215
73–72 C10, C15, Suburban C10, Suburban C15:	.030	11.000	11.060	.030	1.230	1.215
94–92 1500 C/K Pickup: w/1.0" rotor	.030	10.000	10.051	.030	.980	.965
w/1.25" rotor	.030	10.000	10.051	.030	1.230	1.215
94–92 2500 C/K Pickup:	.030	11.150	11.210	.030	1.23	1.215
94–92 3500 C/K Pickup:						
w/1.54" rotor	.030	13.000	13.060	.030	1.480	1.465
w/1.26" rotor	.030	13.000	13.060	.030	1.230	1.215
Chassis Cab F	—	—	—	.030	1.480	1.465
Chassis Cab R	—	—	—	.030	1.480	1.465
94–92 1500 Suburban, 2500 Suburban: w/1.25" rotor	.030	11.150	11.210	.030	1.230	1.215
w/1.26" rotor	.030	13.000	13.060	.030	1.230	1.215
91–88 C20, C25, C30, C35, K20, K25, K30, K35:						
w/11 ⅛" rear brakes	.142	11.150	11.210	.030	1.230	1.215
w/13" rear brakes	.142	13.000	13.060	.030	1.230	1.215
87–76 C20, C25, C30, C35, R20, R25, R30, R35, Suburban R20, Suburban R25:						
under 8,600 lbs. GVW w/11 ⅛" rear brakes...........	.062	11.150	11.210	.030	1.230	1.215
under 8,600 lbs. GVW w/13" rear brakes	.062	13.000	13.060	.030	1.230	1.215
75–74 C20, C25, C30, C35, Suburban C20, Suburban C25:						
w/11 ⅛" rear brakes	.030	11.150	11.210	.030	1.230	1.215
w/13" rear brakes	.030	13.000	13.060	.030	1.230	1.215
73–72 C20, C25:						
w/11 ⅛" rear brakes	.030	11.150	11.210	.030	1.230	1.215
W/12" rear brakes	.030	12.000	12.060	.030	1.230	1.215
w/13" rear brakes	.030	13.000	13.060	.030	1.230	1.215
87–76 C30, C35, G30, G35, R30, R35: over 8,600lbs. GVW	.062	13.000	13.060	.030	1.480	1.465
75–74 C30, C35, G30, G35: over 8,600lbs. GVW	.030	13.000	13.060	.030	1.480	1.465
73–72 C30, C35, G30, G35:						
w/13" rear brakes	.030	13.000	13.060	.030	1.230	1.215
w/15" rear brakes	.030	15.000	15.060	.030	1.230	1.215
88–82 El Camino:	①	9.500	9.560	.030	.980	.965
81–79 El Camino:	①	9.500	9.560	.030	.980	.965
78 El Camino:	①	11.000	11.060	.030	.980	.965
77 El Camino:	①	11.000	11.060	.030	.980	.965
76–73 El Camino:	①	9.500	9.560	.030①	.980	.965
72 El Camino:	①	9.500	9.560	.030①	.980	.965
94–92 G10, G15, G20, G25:	.030	11.150	11.210	.030	1.230	1.215
94–92 G30, G35: exc. Chassis Cab	.030	13.000	13.060	.030	1.230	1.215
91–88 G10, G15: w/11" rear brakes	.030	11.000	11.060	.030	1.230	1.215
w/11 ⅛" rear brakes	.030	11.150	11.210	.030	1.230	1.215
87–76 G10, G15: w/1.0" rotor						
w/11" rear brakes.............	.062	11.000	11.060	.030	.980	.965
w/11 ⅛" rear brakes............	.062	11.150	11.210	.030	.980	.965
w/1.25" rotor						
w/11" rear brakes.............	.062	11.000	11.060	.030	1.230	1.215
w/11 ⅛" rear brakes............	.062	11.150	11.210	.030	1.230	1.215
75–74 G10, G15: w/11" rear brakes	.030	11.000	11.060	.030	1.230	1.215
w/11 ⅛" rear brakes	.030	11.050	11.210	.030	1.230	1.215
73–72 G10, G15:	.030	11.000	11.060	.030	1.230	1.215
91–88 G20, G25:	.030	11.150	11.210	.030	1.230	1.215
87–76 G20, G25, G30, G35:						
under 8,600 lbs. GVW w/11 ⅛" rear brakes...........	.062	11.150	11.210	.030	1.230	1.215
under 8,600 lbs. GVW w/13" rear brakes.............	.062	13.000	13.060	.030	1.230	1.215
75–74 G20, G25, G30, G35:						
w/11 ⅛" rear brakes	.030	11.150	11.210	.030	1.230	1.215
w/13" rear brakes	.030	13.000	13.060	.030	1.230	1.215

① .030" over rivet head; if bonded lining use .062"

(continues)

GENERAL MOTORS TRUCK — Chevrolet, GMC (Continued)

73–72 G20, G25:	.030	11.000	11.060	.030	1.230	1.215
94–92 G30, G35: Chassis Cab	.030	13.000	13.060	.030	1.480	1.465
91–88 G30, G35: w/1.28" rotor	.030	13.000	13.060	.030	1.230	1.215
w/1.54" rotor	.030	13.000	13.060	.030	1.480	1.465
87–76 K10, K15, K20, K25, V10, V15, V20, V25, Suburban:						
w/11 1/8" rear brakes	.062	11.150	11.210	.030	1.230	1.215
w/13" rear brakes	.062	13.000	13.060	.030	1.230	1.215
75–74 K10, K15, K20, K25:						
w/11" rear brakes	.030	11.000	11.060	.030	1.230	1.215
w/11 1/8" rear brakes	.030	11.150	11.210	.030	1.230	1.215
W/13" rear brakes	.030	13.000	13.060	.030	1.230	1.215
73–72 K10, K15, K20, K25:						
w/11" rear brakes	.030	11.000	11.060	.030	1.230	1.215
w/11 1/8" rear brakes	.030	11.150	11.210	.030	1.230	1.215
w/13" rear brakes	.030	13.000	13.060	.030	1.230	1.215
87–77 K30, K35, V30, V35:	.062	13.000	13.060	.030	1.480	1.465
94–93 Lumina APV:	.030	8.863	8.877	.030	1.224	1.209
92–90 Lumina APV:	①	8.863	8.877	.030	.972	.957
82–81 LUV Pickup:	.059	10.000	10.059	.236	.668	.653
80–76 LUV Pickup:	.059	10.000	10.059	.236	.668	.653
75–72 LUV Pickup:	.059	10.000	10.059	—	—	—
91–88 R10, R15, Suburban R10, Suburban R15:	.030	11.150	11.210	.030	1.230	1.215
91–88 R20, R25, R30, R35, Suburban R20, Suburban R25:						
w/1.28" rotor	.030	13.000	13.060	.030	1.230	1.215
w/1.54" rotor	.030	13.000	13.060	.030	1.480	1.465
91–88 V10, V15, Suburban V10, Suburban V15:	.030	11.150	11.210	.030	1.230	1.215
91–88 V20, V25, V30, V35, Suburban V20, Suburban V25:						
w/1.28" rotor	.030	13.000	13.060	.030	1.230	1.215
w/1.54" rotor	.030	13.000	13.060	.030	1.480	1.465

① .030" over rivet head; if bonded lining use .062".

GEO TRUCK

94–91 GEO Tracker:	.039	8.660	8.740	.030	.345	.315
90–89 GEO Tracker:	.039	8.660	8.740	.030	.345	.315

HONDA TRUCK

94 Passport: w/2.6L eng. Front	—	—	—	.039	.826	.811
w/3.2L eng. Front	—	—	—	.039	.983	.969
Rear	—	—	—	.039	.668	.654

HYUNDAI TRUCK

87–86 Pony Pickup:	.039	7.992	①	.060	—	.449
85–83 Pony Pickup:	.098	7.992	①	.060	—	.314

① Machining specification is not available; discard at 8.071".

ISUZU TRUCK

94 Amigo: w/rear drum brakes	.039	10.000	10.059	.039	.985	.970
w/rear disc brakes	—	—	—	.039	.669	.654
93–90 Amigo: w/rear drum brakes	.039	10.000	10.059	.039	.826	.811
w/rear disc brakes	—	—	—	.039	.432	.417
94–88 Pickup: w/rear drum brakes	.039	10.000	10.059	.039	.826	.811
w/rear disc brakes	—	—	—	.039	.432	.417
87–84 Pickup:	.039	10.000	10.039	.039	.658	.654
83–81 Pickup:	.039	10.000	10.039	.039	.453	.437
94–93 Rodeo: w/2.6L eng.	.039	10.000	10.059	.039	.826	.811
w/3.2L eng. Front	—	—	—	.039	.983	.969
Rear	—	—	—	.039	.668	.654
92–91 Rodeo:	.039	10.000	10.059	.039	.826	.811
94–92 Trooper: Front	—	—	—	.039	.983	.969
Rear	—	—	—	.039	.668	.654
91–88 Trooper II: Front	—	—	—	.039	.826	.811
Rear	—	—	—	.039	.432	.417
87 Trooper II:	.039	10.000	10.039	.039	.826	.811
86–84 Trooper II:	.039	10.000	10.039	.039	.668	.654

JEEP TRUCK

Year / Model						
94 Cherokee: Front				.030	—	.890
Rear w/9" drum	.030	9.000	9.060			
Rear w/10" drum	.030	10.000	10.060			
93–91 Cherokee: (4x2)	①	9.000	9.050	.030	—	.866
(4X4)	①	9.000	9.050	.030	—	.890
90 Cherokee: (4x2)	①	9.000	9.050	.030	—	.860
(4x4)	①	9.000	9.050	.030	—	.940
89–84 Cherokee, Wagoneer: (Sportwagons)	.030	10.000	10.060	.030	—	.815
89–82 Cherokee, Wagoneer: (Full Size)	①	11.000	11.060	.062	—	1.215
81–78 Cherokee, Wagoneer:	①	11.000	11.060	.062	—	1.215
77–76 Cherokee, Wagoneer:	①	11.000	11.060	.062	—	1.215
75–74 Cherokee, Wagoneer:	.030	11.000	11.060	.062	1.230	1.215
86–82 CJ, Scrambler:	①	10.000	10.060	.062	—	.815
81–79 CJ:	①	10.000	10.060	.062	—	.815
78–77 CJ:	①	11.000	11.060	.062	—	1.120
76–72 CJ:	.030	11.000	11.060	—	—	—
92–90 Comanche: (4x2) w/9" rear brakes	①	9.000	9.050	.030	—	.860
(4x2) w/10" rear brakes	①	10.000	10.060	.030	—	.860
92–91 Comanche: (4x4) w/9" rear brakes	①	9.000	9.050	.030	—	.890
(4x4) w/10" rear brakes	①	10.000	10.060	.030	—	.890
90 Comanche: (4x4) w/9" rear brakes	①	9.000	9.050	.030	—	.940
(4x4) w/10" rear brakes	①	10.000	10.060	.030	—	.940
89–84 Comanche:	.030	10.000	10.060	.030	—	.815
94 Grand Cherokee:	.030	10.000	10.060	.030	—	.890
w/rear disc brakes	—	—	—	.030	—	—
93 Grand Cherokee, Grand Wagoneer:	.030	10.000	10.060	.030	—	.890
91–90 Grand Wagoneer:	①	11.000	11.060	.030	—	1.215
89–82 J10 Pickup:	①	11.000	11.060	.062	—	1.215
81–78 J10 Pickup:	①	11.000	11.060	.062	—	1.215
77–76 J10 Pickup:	①	11.000	11.060	.062	—	1.215
75–74 J10 Pickup:	.030	11.000	11.060	.062	1.230	1.215
87–82 J20 Pickup:	①	12.000	12.060	.062	—	1.215
81–78 J20 Pickup:	①	12.000	12.060	.062	—	1.215
77–76 J20 Pickup:	①	12.000	12.060	.062	—	1.215
75–74 J20 Pickup:	.030	12.000	12.060	.062	1.230	1.215
94–92 Wrangler:	.030	9.000	9.060	.030	—	.890
91 Wrangler:	①	9.000	9.050	.030	—	.890
90 Wrangler:	①	9.000	9.050	.030	—	.940
89–87 Wrangler:	.030	10.000	10.060	.030	—	.815
73–72 Wagoneer, Pickup: w/11" brakes	.030	11.000	11.060	—	—	—
w/12" brakes	.030	12.000	12.060	—	—	—
73–72 Camper: 8,000lbs. GVW	.030	12.125	12.185	—	—	—

① .030" over rivet head; if bonded lining use .062".

MAZDA TRUCK

Year / Model						
94 B2300 Pickup, B3000 Pickup, B4000 Pickup: (4x2)	.030	9.010	9.040	.118	—	.810
(4x4)	.030	10.000	10.030	.118	—	.810
93–87 B2200 Pickup, B2600 Pickup: (4x2)	.040	10.240	10.300	.118	—	.710
(4x4)	.040	10.240	10.300	.118	—	.790
87–86 B2000 Pickup:	.040	10.240	10.310	.118	—	.710
84 B2200 Pickup:	.040	10.236	10.275	.040	—	.748
84 B2000 Pickup:	.040	10.236	10.275	.040	—	.433
83–82 B2200 Pickup:	.040	10.236	10.275	.276①	—	.748
83–77 B1800 Pickup, B2000 Pickup:	.040	10.236	10.275	.276①	—	.433
76–72 B1600 Pickup:	.040	10.236	10.275	—	—	—
77–74 Pickup: Rotary eng.	.040	10.236	10.275	.276①	—	.433
94 MPV: 2WD Front	—	—	—	.080	—	1.100
4WD Front	—	—	—	.080	—	1.020
w/rear disc brakes	—	—	—	.080	—	.630
93–92 MPV: 2WD	.040	10.240	10.300	.080	—	1.100
4WD	.040	10.240	10.300	.080	—	1.020
91–89 MPV:	.040	10.240	10.300	.080	—	.870
94–91 Navajo: w/9" rear brakes	.030	9.000	9.060	.062	—	.810
w/10" rear brakes	.030	10.000	10.060	.062	—	.810

① Measurement of lining & metal.

MERCURY TRUCK

Year / Model						
94–93 Villager:	.059	9.840	9.900	.080	.974	.945

MITSUBISHI TRUCK

94–92 Expo, Expo LVR:						
w/8″ drum brakes	.040	8.071	8.100	.080	—	.880
w/9″ drum brakes	.040	9.079	9.100	.080	—	.880
w/rear disc brakes	—	—	—	.080	—	.330
94–92 Montero: Front	—	—	—	.079	—	.882
Rear	—	—	—	.079	—	.646
91–88 Montero:	.040	10.000	10.079	.079	—	.803
87 Montero:	.040	10.000	10.040	.040	—	.724
86–83 Montero:	.040	10.000	10.040	.040	—	.720
94–87 Pickup: (4x2)	.040	10.000	10.079	.080	—	.803
(4x4)	.040	10.000	10.079	.080	—	.803
86–83 Pickup: w/9½″ rear brakes	.040	9.500	9.550	.040	—	.720
w/10″ rear brakes	.040	10.000	10.070	.040	—	.720
90–87 Van:	.040	10.000	10.080	.079	—	.803

NAVISTAR — INTERNATIONAL TRUCK

80–72 Scout II:	.030	11.000	11.060	.125	—	1.120
75–74 Travel All 100/150: (½ Ton)	.030	11.000	11.060	.125	—	1.120
75–74 Travel All 200: (¾ Ton)	.030	12.000	12.060	.125	—	1.120

NISSAN (DATSUN) TRUCK

79–78 620 Pickup:	.059	10.000	10.055	.080	—	.413
77–72 620 Pickup:	.059	10.000	10.055	—	—	—
86–84 720 Pickup: w/single wheel	.059	10.000	10.060	.080	—	.787
w/dual wheels	.059	8.660	8.720	.080	—	.787
83–82 720 Pickup:	.059	10.000	10.055	.080	—	.413
81 720 Pickup:	.059	10.000	10.055	.080	—	.413
80 720 Pickup:	.059	10.000	10.055	.080	—	.413
90 Axxess: w/9″ rear brakes	.059	9.000	9.060	.079	—	.787
w/10″ rear brakes	.059	10.240	10.300	.079	—	.787
94–86 D21 Pickup:						
(4x2) w/4 cyl. gas eng.........	.059	10.240	10.300	.079	—	.787
(4x2) exc. 4 cyl. gas eng.						
w/8.66″ rear brakes	.059	8.660	8.720	.079	—	.945
w/10″ rear brakes	.059	10.000	10.060	.079	—	.945
w/10.24″ rear brakes	.059	10.240	10.300	.079	—	.945
(4x4) w/10.24″ rear brakes	.059	10.240	10.300	.079	—	.945
(4x4) w/11.61″ rear brakes	.059	11.610	11.670	.079	—	.945
94–90 Pathfinder: Front	—	—	—	.079	—	.945
Rear w/10.24″ rear brakes	.059	10.240	10.300	—	—	—
Rear w/11.61″ rear brakes	.059	11.610	11.670	—	—	—
Rear w/rear disc brakes	—	—	—	.079	—	.630
89–87 Pathfinder: w/rear drum brakes	.059	10.000	10.060	.079	—	.945
w/rear disc brakes	—	—	—	.079	—	.630
94–93 Quest:	.079	9.840	9.900	.079	—	.945
88–87 Vanette:	.059	10.240	10.300	.079	—	.945

OLDSMOBILE TRUCK

94–91 Bravada:	①	9.500	9.560	.030	.980	.965
94–93 Silhouette APV:	.030	8.863	8.877	.030	1.224	1.209
92–90 Silhouette APV:	①	8.863	8.877	.030	.972	.957

① .030″ over rivet head; if bonded lining use .062″.

PONTIAC TRUCK

94–93 Trans Sport APV:	.030	8.863	8.877	.030	1.224	1.209
92–90 Trans Sport APV:	①	8.863	8.877	.030	.972	.957

① .030″ over rivet head; if bonded lining use .062″.

SUBARU TRUCK

86–85 Brat:	.060	7.090	7.170	.295②	—	.630
w/rear disc brakes	—	—	—	.256②	—	.335
84–83 Brat: w/solid rotor	.060	7.090	7.120	.295②	—	.394
w/vented rotor	.060	7.090	7.120	.295②	—	.610
82–80 Brat:	.060	7.090	7.120	.295②	—	.394
79–77 Brat:	.060	7.090	7.120	.060	—	.330

② Measurement of lining & metal.

SUZUKI TRUCK

94–86 Samurai:	.120①	8.660	8.740	.236①	—	.334
94–89 Sidekick: 2 Door	.120①	8.660	8.740	.315①	—	.315
94–92 Sidekick: 4 Door	.120①	10.000	10.070	.315①	—	.591

① Measurement of lining & metal.

TOYOTA TRUCK

94–92 4 Runner:	.039	11.614	11.693	.039	—	.906
91–89 4 Runner:	.040	11.614	11.693	.040	—	.709
88–87 4 Runner:	.040	11.614	11.693	.040	—	.748
86 4 Runner:	.060	11.614	11.654	.040	—	.748
85–84 4 Runner:	.040	10.000	10.060	.040	—	.453
94–93 Land Cruiser:	.060	11.614	11.693	.157	—	1.181
wrear disc brakes	—	—	—	.039	—	.709
92–91 Land Cruiser:	.060	11.614	11.693	.157	—	.984
w/rear disc brakes	—	—	—	.039	—	.709
90–89 Land Cruiser:	.060	11.614	11.693	.157	—	.748
88–85 Land Cruiser:	.060	11.614	11.693	.040	—	.748
84–75 Land Cruiser:	.060	11.610	11.650	.040	—	.748
74–72 Land Cruiser:	.060	11.400	11.440	—		
94–89 Pickup: (4x2) ½ Ton	.039	10.000	10.079	.039	—	.787
1 Ton exc. H.D. brakes	.039	10.000	10.079	.039	—	.906
1 Ton w/H.D. brakes	.039	10.000	10.079	.039	—	1.102
88–85 Pickup: (4x2) ½ Ton	.040	10.000	10.060	.040	—	.827
1 Ton H.D.	.040	10.000	10.060	.040	—	.945
84 Pickup: (4x2)	.040	10.000	10.060	.040	—	.827
83–75 Pickup: (4x2)	.040	10.000	10.060	.040	—	.453
83–79 Cab & Chassis: (4x2)	.040	10.000	10.060	.040	—	.748
74–72 Pickup: (4x2) w/10" brakes	.040	10.000	10.060	—	—	—
72 Pickup: (4x2) w/9" brakes	.040	9.000	9.060	—	—	—
94–89 Pickup: (4x4) ½ Ton	.039	11.614	11.693	.039	—	.709
88–87 Pickup: (4x4)	.040	11.614	11.693	.040	—	.748
86 Pickup: (4x4)	.060	11.614	11.654	.040	—	.748
85–84 Pickup: (4x4)	.040	10.000	10.060	.040	—	.453
83–79 Pickup: (4x4)	.040	10.000	10.060	.040	—	.453
94–92 Previa: w/rear drum brakes	.039	10.000	10.079	.039	—	.906
w/rear disc brakes Front	—	—	—	.039	—	.787
Rear	—	—	—	.039	—	.669
91 Previa: w/rear drum brakes	.040	10.000	10.080	.040	—	.827
w/rear disc brakes	—	—	—	.040	—	.669
94–93 T100 Pickup: (4x2) ½ Ton	.039	11.610	11.690	.039	—	.906
(4x2) 1 Ton	.039	11.610	11.690	.039	—	.906
(4x4)	.039	11.610	11.690	.039	—	.906
88–87 Van: (4x2)	.040	10.000	10.079	.040	—	.748
(4x4)	.040	10.000	10.079	.118	—	.945
86–84 Van:	.040	10.000	10.060	.040	—	.748

VOLKSWAGEN TRUCK

91–86 Vanagon:	.098	9.921	9.960	.079	—	.512
85–80 Vanagon:	.098	9.921	9.960	.078	.452	.433
79–77 Van, Bus, Wagon, Transporter, Kombi:	.098	9.921	9.960	.080	.492	.453
76–73 Van, Bus, Wagon, Transporter, Kombi:	.100	9.921	9.960	.080	—	.472
72 Van, Bus, Wagon, Transporter, Kombi:	.100	9.921	9.960	.080	.482	.472
84–82 Rabbit Pickup:	.098	7.874	7.894	.250④	.413	.393
81–80 Rabbit Pickup: w/Kelsey-Hayes Caliper	②	7.086	7.105	.080	.413	.393

② .098" riveted; .059" bonded.   ④ Measurement of lining & metal.

INDEX